SymbolicC++: An Introduction to Computer Algebra using Object-Oriented Programming

Springer
London
Berlin
Heidelberg
New York
Barcelona
Hong Kong
Milan
Paris
Singapore
Tokyo

Tan Kiat Shi, Willi-Hans Steeb and Yorick Hardy

SymbolicC++: An Introduction to Computer Algebra using Object-Oriented Programming

2nd extended and revised edition

 Springer

Tan Kiat Shi
ILOG Co. Ltd., Tokyo, Japan

Willi-Hans Steeb
Yorick Hardy
International School for Scientific Computing, Rand Africaans University,
Johannesburg, South Africa

ISBN 1-85233-260-3 Springer-Verlag London Berlin Heidelberg

British Library Cataloguing in Publication Data
Tan, Kiat Shi
  Symbolic C++ : an introduction to computer algebra using
  Object-oriented programming. – 2nd rev.ed.
  1. Object-oriented programming   2. C++ (Computer program
  language)   3. Algebra – Data processing
  I. Title   II. Steeb, W.-H.   III. Hardy, Yorick
  512'.00285'5133
ISBN 1852332603

Library of Congress Cataloging-in-Publication Data
A catalog record for this book is available from the Library of Congress

Typesetting: Camera ready by authors
Printed and bound at the Athenæum Press Ltd., Gateshead, Tyne & Wear
34/3830-543210 Printed on acid-free paper SPIN 10749339

Preface

In this text we will show how object-oriented programming can be used to implement a symbolic algebra system and how the system is applied to different areas in mathematics and physics.

In the most restrictive sense, computer algebra is used for the manipulation of scientific and engineering formulae. Usually, a mathematical formula described in the programming languages such as C, Fortran and Pascal can only be evaluated numerically, by assigning the respective values to each variable. However, the same formula may be treated as a mathematical object in a symbolic algebra system, which allows formal transformation, such as differentiation, integration and series expansion, on top of the numerical manipulations. This is therefore an indispensable tool for research and scientific computation.

Object-oriented programming has created a new era for programming in computer science as it has been suggested as a possible solution to software development. Basically, object-oriented programming is an important approach to analyzing problems, designing systems and building solutions. By applying this method effectively, the software products become less error prone, easier to maintain, more reusable and extensible.

The purpose of this book is to demonstrate how the features of object-oriented programming may be applied to the development of a computer algebra system. Among the many object-oriented programming languages available nowadays, we have selected C++ as our programming language. It is the most widely used object-oriented programming language, which has been successfully utilized by many programmers in various application areas. The design is based partly on acknowledged principles and partly on solid experience and feedback from actual use. Many experienced individuals and organizations in the industry and academia use C++. On top of the reasons stated above, we have selected C++ over other object-oriented languages because of its efficiency in execution speed and its utilization of pointers and templates.

Chapter 1 introduces the general notion of Computer Algebra. We discuss the essential properties and requirements of a computer algebra system. Some pitfalls and limitations are also listed for reference. Finally, we present a computer algebra system — **SymbolicC++**. This new system has many advantages over existing computer algebra systems. A brief summary of features and functionalities is given.

v

Chapter 2 presents the general mathematics for a computer algebra system. We describe how fundamental mathematical quantities are built up to form more complex mathematical structures.

Chapter 3 gives a brief introduction to some computer algebra systems available in the market place, such as Reduce, Maple, Axiom, Mathematica and MuPAD. The basic operations are described for each system. Examples are used to demonstrate the features of these systems.

The basic concepts of object-oriented programming, such as objects, classes, abstract data types, message passing, inheritance, polymorphism and so on are introduced in Chapter 4. Examples are given to assist the readers in understanding these important concepts. The object-oriented language Java is discussed in detail and two classes are developed. Finally, the chapter briefly describes three other object-oriented programming languages — Eiffel, Smalltalk and Oberon.

In Chapter 5, we introduce the language tools in C++. Along with the description, we construct the **String** data type, which serves as a carriage for introducing the facilities available in C++. Some other examples are also given. The Standard Template Library is also introduced together with a large number of examples. Finally we describe recursion. By the end of this chapter, all the features needed in the book will have been introduced.

Chapter 6 gives a collection of useful classes for computer algebra. We investigate very long integers, rational numbers, complex numbers, quaternions, exact derivatives, vectors, matrices, arrays, strings, bit vectors, finite sets and polynomials. They are the building blocks of mathematics as described in Chapter 2. The internal structures and external interfaces of these classes are described in great detail.

In Chapter 7, we describe how a mathematical expression can be constructed using object-oriented techniques. We introduce the new computer algebra system **SymbolicC++** and describe its internal representations and public interfaces. Several examples are also presented to demonstrate the functionalities of the system.

In Chapter 8, we apply the classes developed in Chapters 6 and 7 to problems in mathematics and physics. Applications are categorized according to classes. Several classes may be used simultaneously to solve a particular problem. Many interesting problems are presented, such as ghost solutions, Padé approximant, Lie series techniques, Picard's method, Mandelbrot set, etc.

In Chapter 9, we discuss how the programming language Lisp can be used to implement a computer algebra system. We implement an algebraic simplification and differentiation program. Furthermore, we develop a Lisp system using the object-oriented language C++.

The header files of the classes (abstract data type) introduced in Chapters 6 and 7 are listed in Chapter 10, while in Chapter 11, we show how a PVM (Parallel Virtual Machine) can be used with abstract data types. In Chapter 12, we describe some error handling techniques. We introduce the concept of exception handling. Examples are also given for demonstration purposes. Finally, in Chapter 13 we show how Gnuplot and PostScript can be used to draw figures.

The level of presentation is such that one can study the subject early on in ones education in science. There is a balance between practical programming and the underlying language. The book is ideally suited for use in lectures on symbolic computation and object-oriented programming. The beginner will also benefit from the book.

The reference list gives a collection of textbooks useful in the study of the computer language C++ [6], [13], [23], [29], [31], [37], [45], [55]. For data structures we refer to Budd (1994) [9]. For applications in science we refer to Steeb et al (1993) [48], Steeb (1994) [49] and Steeb (1999) [52].

The C++ programs have been tested with all newer C++ compilers which comply to the C++ Standard and include an implementation of the Standard Template Library.

Without doubt, this book can be extended. If you have comments or suggestions, we would be pleased to have them. The email addresses of the authors are:

```
Willi-Hans Steeb: whs@na.rau.ac.za
                  steeb_wh@yahoo.com
Tan Kiat Shi:     ktan@ilog.co.jp
Yorick Hardy:     yorickhardy@yahoo.com
```

SymbolicC++ was developed by the International School for Scientific Computing. The Web pages of the International School for Scientific Computing are

```
http://zeus.rau.ac.za/
http://issc.rau.ac.za/
```

The second web page also provides the header files for SymbolicC++.

Contents

Chapter 1

Introduction

1.1 What is Computer Algebra ?

Computer algebra [12], [34], [41] is the name of the technology for manipulating mathematical formulae symbolically by digital computers. For example an expression such as

$$x - 2 * x + \frac{d}{dx}(x - a)^2$$

should evaluate symbolically to

$$x - 2 * a.$$

Symbolic simplifications of algebraic expressions are the basic properties of computer algebra. Symbolic differentiation using the sum rule, product rule and division rule has to be part of a computer algebra system. Symbolic integration should also be included in a computer algebra system. Furthermore expressions such as $\sin^2(x) + \cos^2(x)$ and $\cosh^2(x) - \sinh^2(x)$ should simplify to 1. Thus another important ingredient of a computer algebra system is that it should allow one to define rules. Examples are the implementations of the exterior product and Lie algebras. Another important part of a computer algebra system is the symbolic manipulation of polynomials, for example to find the greatest common divisor of two polynomials or finding the coefficients of a polynomial.

The name of this discipline has long hesitated between symbolic and algebraic calculation, symbolic and algebraic manipulations and finally settled down as Computer Algebra. Algebraic computation programs have already been applied to a large number of areas in science and engineering. It is most extensively used in the fields where the algebraic calculations are extremely tedious and time consuming, such as general relativity, celestial mechanics and quantum chromodynamics. One of the first applications was the calculation of the curvature of a given metric tensor field. This involves mainly symbolic differentiation.

1

1.2   Properties of Computer Algebra Systems

What should a computer algebra system be able to do? First of all it should be able to handle the data types such as very long integers, rational numbers, complex numbers, quaternions, etc. The basic properties of the symbolic part should be simplifications of expressions, for example

$$a + 0 = a, \qquad 0 + a = a$$

$$a - a = 0, \qquad -a + a = 0$$

$$a * 0 = 0, \qquad 0 * a = 0$$

$$a * 1 = a, \qquad 1 * a = a$$

$$a^0 = 1.$$

In most systems it is assumed that the symbols are *commutative*, i.e.,

$$a * b = b * a.$$

Thus an expression such as

$$(a + b) * (a - b)$$

should be evaluated to

$$a * a - b * b.$$

If the symbols are not commutative then a special command should be given to indicate so. Furthermore, a computer algebra system should do simplifications of trigonometric and hyperbolic functions such as

$$\sin(0) = 0, \qquad \cos(0) = 1$$

$$\cosh(0) = 1, \qquad \sinh(0) = 0.$$

The expression $\exp(0)$ should simplify to 1 and $\ln 1$ should simplify to 0. Expressions such as

$$\sin^2(x) + \cos^2(x), \qquad \cosh^2(x) - \sinh^2(x)$$

should simplify to 1. Besides symbolic differentiation and integration a computer algebra system should also allow the symbolic manipulation of vectors, matrices and arrays. Thus the scalar product and vector product must be calculated symbolically. For square matrices the trace and determinant have to be evaluated symbolically. Furthermore, the system should also allow numerical manipulations. Thus it should be able to switch from symbolic to numerical manipulations. The computer algebra system should also be a programming language. For example, it should allow if-conditions and for-loops. Moreover, it must allow functions and procedures.

1.3 Pitfalls in Computer Algebra Systems

Although computer algebra systems have been around for many years there are still bugs and limitations in these systems. Here we list a number of typical pitfalls.

One of the typical pitfalls is the evaluation of

$$\sqrt{a^2 + b^2 - 2ab}.$$

Some computer algebra systems indicate that $a - b$ is the solution. Obviously, the result should be

$$\pm|a - b|.$$

As another example consider the *rank* of the matrix

$$A = \begin{pmatrix} 0 & 0 \\ x & 0 \end{pmatrix}.$$

The rank of a matrix is the number of linearly independent columns (which is equal to the number of linearly independent rows). If $x = 0$, then the rank is equal to 0. On the other hand if $x \neq 0$, then the rank of A is 1. Thus computer algebra systems face an ambiguity in the determination of the rank of the matrix. A similar problem arises when we consider the inverse of the matrix

$$B = \begin{pmatrix} 1 & 1 \\ x & 0 \end{pmatrix}.$$

It only exists when $x \neq 0$.

Another problem arises when we ask computer algebra systems to integrate

$$\int x^n dx$$

where n is an integer. If $n \neq -1$ then the integral is given by

$$\frac{x^{n+1}}{n + 1}.$$

If $n = -1$, the integral is
$$\ln(x).$$

Another ambiguity arises when we consider

$$0^0.$$

Consider for example
$$f(x) = x^x \equiv \exp(x \ln(x))$$
for $x > 0$. Applying *L'Hospital's rule* we find $0^0 = 1$ as a possible definition of 0^0. Many computer algebra systems have problems finding

$$\frac{x}{x + \sin(x)}$$

at $x = 0$ using L'Hospital's rule. The result is $1/2$.

We must also be aware that when we solve the equation

$$a * x = 0$$

the computer algebra system has to distinguish between the cases $a = 0$ and $x = 0$.

A large number of pitfalls can arise when we consider complex numbers and branch points in complex analysis. Complex numbers and functions should satisfy the Aslaksen test [3]. Thus
$$\exp(\ln(z))$$
should simplify to z, but
$$\ln(\exp(z))$$
should not simplify for complex numbers. We have to take care of the branch cuts when we consider multiple-valued complex functions. Most computer algebra systems assume by default that the argument is real-valued.

For a more in-depth survey of the pitfalls in computer algebra systems Stoutemeyer [54] may be perused.

1.4 Design of a Computer Algebra System

Most computer algebra systems are based on Lisp. The computer language Lisp takes its name from list processing. The main task of Lisp is the manipulation of quantities called lists, which are enclosed in parentheses. A number of powerful computer algebra systems are based in Lisp, for example Reduce, Macsyma, Derive, Axiom and MuPAD. The design of Axiom is based on object-oriented programming using Lisp. The computer algebra systems Maple and Mathematica are based on C. All of these systems are powerful software systems which can perform symbolic calculations. However, these software systems are independent systems and the transfer of expressions from one of them to another programming environment such as C is rather tedious, time consuming and error prone. It would therefore be helpful to enable a higher level language to manipulate symbolic expressions. On the other hand, the object-oriented programming languages provide all the necessary tools to perform this task elegantly.

Here we show that object-oriented programming using C++ can be used to develop a computer algebra system. Object-oriented programming is an approach to software design that is based on classes rather than procedures. This approach maximizes modularity and information hiding. Object-oriented design provides many advantages. For example, it combines both the data and the functions that operate on that data into a single unit. Such a unit (*abstract data type*) is called a *class*.

We use C++ as our object-oriented programming language for the following reasons. C++ allows the introduction of abstract data types. Thus we can introduce the data types used in the computer algebra system as abstract data types. The language C++ supports the central concepts of object-oriented programming: encapsulation, inheritance, polymorphism (including dynamic binding) and operator overloading. It has good support for dynamic memory management and supports both procedural and object-oriented programming. A less abstract form of polymorphism is provided via template support. We overload the operators

$$+, \quad -, \quad *, \quad /$$

for our abstract data types, such as verylong integers, rational numbers, complex numbers or symbolic data types. The vector and matrix classes are implemented on a template basis so that they can be used with the other abstract data types.

Another advantage of this approach is that, since the system of symbolic manipulations itself is written in C++, it is easy to enlarge it and to fit it to the special problem at hand. The classes (abstract data types) are included in a header file and can be provided in any C++ program by giving the command #include "ADT.h" at the beginning of the program.

For the realization of this concept we need to apply the following features of C++:

(1) the class concept
(2) overloading of operators
(3) overloading of functions
(4) inheritance of classes
(5) virtual functions
(6) function templates
(7) class templates
(8) Standard Template Library.

The developed system **SymbolicC++** includes the following abstract data types (classes):

`Verylong`	handles very long integers
`Rational`	template class that handles rational numbers
`Complex`	template class that handles complex numbers
`Quaternion`	template class that handles quaternions
`Derive`	template class to handle exact differentiation
`Vector`	template class that handles vectors
`Matrix`	template class that handles matrices
`Array`	template class that handles arrays up to four dimensions
`String`	handles strings
`Bitvector`	handles bit vectors
`List`	template class that handles linked lists
`Set`	template class that handles finite sets
`Polynomial`	template class that handles polynomials
`Sum`	template class that handles symbolic manipulations, such as rules, simplifications, differentiation, integration, commutativity and non-commutativity
`Type/Pair`	handles the atom and dotted pair for a Lisp system All the standard Lisp functions are included.

Suitable type conversions between the abstract data types and between abstract data types and basic data types are also provided.

The advantages of SymbolicC++ are as follows:

(1) Object-oriented design and proper handling of basic and abstract data types.
(2) The system is operating system independent, i.e. for all operating systems powerful C++ compilers are available.
(3) The user is provided with the source code.
(4) New classes (abstract data types) can easily be added.
(5) The ANSI C++ standard is taken into account.
(6) The user only needs to learn C++ to apply the computer algebra system.
(7) Assembler code can easily be added to run on a specific CPU.
(8) Member functions and type conversion operators provide a symbolic-numeric interface.
(9) The classes (abstract data types) are included in a header file and can be provided in any C++ program by giving the command
    `#include "ADT.h"`
    at the beginning of the program.
(10) The classes can be linked with Parallel Virtual Machine (PVM).
(11) Standard Template Library can be used with SymbolicC++.

Chapter 2

Mathematics for Computer Algebra

2.1 Rings and Fields

The data types in a computer algebra system include integers, rational numbers, real numbers, complex numbers, quaternions and symbols. On the other hand the basic data types in computer languages such as C and C++ include integers, characters, float (double) numbers and pointers. From these basic data types we can form arrays, for example arrays of integers or arrays of characters (strings). The mathematical structure of these data types (except pointers) are rings and fields [4], [22].

A *ring* is an ordered triple $(R, +, *)$ consisting of a set R with two binary operations $+$ (addition) and $*$ (multiplication), satisfying the following conditions:
(a) the pair $(R, +)$ is a commutative group;
(b) multiplication is associative;
(c) it admits an identity (or unit) element, denoted by I;
(d) multiplication is distributive (on both sides) over addition, i.e.

$$x * (y + z) = x * y + x * z \quad \text{and} \quad (x + y) * z = x * z + y * z, \quad \text{for all } x, y, z \in R.$$

It is common not to include the existence of a multiplicative identity element, amongst the ring axioms. One then distinguishes between rings and rings with an identity (unit) element. Similar differences also apply to the definitions of subring and integral domain given below. If, moreover, multiplication is commutative, then R is said to be a *commutative ring*.

A *subring* of R is a subset S of R satisfying:
(a) S is a subgroup of the additive group R;
(b) $x \in S$ and $y \in S$ together imply $x * y \in S$;
(c) $I \in S$.

Thus a subring is a ring.

If an element $a \in R$ possesses an inverse element with respect to multiplication, i.e. if there exists a (unique) $a^{-1} \in R$ such that

$$a * a^{-1} = a^{-1} * a = I,$$

then we say that a is an *invertible element* of R.

The set of invertible elements of a ring R is denoted by R^*. If every non-zero element of R is invertible, then R is said to be a *division ring*. A commutative division ring is called a *field*. S is a *subfield* of F if S is a subring of F and $x \in S$, $x \neq 0$ together imply $x^{-1} \in S$.

A ring is ordered when there is a non-empty subset $P \subset R$, called the set of positive elements of R, satisfying:
(a) $a \in P$ and $b \in P$ together imply $a + b \in P$ and $a * b \in P$;
(b) for each $a \in R$ exactly one of the following holds:

$$a \in P, \qquad a = 0, \qquad (-a) \in P.$$

A commutative ring is said to be an *integral domain* if $x \in K$, $y \in K$ and $x * y = 0$, together imply $x = 0$ or $y = 0$. If $x * y = 0$, $x \neq 0$ and $y \neq 0$, then x and y are said to be *zero divisors* in K.

The *characteristic* of a ring is defined as the additive order of its (multiplicative) identity element, i.e. R has finite characteristic m if m is the least positive integer for which $m * I = 0$. It has characteristic 0 (or ∞, for usage differs) if no such multiple is zero.

Examples of rings are $(\mathbf{Z}, +, *)$ and $(\mathbf{Q}, +, *)$. $\mathbf{Z}$ is a subring of $\mathbf{Q}$, but it can be shown that $\mathbf{Z}$ has no proper subrings. The only invertible elements of $\mathbf{Z}$ are 1 and -1 whereas all non-zero elements of $\mathbf{Q}$ are invertible. Since $\mathbf{Q}$ is a commutative ring, it follows that $\mathbf{Q}$ is a field. $\mathbf{Q}$ is a subfield of $\mathbf{R}$. $\mathbf{Z}$, $\mathbf{Q}$ and $\mathbf{R}$ can be ordered by $<$ used in its usual sense. They are all integral domains and have characteristic 0.

A two-sided *ideal*, I, of a ring R is a non-empty subset of R satisfying:
(a) I is a subgroup of the additive group R;
(b) for all $x \in I$ and all $a, b \in R$ we have

$$\text{(i)} \quad a * x \in R \quad \text{and} \quad \text{(ii)} \quad x * b \in R.$$

I is a left ideal if it satisfies axioms (a) and (b)(i) and a right ideal if it satisfies (a) and (b)(ii). When R is commutative all ideals are two-sided.

An ideal, $I \neq K$, of a commutative ring K is said to be a maximal ideal of K if, whenever M is an ideal of K satisfying $I \subset M \subset K$, either $M = I$ or $M = K$. Given two ideals I and J of a commutative ring K, we define their sum, $I + J$, to be the set of all elements of K of the form $x + y$ where $x \in I$ and $y \in J$, and their product, IJ, to be all elements of K which can be written in the form

$$x_1 * y_1 + x_2 * y_2 + \ldots + x_n * y_n$$

where $n \in \mathbf{N}$, $x_i \in I$ and $y_i \in J$ for all $i = 1, \ldots, n$. $I + J$ and IJ are again ideals.

If K is a commutative ring and $I = \{x * b \mid x \in K\}$ where b is a fixed element of K, then I is an ideal, called a principal ideal. b is then said to generate I. An integral domain all of whose ideals are principal is called a principal ideal domain (or principal ideal ring). If I is an ideal of a principal ideal domain D and is generated by the elements $x_1, \ldots, x_n$ i.e. I is the set of all elements of the form

$$u_1 * x_1 + u_2 * x_2 + \ldots + u_n * x_n$$

where $u_i \in D$, then I is generated by some single element d. We say that d is a *greatest common divisor* (g.c.d.) of $x_1, \ldots, x_n$. $x_1, \ldots, x_n$ are said to be mutually or relatively prime if they have g.c.d. 1.

The intersection of the n ideals generated by $x_1, \ldots, x_n$ taken one at a time is also an ideal and so will be generated by a single element m. m is called a least common multiple (l.c.m.) of $x_1, \ldots, x_n$. An element p of D is said to be prime or irreducible if it is not an invertible element of D and if $p = a * b$ ($a, b \in D$) implies that a or b is invertible.

An integral domain D is a unique factorization domain if

(a) for all $a \in D \setminus \{0\}$, a is either invertible or can be written as the product of a finite number of irreducible elements of D, and
(b) the decomposition in (a) is unique up to the ordering of the irreducible elements and substitution by associates.

Example. The sets

$$I_2 = \{\ldots, -4, -2, 0, 2, 4, \ldots\} \quad \text{and} \quad I_3 = \{\ldots, -6, -3, 0, 3, 6, \ldots\}$$

are maximal ideals of $\mathbf{Z}$.

$$I_{12} = \{\ldots, -24, -12, 0, 12, 24, \ldots\} \quad \text{and} \quad I_8 = \{\ldots, -16, -8, 0, 8, 16, \ldots\}$$

are ideals but not maximal ideals.

2.2   Integers

In this section we introduce integers. For the proof of the following theorems we refer to the literature [4], [22]. Let $\mathbf{N}$ be the set of natural numbers

$$\mathbf{N} := \{\, 0, 1, 2, \ldots \,\}.$$

The set of *integers*

$$\mathbf{Z} := \{\ldots, -3, -2, -1, 0, 1, 2, 3, \ldots\}$$

can be constructed from $\mathbf{N}$ in the following way: define an equivalence relation E on $\mathbf{N} \times \mathbf{N}$ by

$$(x, y)E(x', y') \Leftrightarrow x + y' = x' + y.$$

We want the formula $x - y = x' - y'$ to hold once 'minus' has been defined. The set $\mathbf{Z}$ is then defined as $(\mathbf{N} \times \mathbf{N})/E$. Given elements z and z' of $\mathbf{Z}$ such that

$$z = p_E(x, y), \qquad z' = p_E(x', y')$$

where p_E denotes the canonical mapping from $\mathbf{N} \times \mathbf{N}$ onto $\mathbf{Z}$, we define the sum and product of z and z' by

$$z + z' = p_E(x + x', y + y'), \qquad z * z' = p_E(x * x' + y * y', x * y' + x' * y).$$

This definition is chosen because we want the formulae

$$(x - y) + (x' - y') = (x + x') - (y + y')$$

and

$$(x - y) * (x' - y') = (x * x' + y * y') - (x * y' + x' * y)$$

to hold. With these two operations $\mathbf{Z}$ is a *commutative ring*. The mapping $n \mapsto p_E(n, 0)$ is an injection of $\mathbf{N}$ into $\mathbf{Z}$ which preserves addition and multiplication. We can, therefore, identify $\mathbf{N}$ with a subset of $\mathbf{Z}$. If we define the negative of z, written $-z$, as the inverse element of z under addition, then it can be shown that either $z \in \mathbf{N}$ or $-z \in \mathbf{N}$. If $z \in \mathbf{N} \setminus \{0\}$ we say that z is a positive integer, if $-z \in \mathbf{N} - \{0\}$ we say that z is a negative integer. Note that $a - b = a + (-b)$ if $a, b \in \mathbf{N}$. An integer $a \neq 0$ is called a *divisor* of an integer b (written $a|b$) if there exists an integer c such that $b = a * c$. When $a|b$ we also say that b is an integral multiple of a. To show that the restriction $a \neq 0$ is necessary, suppose $0|b$. If $b \neq 0$, we must have $b = 0 * c$ for some $c \in \mathbf{Z}$, which is impossible; while if $b = 0$, we would have $0 = 0 * c$, which is true for every $c \in \mathbf{Z}$. When $b, c, x, y \in \mathbf{Z}$, the integer $b * x + c * y$ is called a linear combination of b and c. We have

Theorem. If $a|b$ and $a|c$ then $a|(b * x + c * y)$ for all $x, y \in \mathbf{Z}$.

An integer $p \neq 0, \pm 1$ is called a *prime* if and only if its only divisors are ± 1 and $\pm p$. It is clear that $-p$ is a prime if and only if p is a prime. Hereafter, we restrict our attention mainly to positive primes. The number of positive primes is infinite. When $a = b * c$ with $|b| > 1$ and $|c| > 1$, we call the integer a composite. Thus every integer $a \neq 0, \pm 1$ is either a prime or a composite. When $a|b$ and $a|c$, we call a a *common divisor* of b and c. When, in addition, every common divisor of b and c is also a divisor of a, we call a a *greatest common divisor* of b and c. Suppose c and d are two different greatest common divisors of $a \neq 0$ and $b \neq 0$. Then $c|d$ and $d|c$; hence c and d differ only in sign. We limit our attention to the positive greatest common divisor of two integers a and b and use either d or (a, b) to denote it. Thus, d is truly the largest (greatest) integer which divides both a and b.

We have assumed (a) that every two non-zero integers have a positive greatest common divisor and (b) that any integer $a > 1$ has a unique factorization, except for the order of the factors, as a product of positive primes. Of course, in (b) it must be understood that when a is itself a prime, "a product of positive primes" consists of a single prime.

Division Algorithm. For any two non-zero integers a and b, there exist unique integers q and r, called respectively *quotient* and *remainder*, such that

$$a = b * q + r, \qquad 0 \leq r < |b|.$$

It follows that $b|a$ and $(a, b) = b$ if and only if $r = 0$. When $r \neq 0$ it is easy to show that a common divisor of a and b also divides r and a common divisor of b and r also divides a. Then $(a, b)|(b, r)$ and $(b, r)|(a, b)$ so that $(a, b) = (b, r)$. Now either $r|b$ or $r \nmid b$. In the latter case, we use the division algorithm to obtain

$$b = r * q_1 + r_1, \qquad 0 < r_1 < r.$$

Again, either $r_1|r$ and $(a, b) = r_1$ or, using the division algorithm,

$$r = r_1 * q_2 + r_2, \qquad q < r_2 < r_1$$

and $(a, b) = (b, r) = (r, r_1) = (r_1, r_2)$. Since the remainders $r_1, r_2, \ldots$, assuming the process to continue, constitute a set of decreasing non-negative integers there must eventually be one which is zero. Suppose the process terminates with

$$
\begin{array}{llll}
(k) & r_{k-3} & = & r_{k-2} * q_{k-1} + r_{k-1} \quad 0 < r_{k-1} < r_{k-2} \\
(k+1) & r_{k-2} & = & r_{k-1} * q_k + r_k \qquad\ \ 0 < r_k < r_{k-1} \\
(k+2) & r_{k-1} & = & r_k * q_{k+1} + 0 \quad .
\end{array}
$$

Then

$$(a, b) = (b, r) = (r, r_1) = \ldots = (r_{k-2}, r_{k-1}) = (r_{k-1}, r_k) = r_k.$$

Since $r = a - b * q = a + (-q) * b = m_1 * a + n_1 * b$ we find

$$
\begin{aligned}
r_1 &= b - r * q_1 \\
&= b - (m_1 * a + n_1 * b) * q_1 \\
&= -m_1 * q_1 * a + (1 - n_1 * q_1) * b \\
&= m_2 * a + n_2 * b
\end{aligned}
$$

and

$$
\begin{aligned}
r_2 &= r - r_1 * q_2 \\
&= (m_1 * a + n_1 * b) - (m_2 * a + n_2 * b) * q_2 \\
&= (m_1 - q_2 * m_2) * a + (n_1 - q_2 * n_2) * b \\
&= m_3 * a + n_3 * b
\end{aligned}
$$

and continuing, we obtain finally

$$
r_k = m_{k+1} * a + n_{k+1} * b.
$$

Thus, we have

Theorem. When $d = (a, b)$, there exist $m, n \in \mathbf{Z}$ such that $d = (a, b) = m * a + n * b$.

In $(a, b) = m * a + n * b$, the integers m and n are not unique; in fact, $(a, b) = (m + k * b) * a + (n - k * a) * b$ for every $k \in \mathbf{N}$. The importance of this theorem is as follows: if $a|c$, if $b|c$, and if $(a, b) = d$, then $a * b|c * d$. Since $a|c$ and $b|c$, there exist integers s and t such that $c = a * s = b * t$. There exist $m, n \in \mathbf{Z}$ such that $d = m * a + n * b$. Then

$$
c * d = c * m * a + c * n * b = b * t * m * a + a * s * n * b = a * b * (t * m + s * n)
$$

and $a * b|c * d$.

A second consequence of the Division Algorithm is

Theorem. Any non-empty set K of integers which is closed under the binary operations addition and subtraction is either $\{0\}$ or consists of all multiples of its least positive element.

For given $a, b \in \mathbf{Z}$, suppose there exist $m, n \in \mathbf{Z}$ such that $a * m + b * n = 1$. Now every common factor of a and b is a factor of the right member 1; hence, $(a, b) = 1$. Two integers a and b for which $(a, b) = 1$ are said to be *relatively prime*.

Unique Factorization Theorem. Every integer $a > 1$ has a unique factorization, except for order,

$$
a = p_1 * p_2 * p_3 * \ldots * p_n
$$

into a product of positive primes. Evidently,

$$-a = -(p_1 * p_2 * p_3 * \ldots * p_n).$$

Moreover, since the p_i's are not necessarily distinct, we may write

$$a = p_1^{\alpha_1} * p_2^{\alpha_2} * p_3^{\alpha_3} * \ldots * p_s^{\alpha_s}$$

where each $\alpha_i \geq 1$ and the primes $p_1, p_2, p_3, \ldots, p_s$ are distinct.

Let m be a positive integer. The relation congruent modulo m ($\equiv$ (mod m)) is defined on all $a, b \in \mathbf{Z}$ by $a \equiv b \,(\text{mod } m)$ if and only if $m|(a - b)$. An alternate definition, often more useful than the original, is $a \equiv b \,(\text{mod } m)$ if and only if a and b have the same remainder when divided by m. As immediate consequences of these definitions, we have

Theorem. If $a \equiv b \,(\text{mod } m)$, then for any $n \in \mathbf{Z}$, $m*n+a \equiv b \,(\text{mod } m)$ and conversely.

Theorem. If $a \equiv b \,(\text{mod } m)$, then for all $x \in \mathbf{Z}$, $a + x \equiv b + x \,(\text{mod } m)$ and $a * x \equiv b * x \,(\text{mod } m)$.

Theorem. If $a \equiv b \,(\text{mod } m)$ and $c \equiv e \,(\text{mod } m)$, then $a + c \equiv b + e \,(\text{mod } m)$, $a - c \equiv b - e \,(\text{mod } m)$, $a * c \equiv b * e \,(\text{mod } m)$.

Theorem. Let $(c, m) = d$ and write $m = m_1 * d$. If $c * a \equiv c * b \,(\text{mod } m)$, then $a \equiv b \,(\text{mod } m_1)$ and conversely.

The relation $\equiv$ (mod m) on $\mathbf{Z}$ is an equivalence relation and separates the integers into m equivalence classes, $[0], [1], [2], \ldots, [m-1]$, called *residue classes modulo m*, where

$$[r] := \{\, a : \ a \in \mathbf{Z}, \ a \equiv r \,(\text{mod } m) \,\}.$$

We denote the set of all residue classes modulo m by $\mathbf{Z}/(m)$. Two basic properties of the residue classes modulo m are: If a and b are elements of the same residue class $[s]$, then $a \equiv b \,(\text{mod } m)$. If $[s]$ and $[t]$ are distinct residue classes with $a \in [s]$ and $b \in [t]$, then $a \not\equiv b \,(\text{mod } m)$.

Consider the linear congruence

$$a * x \equiv b \,(\text{mod } m)$$

in which a, b, m are fixed integers with $m > 0$. By a solution of the congruence we mean an integer $x = x_1$ for which $m|(a * x_1 - b)$. Now if x_1 is a solution so that $m|(a * x_1 - b)$, then for any $k \in \mathbf{Z}$, $m|(a * (x_1 + k * m) - b)$ and $x_1 + k * m$ is another solution. Thus, if x_1 is a solution so is every other element of the residue class $[x_1]$ modulo m. If the linear congruence has solutions, they consists of all the elements of one or more of the residue classes of $\mathbf{Z}/(m)$.

Suppose $(a, m) = 1 = s*a+t*m$. Then $b = b*s*a+b*t*m$ and $x_1 = b*s$ is a solution. Now assume $x_2 \not\equiv x_1 \,(\mathrm{mod}\ m)$ to be another solution. Since $a*x_1 \equiv b(\mathrm{mod}\ m)$ and $a*x_2 \equiv b\,(\mathrm{mod}\ m)$, it follows from the transitive property of $\equiv$ (mod m) that $a*x_1 \equiv a*x_2(\mathrm{mod}\ m)$. Then $m|a*(x_1-x_2)$ and $x_1 \equiv x_2\,(\mathrm{mod}\ m)$, contrary to the assumption. Thus, one has just one incongruent solution, say x_1, and the residue class $[x_1] \in \mathbf{Z}/(m)$, also called a *congruence class*, includes all solutions.

Theorem. The congruence $a*x \equiv b\,(\mathrm{mod}\ m)$ has a solution if and only if $d = (a, m)$ is a divisor of b. When $d|b$, the congruence has exactly d incongruent solutions (d congruence classes of solutions).

The number 827 016 can be written as

$$827016 = 8*10^5 + 2*10^4 + 7*10^3 + 0*10^2 + 1*10 + 6.$$

This representation is an application of the congruence properties of integers. For suppose a is a positive integer. By the division algorithm,

$$a = 10*q_0 + r_0, \qquad 0 \le r_0 < 10.$$

If $q_0 = 0$, we write $a = r_0$. If $q_0 > 0$, then $q_0 = 10*q_1 + r_1$, $0 \le r_1 < 10$. Now if $q_1 = 0$, then $a = 10*r_1 + r_2$ and we write $a = r_1r_0$; if $q_1 > 0$, then $q_1 = 10*q_2 + r_2$, $0 \le r_2 < 10$. Again, if $q_2 = 0$, then $a = 10^2*r_2 + 10*r_1 + r_0$ and we write $a = r_2r_1r_0$; if $q_2 > 0$, we repeat the process. This must end eventually and we have

$$a = 10^s*r_s + 10^{s-1}*r_{s-1} + \cdots + 10*r_1 + r_0 = r_sr_{s-1}\cdots r_1r_0.$$

This follows from the fact that the q_i's constitute a set of decreasing non-negative integers. Note that in this representation the symbols r_i used are from the set $\{0, 1, 2, 3, \ldots, 9\}$ of remainders modulo 10. The representation is unique. The process is independent of the base and any other positive integer may be used. Thus, if 4 is taken as base, any positive integer will be represented by a sequence of the symbols $0, 1, 2, 3$. For example, the integer (base 10) 155 is given in base 4 as

$$155 = 4^3*2 + 4^2*1 + 4*2 + 3 = 2123 \quad \text{base 4}.$$

Next, we describe an algorithm which generates the prime number sequence called the "sieve of Eratosthenes". It was worked out in the third century B.C. This algorithm discovers all the prime numbers less than a given integer N. It works by removing all the non-prime numbers, leaving the prime number sequence.

Consider the following sequence of integers > 1:

$$2 \quad 3 \quad 4 \quad 5 \quad 6 \quad 7 \quad 8 \quad 9 \quad 10 \quad 11 \quad 12 \quad 13 \quad 14 \quad 15 \quad 16 \quad 17 \quad 18 \quad 19 \quad 20 \quad 21 \quad \ldots$$

As we all know, the first prime number is 2, and all the multiples of 2 are not prime. Thus we cross out all the even numbers greater than 2:

2 | 3 4 5 ~~6~~ 7 ~~8~~ 9 ~~10~~ 11 ~~12~~ 13 ~~14~~ 15 ~~16~~ 17 ~~18~~ 19 ~~20~~ 21 ...

The next number on the sequence which has not been crossed out is 3. Thus we know it is the next prime number. Similarly, we cross out all the multiples of 3. Note that we attempt to cross out the numbers $3 \times 2 = 6$, $3 \times 4 = 12$, $3 \times 6 = 18$ which have already been removed as they are also multiples of 2. In fact, we could save some operations by just removing $3(3+0) = 9$, $3(3+2) = 15$, $3(3+4) = 21$, ...

2 3 | 5 7 ~~9~~ 11 13 ~~15~~ 17 19 ~~21~~ 23 25 ~~27~~ 29 31 ~~33~~ 35 37 ...

In general, starting with a prime number p, we successively cross out the multiples p^2, $p(p+2)$, $p(p+4)$, ... We start the crossing out process from p^2 because all the multiples smaller than that would have been removed in the earlier stages of the process. For example, starting with the prime number 5, we cross out $5(5+0) = 25$, $5(5+2) = 35$, $5(5+4) = 45$, ... We do not need to cross out 5×2 or 5×3 as they have been removed for $p = 2$ or $p = 3$, respectively.

Note that with this process, we may still end up crossing out numbers more than once. For example, $5(5+4) = 45$ has already been crossed out as a multiple of 3. The sequence for $p = 5$ looks like

2 3 5 | 7 11 13 17 19 23 ~~25~~ 29 31 ~~35~~ 37 41 43 47 53 ~~55~~ ...

We continue this process until we reach a prime p with $p^2 > N$, where N is the largest number we wish to consider. At this point, all the non-prime numbers $\leq N$ would have been crossed out. What remains is the prime number sequence $\leq N$. Below we list the prime numbers less than 100 that were generated using the algorithm described above:

2 3 5 7 11 13 17 19 23 29 31 37 41
43 47 53 59 61 67 71 73 79 83 89 97

2.3   Rational Numbers

The set of *rational numbers*, $\mathbf{Q}$, can be constructed in a similar manner to $\mathbf{Z}$, as follows. Let E be the equivalence relation on $\mathbf{Z} \times (\mathbf{Z} - \{0\})$ defined by

$$(a, b)E(c, d) \Leftrightarrow a * d = b * c,$$

and define $\mathbf{Q}$ as

$$\mathbf{Z} \times (\mathbf{Z} \setminus \{0\})/E.$$

Addition and multiplication are defined on $\mathbf{Q}$ in terms of the canonical mapping, p_E, by

$$
\begin{aligned}
p_E(a, b) + p_E(c, d) &= p_E(a * d + b * c, b * d) \\
p_E(a, b) \times p_E(c, d) &= p_E(a * c, b * d).
\end{aligned}
$$

It can be shown that with these operations $\mathbf{Q}$ is a *field* and that there is a natural injection which maps $\mathbf{Z}$ into $\mathbf{Q}$ and preserves the operations of multiplication and addition. We can, therefore, consider $\mathbf{Z}$ to be a subset of $\mathbf{Q}$.

Corresponding to each ordered pair (a, b) of $\mathbf{Z} \times (\mathbf{Z} \setminus \{0\})$ is the *fraction* a/b with *numerator* a and non-zero *denominator* b (the need to make b non-zero accounts for the use of $\mathbf{Z} \setminus \{0\}$, rather than $\mathbf{Z}$, in the definition). Two fractions are then equivalent if the corresponding ordered pairs are equivalent in the sense defined above, and a rational number is an equivalence class of fractions.

We now define two special rational numbers

$$\text{zero} \rightarrow p_E(0, b), \qquad \text{one} \rightarrow p_E(a, a)$$

and the inverse

$$(\text{additive}) : \quad -p_E(a, b) = p_E(-a, b)$$

$$(\text{multiplicative}) : \quad p_E(a, b)^{-1} = p_E(b, a)$$

In the following we use the notation a/b. Addition and multiplication obey the distributive and associative law. Moreover, for every a/b $(a, b \neq 0)$ there exists a multiplicative inverse b/a such that $(a/b) * (b/a) = 1$.

Subtraction and division are defined by

$$\frac{a}{b} - \frac{c}{d} = \frac{a * d - b * c}{b * d}, \qquad b, d \neq 0$$

and

$$\frac{\frac{a}{b}}{\frac{c}{d}} = \frac{a*d}{b*c}.$$

There is an order relation for rational numbers. An element $a/b \in \mathbf{Q}$ is called positive if and only if $a*b > 0$. Similarly, a/b is called negative if and only if $a*b < 0$. Since, by the *Trichotomy Law*, either $a*b > 0$, $a*b < 0$ or $a*b = 0$, it follows that each element of $\mathbf{Q}$ is either positive, negative or zero. The order relations $<$ and $>$ on $\mathbf{Q}$ are defined as follows. For each $x, y \in \mathbf{Q}$,

$$x < y \text{ if and only if } x - y < 0$$

$$x > y \text{ if and only if } x - y > 0.$$

These relations are transitive but neither reflexive nor symmetric. $\mathbf{Q}$ also satisfies the Trichotomy Law. If $x, y \in \mathbf{Q}$, one and only one of

(a) $x = y$,   (b) $x < y$,   (c) $x > y$

holds.

Consider any arbitrary $s/m \in \mathbf{Q}$ with $m \neq 0$. Let the (positive) greatest common divisor of s and m be d and write $s = d*s_1$, $m = d*m_1$. Since $(s, m) \sim (s_1, m_1)$, it follows that $s/m = s_1/m_1$. Thus, any rational number $\neq 0$ can be written uniquely in the form a/b where a and b are relatively prime integers. Whenever s/m has been replaced by a/b, we say that s/m has been reduced to lowest terms. Hereafter, any arbitrary rational number introduced is assumed to have been reduced to lowest terms.

Theorem. If x and y are positive rationals with $x < y$, then $1/x > 1/y$.

Density Property. If x and y, with $x < y$, are two rational numbers, there exists a rational number z such that $x < z < y$.

Archimedean Property. If x and y are positive rational numbers, there exists a positive integer p such that $p*x > y$.

Consider the positive rational number a/b in which $b > 1$. Now

$$a = q_0 * b + r_0, \qquad 0 \leq r_0 < b$$

and

$$10 * r_0 = q_1 * b + r_1, \qquad 0 \leq r_1 < b.$$

Since $r_0 < b$ and, hence, $q_1 * b + r_1 = 10 * r_0 < 10 * b$, it follows that $q_1 < 10$. If $r_1 = 0$, then

$$r_0 = \frac{q_1}{10} * b, \qquad a = q_0 * b + \frac{q_1}{10} * b, \qquad \frac{a}{b} = q_0 + \frac{q_1}{10}.$$

We write $a/b = q_0 * q_1$ and call $q_0 q_1$ the *decimal representation* of a/b. If $r_1 \neq 0$, we have

$$10 * r_1 = q_2 * b + r_2, \qquad 0 \leq r_2 \leq b$$

in which $q_2 < 10$. If $r_2 = 0$, then $r_1 = \frac{q_2}{10} * b$ so that $r_0 = \frac{q_1}{10} * b + \frac{q_2}{10^2} * b$ and the decimal representation of a/b is $q_0.q_1 q_2$. If $r_2 = r_1$, the decimal representation of a/b is the repeating decimal

$$q_0.q_1 q_2 q_2 q_2 \cdots$$

If $r_2 \neq 0$, we repeat the process. Now the distinct remainders $r_0, r_1, r_2, \ldots$ are elements of the set $\{0, 1, 2, 3, \ldots, b-1\}$ of residues modulo b so that, in the extreme case, r_b must be identical with some one of $r_0, r_1, r_2, \ldots, r_{b-1}$, say r_c, and the decimal representation of a/b is the repeating decimal

$$q_0.q_1 q_2 q_3 \cdots q_{b-1} q_{c+1} q_{c+2} \cdots q_{b-1} q_{c+1} q_{c+2} \cdots q_{b-1} \cdots$$

Thus, every rational number can be expressed as either a terminating or a repeating decimal.

Example. For 11/6, we find

$$
\begin{aligned}
11 &= 1 * 6 + 5; & q_0 &= 1, \; r_0 = 5 \\
10 * 5 &= 8 * 6 + 2; & q_1 &= 8, \; r_1 = 2 \\
10 * 2 &= 3 * 6 + 2; & q_2 &= 3, \; r_2 = 2 = r_1
\end{aligned}
$$

and $11/6 = 1.833333\ldots$

Conversely, it is clear that every terminating decimal is a rational number. For example, $0.17 = 17/100$ and $0.175 = 175/1000 = 7/40$.

Theorem. Every repeating decimal is a rational number.

The proof makes use of two preliminary theorems:

(i) Every repeating decimal may be written as the sum of an infinite geometric progression.

(ii) The sum of an infinite geometric progression whose common ratio r satisfies $|r| < 1$ is a finite number.

2.4 Real Numbers

Beginning with $\mathbf{N}$, we can construct $\mathbf{Z}$ and $\mathbf{Q}$ by considering quotient sets of suitable Cartesian products. It is not possible to construct $\mathbf{R}$, the set of all real numbers, in a similar fashion and other methods must be employed.

We first note that the order relation $<$ defined on $\mathbf{Q}$ has the property that, given $a, b \in \mathbf{Q}$ such that $a < b$, there exists $c \in \mathbf{Q}$ such that $a < c$ and $c < b$. Let $\mathcal{P}(\mathbf{Q})$ be the power set of $\mathbf{Q}$, i.e. $\mathcal{P}(\mathbf{Q})$ denotes the set whose elements are the subsets of $\mathbf{Q}$. Now consider ordered pairs of elements of $\mathcal{P}(\mathbf{Q})$, (A, B) say, satisfying:

(i) $A \cup B = \mathbf{Q}$, $A \cap B = \emptyset$,

(ii) A and B are both non-empty,

(iii) $a \in A$ and $b \in B$ together imply $a < b$.

Such a pair of sets (A, B) is known as a *Dedekind cut*. An equivalence relation R is defined upon the set of cuts by

$$(A, B)R(C, D)$$

if and only if there is at most one rational number which is either in both A and D or in both B and C. This ensures that the cuts $(\{x|x \leq q\}, \{x|x > q\})$ and $(\{x|x < q\}, \{x|x \geq q\})$ are equivalent for all $q \in \mathbf{Q}$. Each equivalence class under this relation is defined as a *real number*. The set of all real numbers, denoted by $\mathbf{R}$ is then the set of all such equivalence classes. If the class contains a cut (A, B) such that A contains positive rationals, then the class is a positive real number, whereas if B should contain negative rationals then the class is a negative real number. Thus, for example, $\sqrt{2}$ which contains the cut $(\{x|x^2 < 2\}, \{x|x^2 > 2\})$ is positive since $1 \in A = \{x|x^2 < 2\}$. To define addition of real numbers we must consider cuts (A_1, B_1) and (A_2, B_2) representing the real numbers α_1 and α_2. We define $\alpha_1 + \alpha_2$ to be the class containing the cut (A_3, B_3) where A_3 consists of all the sums $a = a_1 + a_2$ obtained by selecting a_1 from A_1 and a_2 from A_2. Given the real number α, represented by the cut (A_1, B_1), we define $-\alpha$, negative α, to be the class containing the cut $(-B_1, -A_1)$ defined by $a \in A_1 \Leftrightarrow -a \in -A_1$ and $b \in B_1 \Leftrightarrow -b \in -B_1$. It will be observed that $\alpha + (-\alpha) = 0$, and that subtraction can now be defined by $\alpha - \beta = \alpha + (-\beta)$. Of two non-zero numbers α and $-\alpha$, one is always positive. The one which is positive is known as the *absolute value* or *modulus* of α and is denoted by $|\alpha|$. Thus $|\alpha| = \alpha$ if α is positive and $|\alpha| = -\alpha$ if α is negative. $|0|$ is defined to be 0. If α_1 and α_2 are two positive real numbers, then the product $\alpha_1 * \alpha_2$ is the class containing the cut (A_4, B_4) where A_4 consists of the negative rationals, zero, and all the products $a = a_1 * a_2$ obtained by selecting a positive a_1 from A_1 and a positive a_2 from A_2. The definition is extended to negative numbers by agreeing that if α_1 and α_2 are positive, then

$$(-\alpha_1) * \alpha_2 = \alpha_1 * (-\alpha_2) = -\alpha_1 * \alpha_2, \qquad (-\alpha_1) * (-\alpha_2) = \alpha_1 * \alpha_2.$$

Finally, we define
$$0 * \alpha = \alpha * 0 = 0 \quad \text{for all } \alpha.$$
With these definitions it can be shown that the real numbers $\mathbf{R}$ form an *ordered field*.

By associating the element $q \in \mathbf{Q}$ with the class containing the cut

$$(\{x|x \leq q\}, \ \{x|x > q\})$$

one can define a monomorphism (of fields) $\mathbf{Q} \to \mathbf{R}$. We can, therefore, consider $\mathbf{Q}$ to be a subfield of $\mathbf{R}$ (i.e. identify $\mathbf{Q}$ with a subfield of $\mathbf{R}$). Those elements of $\mathbf{R}$ which do not belong to $\mathbf{Q}$ are known as *irrational numbers*.

An important property of $\mathbf{R}$ which can now be established is that given any non-empty subset $V \subset \mathbf{R}$ for which there exists an upper bound, M, i.e. an element $M \in \mathbf{R}$ such that $v \leq M$ for all $v \in V$, then there exists a *supremum (sup)* L such that if M is any upper bound of V, then $L \leq M$. In a similar manner, we can define an *infimum (inf)* for any non-empty subset V or $\mathbf{R}$ which possesses a lower bound.

Not every subset in $\mathbf{R}$ possesses an upper (or lower) bound in $\mathbf{R}$, for example, $\mathbf{N}$. In order to overcome certain consequences of this, one often makes use in analysis of the *extended real number system*, $\overline{\mathbf{R}}$, consisting of $\mathbf{R}$ together with the two symbols $-\infty$ and $+\infty$ having the properties:

(a) If $x \in \mathbf{R}$, then
$$-\infty < x < +\infty,$$
and
$$x + \infty = +\infty, \qquad x - \infty = -\infty, \qquad \frac{x}{+\infty} = \frac{x}{-\infty} = 0.$$

(b) If $x > 0$, then
$$x * (+\infty) = +\infty, \qquad x * (-\infty) = -\infty.$$

(c) If $x < 0$, then
$$x * (+\infty) = -\infty, \qquad x * (-\infty) = +\infty.$$

Note that $\overline{\mathbf{R}}$ does not possess all the algebraic properties of $\mathbf{R}$.

In the following, we list the basic properties of the system $\mathbf{R}$ of all real numbers.

Addition

A_1.	Closure Law	$r + s \in \mathbf{R}$, for all $r, s \in \mathbf{R}$.
A_2.	Commutative Law	$r + s = s + r$, for all $r, s \in \mathbf{R}$.
A_3.	Associative Law	$r + (s + t) = (r + s) + t$, for all $r, s, t \in \mathbf{R}$.
A_4.	Cancellation law	If $r + t = s + t$, then $r = s$ for all $r, s, t \in \mathbf{R}$.
A_5.	Additive Identity	There exists a unique additive identity element $0 \in \mathbf{R}$. such that $r + 0 = 0 + r = r$, for every $r \in \mathbf{R}$.
A_6.	Additive Inverses	For each $r \in \mathbf{R}$, there exists a unique additive inverse $-r \in \mathbf{R}$ such that $r + (-r) = (-r) + r = 0$.

Multiplication

M_1.	Closure Law	$r * s \in \mathbf{R}$, for all $r, s \in \mathbf{R}$.
M_2.	Commutative Law	$r * s = s * r$, for all $r, s \in \mathbf{R}$.
M_3.	Associative Law	$r * (s * t) = (r * s) * t$, for all $r, s, t \in \mathbf{R}$.
M_4.	Cancellation Law	If $m * p = n * p$, then $m = n$ for all $m, n \in \mathbf{R}$ and $p \neq 0 \in \mathbf{R}$.
M_5.	Multiplicative Identity	There exists a unique multiplicative identity element $1 \in \mathbf{R}$ such that $1 * r = r * 1 = r$ for every $r \in \mathbf{R}$.
M_6.	Multiplicative Inverses	For each $r \neq 0 \in \mathbf{R}$, there exists a unique multiplicative inverse $r^{-1} \in \mathbf{R}$ such that $r * r^{-1} = r^{-1} * r = 1$.

Distributive Laws D_1. D_2.	For every $r, s, t \in \mathbf{R}$, $r * (s + t) = r * s + r * t$ $(s + t) * r = s * r + t * r$
Density Property	For each $r, s \in \mathbf{R}$, with $r < s$, there exists $t \in \mathbf{Q}$ such that $r < t < s$.
Archimedean Property	For each $r, s \in \mathbf{R}^+$, with $r < s$, there exists $n \in \mathbf{N}$ such that $n * r > s$.
Completeness Property	Every non-empty subset of $\mathbf{R}$ having a lower bound (upper bound) has a greatest lower bound (least upper bound).

2.5   Complex Numbers

The field of *complex numbers*, $\mathbf{C}$, can be defined in several ways. Consider $\mathbf{R} \times \mathbf{R}$ and take the elements of $\mathbf{C}$ to be ordered pairs $(x, y) \in \mathbf{R} \times \mathbf{R}$. The operations of addition and multiplication of elements of $\mathbf{C}$ are defined by

$$(x_1, y_1) + (x_2, y_2) := (x_1 + x_2, y_1 + y_2)$$

and

$$(x_1, y_1) * (x_2, y_2) := (x_1 x_2 - y_1 y_2, x_1 y_2 + x_2 y_1).$$

Then we can show that $\mathbf{C}$ is a field. We can define a monomorphism (of fields),

$$r : \mathbf{R} \to \mathbf{C}, \quad \text{by} \quad r(x) = (x, 0).$$

This enables us to regard $\mathbf{R}$ as a subfield of $\mathbf{C}$. It can, moreover, be checked that

$$(x, y) = (x, 0) + (0, 1) * (y, 0).$$

Thus making use of the monomorphism defined above, one can write

$$(x, y) = x + i * y$$

where $x, y \in \mathbf{R}$ and $i = (0, 1)$. It is seen that

$$i^2 = (0, 1) * (0, 1) = (-1, 0) = -1.$$

Given a complex number

$$z = x + i * y, \qquad \text{where } x, y \in \mathbf{R},$$

we say that x is the *real part*, $\Re(z)$, and y is the *imaginary part*, $\Im(z)$, of z. The number

$$x - i * y$$

is known as the *complex conjugate* of z and is denoted by $\bar{z}$ (or z^*). An obvious geometrical representation of the complex numbers are points in the Cartesian plane $\mathbf{R} \times \mathbf{R}$. This representation is known as an *Argand diagram*. The diagram is based on a pair of perpendicular coordinate axes in the plane. The number $z = x + i * y$ is associated with the point with coordinates (x, y). With this representation, the addition of complex numbers is interpreted as the addition of vectors in the plane. The length, r, of the segment

$$(0, 0) - (x, y)$$

is known as the *absolute value* or *modulus* of z and is denoted by $|z|$. We therefore have

$$|z| = r = \sqrt{x^2 + y^2} = \sqrt{z\bar{z}}.$$

The angle which the segment Oz makes with the Ox axis is known as the *argument* (amplitude or angle) of z and is denoted by arg z. We therefore have

$$\tan(\arg z) = \tan \theta = \frac{y}{x}$$

and arg z is defined as a real number modulo 2π (provided $z \neq 0$, for arg 0 is not defined). Some authors take

$$0 \leq \arg z < 2\pi$$

while others opt for

$$-\pi < \arg z \leq \pi.$$

This restricted value of the argument is often known as the *principal argument*.

The coordinates r and θ are known as *polar coordinates*. The connections between the polar coordinates of a point and the complex number $z = x + i * y$ which it represents are

$$x = r \cos \theta, \qquad y = r \sin \theta$$

$$r = |z| = \sqrt{x^2 + y^2}$$
$$\theta = \arg z$$
$$z = r(\cos \theta + i \sin \theta).$$

The geometry of the triangle provides the inequality

$$|z_1 + z_2| \leq |z_1| + |z_2|$$

known as the *triangle inequality*. The two formulae

$$|z_1 z_2| = |z_1||z_2|$$

and

$$\arg(z_1 z_2) \equiv \arg(z_1) + \arg(z_2) \pmod{2\pi}$$

allow one to give a geometrical interpretation of the multiplication of complex numbers.

Alternative constructions of **C** are:

(a) Consider the set, M, containing all matrices of $M_2(\mathbf{R})$ of the form

$$\begin{pmatrix} a & -b \\ b & a \end{pmatrix}.$$

Under matrix addition and multiplication M forms a field with zero O_2 and identity element I_2. We have

$$\begin{pmatrix} a & -b \\ b & a \end{pmatrix} \begin{pmatrix} c & -d \\ d & c \end{pmatrix} = \begin{pmatrix} ac - bd & -(ad + bc) \\ ad + bc & ac - bd \end{pmatrix}.$$

Moreover, $p : \mathbf{R} \to M$ defined by

$$p(c) = \begin{pmatrix} c & 0 \\ 0 & c \end{pmatrix}$$

is a field isomorphism between $\mathbf{R}$ and a subfield of M. Mapping the matrix

$$\begin{pmatrix} a & -b \\ b & a \end{pmatrix}$$

onto $a + i * b$, we obtain a field isomorphism between M and $\mathbf{C}$.

(b) We define $\mathbf{C}$ to be the quotient ring of $\mathbf{R}[t]$ by the principal ideal $(t^2 + 1)$, i.e.

$$\mathbf{C} = \mathbf{R}[t]/(t^2 + 1).$$

If we denote the image of the polynomial $t \in \mathbf{R}[t]$ under the canonical mapping by i, then every element of $\mathbf{C}$ can be written uniquely as $x + iy$ where $x, y \in \mathbf{R}$. The operations of addition and multiplication on $\mathbf{C}$ are the natural operations of quotient algebra.

To facilitate the study of certain curves (e.g. the equiangular spiral) one frequently relaxes the conditions $r \geq 0$ and $0 \leq \theta < 2\pi$ on polar coordinates. One then has an extended system of polar coordinates in which r and θ can take all real values. In the extended system any pair (ρ, ω) will determine a unique point of the plane, yet every point in the plane will possess an infinite number of polar coordinates, namely

$$(\rho, \omega + 2n\pi), \quad (-\rho, \omega + (2n + 1)\pi)$$

for all $n \in \mathbf{Z}$. In the extended system the equiangular spiral is described by the single equation $r = e^{a\theta}$, whereas in the restricted system a whole set of equations would be required to define it.

2.6 Vectors and Matrices

Let F be a field. A *vector space* (also called a *linear space*), V, over F is defined as an additive Abelian group V together with a function $F \times V \to V$, $(\lambda, \mathbf{v}) \mapsto \lambda * \mathbf{v}$, satisfying

$$
\begin{aligned}
\lambda * (\mathbf{a} + \mathbf{b}) &= \lambda * \mathbf{a} + \lambda * \mathbf{b} \\
(\lambda + \mu) * \mathbf{a} &= \lambda * \mathbf{a} + \mu * \mathbf{a} \\
(\lambda * \mu) * \mathbf{a} &= \lambda * (\mu * \mathbf{a}) \\
1 * \mathbf{a} &= \mathbf{a}
\end{aligned}
$$

for all $\lambda, \mu \in F$ and $\mathbf{a}, \mathbf{b} \in V$. F is called the ground field of the vector space, its elements are called *scalars* and those of V are called *vectors*. Letters representing vectors are printed in bold type. A *vector subspace* of a vector space V is defined as any subset V' of V for which

(a) V' is a subgroup of the additive group V;
(b) for all $\mathbf{a} \in V'$ and for all $\lambda \in F$, $\lambda * \mathbf{a} \in V'$.

The set of all n-tuples $(x_1, \ldots, x_n)$ where $x_i \in \mathbf{R}$, $i = 1, \ldots, n$, forms a vector space, which we denote by $\mathbf{R}^n$, over the ground field $\mathbf{R}$ when we define

$$(x_1, \ldots, x_n) + (y_1, \ldots, y_n) = (x_1 + y_1, \ldots, x_n + y_n)$$

and

$$\lambda * (x_1, \ldots, x_n) = (\lambda * x_1, \ldots, \lambda * x_n)$$

where $\lambda \in \mathbf{R}$.

A vector space V over a field F which satisfies the ring axioms in such a way that addition in the ring is addition in the vector space and such that

$$\lambda * (\mathbf{v}_1 * \mathbf{v}_2) = (\lambda * \mathbf{v}_1) * \mathbf{v}_2 = \mathbf{v}_1 * (\lambda * \mathbf{v}_2)$$

for all $\mathbf{v}_1, \mathbf{v}_2 \in V$ and $\lambda \in F$ is said to be an *algebra* over F. If, in addition, it forms a commutative ring, then we say it is a *commutative algebra*. A subset of an algebra V is termed a *sub-algebra* if it is both a vector subspace and a subring of V.

Given two vectors $\mathbf{x} = (x_1, x_2, x_3)$, $\mathbf{y} = (y_1, y_2, y_3) \in \mathbf{R}^3$ we define their *vector product* (also called *cross product*) denoted by $\mathbf{x} \times \mathbf{y}$ to be the vector

$$\mathbf{x} \times \mathbf{y} := (x_2 * y_3 - x_3 * y_2, x_3 * y_1 - x_1 * y_3, x_1 * y_2 - x_2 * y_1).$$

The vector product is not commutative. It is an example of a *Lie algebra*. We have

$$\mathbf{x} \times (\mathbf{y} + \mathbf{z}) = \mathbf{x} \times \mathbf{y} + \mathbf{x} \times \mathbf{z}$$

$$\mathbf{x} \times \mathbf{y} = -\mathbf{y} \times \mathbf{x}$$

$$\mathbf{x} \times (\lambda * \mathbf{y}) = \lambda * (\mathbf{x} \times \mathbf{y})$$

$$\mathbf{x} \times (\mathbf{y} \times \mathbf{z}) + \mathbf{z} \times (\mathbf{x} \times \mathbf{y}) + \mathbf{y} \times (\mathbf{z} \times \mathbf{x}) = \mathbf{0}.$$

The last equation is the *Jacobi identity*.

Given two vector spaces U and V over a field F, a *homomorphism* of U to V is a function $t : U \to V$ satisfying

$$
\begin{aligned}
t(\mathbf{a} + \mathbf{b}) &= t(\mathbf{a}) + t(\mathbf{b}) \\
t(\lambda * \mathbf{a}) &= \lambda * t(\mathbf{a})
\end{aligned}
$$

for all $\mathbf{a}, \mathbf{b} \in U$ and $\lambda \in F$. Homomorphisms of vector spaces therefore preserve linear combinations of the type

$$\lambda_1 * \mathbf{a}_1 + \lambda_2 * \mathbf{a}_2 + \ldots + \lambda_n * \mathbf{a}_n$$

where $\lambda_1, \ldots, \lambda_n \in F$, $\mathbf{a}_1, \ldots, \mathbf{a}_n \in U$. For this reason a homomorphism of vector spaces is called a *linear transformation* or *linear mapping*. The set of all linear combinations of $a_1, \ldots, a_n \in U$ forms a vector subspace of U, called the subspace generated or spanned by $\mathbf{a}_1, \ldots, \mathbf{a}_n$. A vector space, V, is said to be finitely generated if there exists a finite set of elements $\mathbf{a}_1, \ldots, \mathbf{a}_n$ which generate V.

The vectors $\mathbf{a}_1, \ldots, \mathbf{a}_n$ are said to be *linearly independent* if the only choice of $\lambda_1, \ldots, \lambda_n$ satisfying the relation

$$\lambda_1 * \mathbf{a}_1 + \lambda_2 * \mathbf{a}_2 + \ldots + \lambda_n * \mathbf{a}_n = 0, \qquad 0 \in V$$

is

$$\lambda_1 = \lambda_2 = \ldots = \lambda_n = 0, \qquad 0 \in F.$$

The vectors $\mathbf{a}_1, \ldots, \mathbf{a}_n$ are linearly dependent if and only if there exist $\lambda_1, \ldots, \lambda_n$ not all zero, for which

$$\lambda_1 * \mathbf{a}_1 + \lambda_2 * \mathbf{a}_2 + \ldots + \lambda_n * \mathbf{a}_n = \mathbf{0}.$$

If $\mathbf{a}_1, \ldots, \mathbf{a}_n$ generate a vector space, V, and are linearly independent, then we say that $\mathbf{a}_1, \ldots, \mathbf{a}_n$ form a *basis* of V. If $\mathbf{a}_1, \ldots, \mathbf{a}_n$ form a basis of V and $\mathbf{b}$ is any vector of V, then there exists a unique n-tuple $(\lambda_1, \ldots, \lambda_n)$ such that

$$\mathbf{b} = \lambda_1 * \mathbf{a}_1 + \lambda_2 * \mathbf{a}_2 + \ldots + \lambda_n * \mathbf{a}_n.$$

The scalars $\lambda_1, \ldots, \lambda_n$ are then called the coordinates or components of $\mathbf{b}$ with respect to the basis $\mathbf{a}_1, \ldots, \mathbf{a}_n$. It can be shown that every finitely-generated vector space,

V, has a basis and that, in particular, any two bases of V contain the same number of elements, say n. The number n is called the *dimension* of the vector space V over F and we write $\dim V = n$. The vector space V is then isomorphic to F^n, the vector space of all n-tuples $(x_1, \ldots, x_n)$ with $x_i \in F$, $i = 1, \ldots, n$. By definition, $\dim\{0\} = 0$.

If U and V are finite-dimensional vector spaces over the same field F, and if 'addition' of linear transformations and 'multiplication of linear transformations by a scalar' are defined by

$$(t_1 + t_2)(\mathbf{a}) = t_1(\mathbf{a}) + t_2(\mathbf{a})$$
$$(\lambda * t)(\mathbf{a}) = \lambda * (t(\mathbf{a})), \quad \text{for all } \mathbf{a} \in U$$

then it can easily be shown that the set of linear transformations itself forms a finite-dimensional vector space over F (having dimension $(\dim U) \times (\dim V)$). This vector space is denoted by $\mathrm{Hom}(U, V)$. Let $t \in \mathrm{Hom}(U, V)$, $\mathbf{u}_1, \ldots, \mathbf{u}_n$ be a basis for U and $\mathbf{v}_1, \ldots, \mathbf{v}_m$ be a basis for V. Then t is completely determined by the formulae which tell how the components $(x'_1, \ldots, x'_m)$ of the vector $t(\mathbf{x})$, with respect to the basis $\mathbf{v}_1, \ldots, \mathbf{v}_m$ of V can be obtained in terms of $(x_1, \ldots, x_n)$, the components of the vector $\mathbf{x}$ with respect to the basis $\mathbf{u}_1, \ldots, \mathbf{u}_n$ of U. We have

$$\begin{aligned}
x'_1 &= a_{11}x_1 + a_{12}x_2 + \ldots + a_{1n}x_n \\
x'_2 &= a_{21}x_1 + a_{22}x_2 + \ldots + a_{2n}x_n \\
&\ \ \vdots \qquad\qquad \vdots \\
x'_m &= a_{m1}x_1 + a_{m2}x_2 + \ldots + a_{mn}x_n, \qquad a_{ij} \in F
\end{aligned}$$

and the coefficients a_{ij} determine the homomorphism t uniquely with respect to the chosen bases. The rectangular array of coefficients

$$\begin{pmatrix}
a_{11} & a_{12} & \cdots & a_{1n} \\
a_{21} & a_{22} & \cdots & a_{2n} \\
\vdots & \vdots & & \vdots \\
a_{m1} & a_{m2} & \cdots & a_{mn}
\end{pmatrix}$$

is said to form a *matrix*, A, having m *rows* and n *columns*. More abstractly we can think of the matrix as a function

$$A : \{1, 2, \ldots, m\} \times \{1, 2, \ldots, n\} \to F.$$

The matrix A is often abbreviated to (a_{ij}), and to denote that this represents the linear transformation t we write $(t) = (a_{ij})$. The matrix corresponding to the zero mapping, $t : \mathbf{x} \mapsto 0$ (all $\mathbf{x} \in U$), is called the *zero matrix* and is denoted by 0.

The sum of two $m \times n$ matrices, $A = (a_{ij})$ and $B = (b_{ij})$ is defined as the $m \times n$ matrix

$$A + B = (a_{ij} + b_{ij}).$$

The product of a $m \times n$ matrix $A = (a_{ij})$ with the scalar λ is defined to be the $m \times n$ matrix $\lambda * A = (\lambda * a_{ij})$. With these definitions the set of all $m \times n$ matrices with coefficients in F becomes a vector space of dimension $m * n$ and is isomorphic to $\text{Hom}(U, V)$. We define the product matrix, $C = A * B$, of two matrices A $(a_{ij}$ an $m \times n$ matrix), and B $(b_{jk}$ an $n \times p$ matrix), as the $m \times p$ matrix (c_{ik}) where

$$c_{ik} = \sum_{j=1}^{n} a_{ij} b_{jk}.$$

This definition is motivated by the need to form the composite linear transformation $s \circ t \in \text{Hom}(U, W)$, given $t \in \text{Hom}(U, V)$ corresponding to the matrix B, and $s \in \text{Hom}(V, W)$ corresponding to the matrix A. The matrix product $A * B$ of the matrices A and B is not defined unless the number of columns of A is equal to the number of rows of B (is equal to the dimension of V). The existence of the product $A * B$ will not therefore imply the existence of the product $B * A$. With this definition of multiplication, the set of square matrices of order n, i.e. those matrices having n rows and n columns with coefficients in a ring K, form a non-commutative ring (provided $n > 1$ and $K \neq \{0\}$) denoted by $M_n(K)$. A matrix $A \in M_n(K)$ is said to be *non-singular* or *invertible* if there exists $B \in M_n(K)$ such that

$$A * B = B * A = I_n$$

where I_n is the *identity matrix* (also called the *unit matrix*) of order n having coefficients δ_{ij} (called the *Kronecker delta*), where

$$\delta_{ij} := \begin{cases} 1 & \text{if } \; i = j \\ 0 & \text{otherwise.} \end{cases}$$

The matrix B is then unique and is known as the *inverse* of A; it is denoted by A^{-1}. If there is no matrix B in $M_n(K)$ such that $A * B = B * A = I$, then A is said to be *singular* or *non-invertible* in $M_n(K)$. A linear mapping $U \to V$ will be an isomorphism of vector spaces if and only if it can be represented by an invertible (square) matrix. The *transpose* of $m \times n$ matrix $A = (a_{ij})$ is defined as the $n \times m$ matrix (a_{ij}) obtained from A by interchanging rows and columns. The transpose of A is denoted by A^T. When $A = A^T$ the matrix A is said to be *symmetric*.

For each $n > 0$ the set of non-singular $n \times n$ matrices over the field F forms a multiplicative group, called the *general linear group* GL(n, F). The elements of GL(n, F) having determinant 1 form a subgroup denoted by SL(n, F) and known as the *special linear group*. An $m \times 1$ matrix having only one column is known as a column matrix or column vector, a $1 \times n$ matrix is known as a row matrix or row vector. The equations which determine the homomorphism t with matrix A can therefore be written as

$$\mathbf{x}' = A * \mathbf{x}$$

where $\mathbf{x}'$ is a column vector whose m elements are the components of $t(\mathbf{x})$ with respect to $\mathbf{v}_1, \ldots, \mathbf{v}_m$ and $\mathbf{x}$ is a column vector having n elements, the components of $\mathbf{x}$ with respect to $\mathbf{u}_1, \ldots, \mathbf{u}_n$.

In particular, a $1 \times n$ matrix will describe a homomorphism from an n-dimensional vector space $\mathbf{u}$ with basis $\mathbf{u}_1, \ldots, \mathbf{u}_n$ to a one-dimensional vector space V with basis $\mathbf{v}_1$. The set of all $1 \times n$ matrices will form a vector space isomorphic to the vector space Hom(U, F) of all homomorphisms mapping U onto its ground field F, which can be regarded as a vector space of dimension 1 over itself. In general, if V is any vector space over a field F, then the vector space Hom(V, F) is called the *dual space* of V and is denoted by V^* or $\hat{V}$. Elements of Hom(V, F) are known as *linear functionals*. It can be shown that every finite-dimensional vector space is isomorphic to its dual.

Example. The *trace* of a square matrix is the sum of the diagonal elements. Thus the trace of a square matrix is a linear functional. This means

$$\begin{aligned} \text{tr}(A + B) &= \text{tr}(A) + \text{tr}(B) \\ \text{tr}(c * A) &= c * \text{tr}(A). \end{aligned}$$

Given two subspaces S and T of a vector space V, we define

$$S + T := \{ x + y \mid x \in S, y \in T \}.$$

Then $S + T$ is a subspace of V. We say that V is the *direct sum* of S and T, written $S \oplus T$, if and only if $V = S + T$ and $S \cap T = \{0\}$. S and T are then called direct summands of V. Any subspace S of a finite-dimensional vector space V is a direct summand of V. Moreover, if $V = S \oplus T$, then

$$\dim T = \dim V - \dim S.$$

If S and T are any finite-dimensional subspaces of a vector space V, then

$$\dim S + \dim T = \dim(S \cap T) + \dim(S + T).$$

2.7   Quaternions

By defining multiplication suitably on $\mathbf{R} \times \mathbf{R}$, it is possible to construct a field $\mathbf{C}$ which is an extension of $\mathbf{R}$. Indeed, since $\mathbf{C}$ is a vector space of dimension 2 over $\mathbf{R}$, $\mathbf{C}$ is a commutative algebra with unity element over $\mathbf{R}$. It is natural to attempt to repeat this process and to try to embed $\mathbf{C}$ in an algebra defined upon $\mathbf{R}^n$ $(n > 2)$. It is, in fact, impossible to find such an extention satisfying the field axioms, but, as the following construction shows, some measure of success can be attained.

Consider an associative algebra of rank 4 with the basis elements

$$1, \quad I, \quad J, \quad K$$

where 1 is the identity element, i.e.

$$1 * I = I, \qquad 1 * J = J, \qquad 1 * K = K.$$

The compositions are

$$I * I = J * J = K * K = -1$$

and

$$I * J = K, \quad J * K = I, \quad K * I = J, \quad J * I = -K, \quad K * J = -I, \quad I * K = -J.$$

This is the so-called *quaternion algebra*. Multiplication, as thus defined, is clearly non-commutative, so the resulting structure cannot be a field. It is a division ring and is known as the quaternion algebra. The algebra is associative.

Any quaternion q can be represented in the form

$$q := a_1 * 1 + a_I * I + a_J * J + a_K * K, \qquad \text{where} \quad a_1, a_I, a_J, a_K \in F.$$

The sum, difference, product and division of two quaternions

$$q := a_1 * 1 + a_I * I + a_J * J + a_K * K, \qquad p := b_1 * 1 + b_I * I + b_J * J + b_K * K$$

are defined as

$$
\begin{aligned}
q + p \; &:= \; (a_1 + b_1) * 1 + (a_I + b_I) * I + (a_J + b_J) * J + (a_K + b_K) * K \\
q - p \; &:= \; (a_1 - b_1) * 1 + (a_I - b_I) * I + (a_J - b_J) * J + (a_K - b_K) * K \\
q * p \; &:= \; (a_1 * 1 + a_I * I + a_J * J + a_K * K) * (b_1 * 1 + b_I * I + b_J * J + b_K * K) \\
&= \; (a_1 * b_1 - a_I * b_I - a_J * b_J - a_K * b_K) * 1 \\
&\quad + (a_1 * b_I + a_I * b_1 + a_J * b_K - a_K * b_J) * I \\
&\quad + (a_1 * b_J + a_J * b_1 + a_K * b_I - a_I * b_K) * J \\
&\quad + (a_1 * b_K + a_K * b_1 + a_I * b_J - a_J * b_I) * K \\
q/p \; &:= \; q * p^{-1} \text{ where } p^{-1} \text{ is the inverse of } p.
\end{aligned}
$$

The negate of q is

$$-q = -a_1 * 1 - a_I * I - a_J * J - a_K * K.$$

The conjugate of q, say q^*, is defined as

$$q^* := a_1 * 1 - a_I * I - a_J * J - a_K * K.$$

The inverse of q is

$$q^{-1} := \frac{q^*}{|q|^2}, \qquad q \neq 0.$$

The magnitude of q is

$$|q|^2 = a_1^2 + a_I^2 + a_J^2 + a_K^2.$$

The normalization of q is defined as

$$q/|q|.$$

A *matrix representation* of the quaternions is given by

$$1 \to \begin{pmatrix} 1 & 0 \\ 0 & 1 \end{pmatrix}$$

$$I \to -i \begin{pmatrix} 0 & 1 \\ 1 & 0 \end{pmatrix} \equiv -i\sigma_x$$

$$J \to -i \begin{pmatrix} 0 & -i \\ i & 0 \end{pmatrix} \equiv -i\sigma_y$$

$$K \to -i \begin{pmatrix} 1 & 0 \\ 0 & -1 \end{pmatrix} \equiv i\sigma_z$$

where $i := \sqrt{-1}$ and σ_x, σ_y, σ_z are the *Pauli spin matrices*. This also shows that the quaternion algebra is associative.

The quaternion algebra can also be obtained as the subring of $M_4(\mathbf{R})$ consisting of matrices of the form

$$\begin{pmatrix} x & -y & -z & -t \\ y & x & -t & z \\ z & t & x & -y \\ t & -z & y & x \end{pmatrix}.$$

2.8   Polynomials

In this section we introduce polynomials. For the proof of the theorems we refer to the literature [4], [22], [30]. Functions of the form

$$1 + 2 * x + 3 * x^2, \qquad x + x^5, \qquad \frac{1}{5} - 4 * x^2 + \frac{3}{2} * x^{10}$$

are called *polynomials* in x. The coefficients in these examples are integers and rational numbers. In elementary calculus, the range of values of x (domain of definition of the function) is $\mathbf{R}$. In algebra, the range is $\mathbf{C}$. Consider, for instance, the polynomial $p(x) = x^2 + 1$. The solution of $p(x) = 0$ is given by $\pm i$. Any polynomial in x can be thought of as a mapping of a set S (range of x) onto a set T (range of values of the polynomial). Consider, for example, the polynomial $1 + \sqrt{2} * x - 3 * x^2$. If $S = \mathbf{Z}$, then $T \subset \mathbf{R}$ and the same is true if $S = \mathbf{Q}$ or $S = \mathbf{R}$; if $S = \mathbf{C}$, then $T \subset \mathbf{C}$. Two polynomials in x are equal if they have identical form. For example, $a + b * x = c + d * x$ if and only if $a = c$ and $b = d$.

Let R be a ring and let x, called an *indeterminate*, be any symbol not found in R. By a polynomial in x over R will be meant any expression of the form

$$\alpha(x) = a_0 * x^0 + a_1 * x^1 + a_2 * x^2 + \cdots = \sum_{k=0} a_k * x^k, \qquad a_k \in R$$

in which only a finite number of the a_k's are different from z, the zero element of R. Two polynomials in x over R, $\alpha(x)$ defined above, and

$$\beta(x) = b_0 * x^0 + b_1 * x^1 + b_2 * x^2 + \cdots = \sum_{k=0} b_k * x^k, \qquad b_k \in R$$

are equal $\alpha(x) = \beta(x)$, provided $a_k = b_k$ for all values of k.

In any polynomial, as $\alpha(x)$, each of the components $a_0 * x^0$, $a_1 * x^1$, $a_2 * x^2, \ldots$ is called a *term*; in any term such as $a_i * x^i$, a_i is called the *coefficient* of the term. The terms of $\alpha(x)$ and $\beta(x)$ have been written in a prescribed (but natural) order. The i, the superscript of x, is merely an indicator of the position of the term $a_i * x^i$ in the polynomial. Likewise, juxtaposition of a_i and x^i in the term $a_i * x^i$ is not to be construed as indicating multiplication and the plus signs between terms are to be thought of as helpful connectives rather than operators. Let z be the zero element of the ring. If in a polynomial such as $\alpha(x)$, the coefficient $a_n \neq z$ while all coefficients of terms which follow are z, we say that $\alpha(x)$ is of *degree* n and call a_n its leading coefficient. In particular, the polynomial $a_0 * x^0 + z * x^1 + z * x^2 + \cdots$ is of degree zero with leading coefficient a_0 when $a_0 \neq z$ and it has no degree (and no leading coefficient) when $a_0 = z$.

Denote by $R[x]$ the set of all polynomials in x over R and, for arbitrary $\alpha(x), \beta(x) \in R[x]$, define addition $(+)$ and multiplication $*$ on $R[x]$ by

$$\alpha(x) + \beta(x) := (a_0 + b_0) * x^0 + (a_1 + b_1) * x^1 + (a_2 + b_2) * x^2 + \cdots = \sum_{k=0} (a_k + b_k) * x^k$$

and

$$\alpha(x) * \beta(x) := a_0 * b_0 * x^0 + (a_0 * b_1 + a_1 * b_0) * x^1 + (a_0 * b_2 + a_1 * b_1 + a_2 * b_0) * x^2 + \cdots = \sum_{k=0} c_k * x^k$$

where

$$c_k := \sum_{i=0}^{k} a_i * b_{k-i}.$$

The sum and product of elements of $R[x]$ are elements of $R[x]$; there are only a finite number of terms with non-zero coefficients $\in R$. Addition on $R[x]$ is both associative and commutative and multiplication is associative and distributive with respect to addition. Moreover, the *zero polynomial*

$$z * x^0 + z * x^1 + z * x^2 + \cdots = \sum_{k=0} z * x^k \in R[x]$$

is the additive identity or zero element of $R[x]$ while

$$-\alpha(x) = -a_0 * x^0 + (-a_1) * x^1 + (-a_2) * x^2 + \cdots = \sum_{k=0} (-a_k) * x^k \in R[x]$$

is the additive inverse of $\alpha(x)$. Thus,

Theorem. The set of all polynomials R in x over R is a ring with respect to addition and multiplication as defined above.

Let $\alpha(x)$ and $\beta(x)$ have respective degrees m and n. If $m \neq n$, the degree of $\alpha(x) + \beta(x)$ is the larger of m, n; if $m = n$, the degree of $\alpha(x) + \beta(x)$ is at most m. The degree of $\alpha(x) * \beta(x)$ is at most $m + n$ since $a_m b_n$ may be z. However, if R is free of divisors of zero, the degree of the product is $m + n$.

Consider now the subset $S := \{ r * x^0 : r \in R \}$ of $R[x]$ consisting of the zero polynomial and all polynomials of degree zero. The mapping

$$R \to S : r \to r * x^0$$

is an isomorphism. As a consequence, we may hereafter write a_0 for $a_0 * x^0$ in any polynomial $\alpha(x) \in R[x]$.

Let R be a ring with unity u. Then $u = u * x^0$ is the unity of $R[x]$ since $ux^0 * \alpha(x) = \alpha(x)$ for every $\alpha(x) \in R[x]$. Also, writing $x = u*x^1 = z*x^0 + u*x^1$, we have $x \in R[x]$. Now $a_k(x \cdot x \cdot x \cdot$ to k factors$) = a_k * x^k \in R[x]$ so that in $\alpha(x) = a_0 + a_1 * x + a_2 * x^2 + \cdots$ we may consider the superscript i and $a_i x^i$ as truly an exponent, juxtaposition in any term $a_i * x^i$ as (polynomial) ring multiplication, and the connective $+$ as (polynomial) ring addition. Any polynomial $\alpha(x)$ of degree m over R with leading coefficient u, the unity of R, will be called monic.

Theorem. Let R be a ring with unity u, $\alpha(x) = a_0 + a_1 * x + \cdots + a_m * x^m \in R[x]$ be either the zero polynomial or a polynomial of degree m, and $\beta(x) = b_0 + b_1 * x + \cdots + u * x^n \in R[x]$ be a monic polynomial of degree n. Then there exist unique polynomials $q_R(x)$, $r_R(x)$, $q_L(x)$, $r_L(x) \in R[x]$ with $r_R(x), r_L(x)$ either the zero polynomial or of degree $< n$ such that

$$\text{(i)}\quad \alpha(x) = q_R(x) * \beta(x) + r_R(x)$$

and

$$\text{(ii)}\quad \alpha(x) = \beta(x) * q_L(x) + r_L(x).$$

In (i) of the theorem we say that $\alpha(x)$ has been divided on the right by $\beta(x)$ to obtain the right quotient $q_R(x)$ and right remainder $r_R(x)$. Similarly, in (ii) we say that $\alpha(x)$ has been divided on the left by $\beta(x)$ to obtain the left quotient $q_L(x)$ and left remainder $r_L(x)$. When $r_R(x) = z$ $(r_L(x) = z)$, we call $\beta(x)$ a right (left) divisor of $\alpha(x)$.

We consider now commutative polynomial rings with unity. Let R be a commutative ring with unity. Then $R[x]$ is a commutative ring with unity and the theorem may be restated without distinction between right and left quotients (we replace $q_R(x) = q_L(x)$ by $q(x)$), remainders (we replace $r_R(x) = r_L(x)$ by $r(x)$), and divisors. Thus (i) and (ii) of the theorem may be replaced by

$$\text{(iii)}\quad \alpha(x) = q(x) * \beta(x) + r(x)$$

and, in particular, we have

Theorem. In a commutative polynomial ring with unity, a polynomial $\alpha(x)$ of degree m has $x - b$ as divisor if and only if the remainder

$$r = a_0 + a_1 * b + a_2 * b^2 + \cdots + a_m * b^m = z.$$

When $r = z$ then b is called a zero (root) of the polynomial $\alpha(x)$.

When R is without divisors of zero so is $R[x]$. For suppose $\alpha(x)$ and $\beta(x)$ are elements of $R[x]$, of respective degrees m and n, and that

$$\alpha(x) * \beta(x) = a_0 * b_0 + (a_0 * b_1 + a_1 * b_0)x + \cdots + a_m * b_n x^{m+n} = z.$$

Then each coefficient in the product and, in particular $a_m * b_n$ is z. But R is without divisors of zero; hence $a_m b_n = z$ if and only if $a_m = z$ or $b_n = z$. Since this contradicts the assumption that $\alpha(x)$ and $\beta(x)$ have degrees m and n, $R[x]$ is without divisors of zero.

Theorem. A polynomial ring $R[x]$ is an integral domain if and only if the coefficient ring R is an integral domain.

An examination of the remainder

$$r = a_0 + a_1 * b + a_2 * b^2 + \cdots + a_m * b^m$$

shows that it may be obtained mechanically by replacing x by b throughout $\alpha(x)$ and, of course, interpreting juxtaposition of elements as indicating multiplication in R. Thus, by defining $f(b)$ to mean the expression obtained by substituting b for x throughout $f(x)$, we may replace r by $\alpha(b)$. This is the familiar substitution process in elementary algebra where x is considered as a variable rather than an indeterminate. For a given $b \in R$, the mapping

$$f(x) \rightarrow f(b) \qquad \text{for all } f(x) \in R[x]$$

is a homomorphism of $R[x]$ onto R.

The most important polynomial domains arise when the coefficient ring is a field F. Every non-zero element of a field F is a unit of F. For the integral domain $F[x]$ the principal results are as follows:

Division Algorithm. If $\alpha(x)$, $\beta(x) \in F[x]$ where $\beta(x) \neq z$, there exist unique polynomials $q(x)$, $r(x)$ with $r(x)$ either the zero polynomial or of degree less than that of $\beta(x)$, such that

$$\alpha(x) = q(x) * \beta(x) + r(x).$$

When $r(x)$ is the zero polynomial, $\beta(x)$ is called a divisor of $\alpha(x)$ and we write $\beta(x)|\alpha(x)$.

Remainder Theorem. If $\alpha(x)$, $x - b \in F[x]$, the remainder when $\alpha(x)$ is divided by $x - b$ is $\alpha(b)$.

Factor Theorem. If $\alpha(x) \in F[x]$ and $b \in F$, then $x - b$ is a *factor* of $\alpha(x)$ if and only if $\alpha(b) = z$, that is, $x - b$ is a factor of $\alpha(x)$ if and only if b is a zero of $\alpha(x)$. This leads to the following theorem.

Theorem. Let $\alpha(x) \in F[x]$ have degree $m > 0$ and leading coefficient a. If the distinct elements $b_1, b_2, \ldots, b_m$ of F are zeros of $\alpha(x)$, then

$$\alpha(x) = a * (x - b_1) * (x - b_2) * \cdots * (x - b_m).$$

Theorem. Every polynomial $\alpha(x) \in F[x]$ of degree $m > 0$ has at most m distinct zeros in F.

Theorem. Let $\alpha(x)$, $\beta(x) \in F[x]$ be such that $\alpha(s) = \beta(s)$ for every $s \in F$. Then, if the number of elements in F exceeds the degrees of both $\alpha(x)$ and $\beta(x)$, we have necessarily $\alpha(x) = \beta(x)$.

The only units of a polynomial domain $F[x]$ are the non-zero elements (i.e., the units) of the coefficient ring F. Thus the only associates of $\alpha(x) \in F[x]$ are the elements $v * \alpha(x)$ of $F[x]$ in which v is any unit of F. Since for any $v \neq z \in F$ and any $\alpha(x) \in F[x]$,

$$\alpha(x) = v^{-1} * \alpha(x) * v$$

while, whenever $\alpha(x) = q(x) * \beta(x)$,

$$\alpha(x) = \left(v^{-1} * q(x)\right) * (v * \beta(x))$$

it follows that (a) every unit of F and every associate of $\alpha(x)$ is a divisor of $\alpha(x)$ and (b) if $\beta(x)|\alpha(x)$ so also does every associate of $\beta(x)$. The units of F and the associates of $\alpha(x)$ are called trivial divisors of $\alpha(x)$. Other divisors of $\alpha(x)$, if any, are called non-trivial divisors. A polynomial $\alpha(x) \in F[x]$ of degree $m \geq 1$ is called a prime (irreducible) polynomial over F if its divisors are all trivial.

Next we consider the polynomial domain $C[x]$. Consider an arbitrary polynomial

$$\beta(x) = b_0 + b_1 * x + b_2 * x^2 + \cdots + b_m * x^m \in C[x]$$

of degree $m \geq 1$. We give a number of elementary theorems related to the zeros of such polynomials and, in particular, with the subset of all polynomials of $C[x]$ whose coefficients are rational numbers. Suppose $r \in C$ is a zero of $\beta(x)$, i.e., $\beta(r) = 0$ and, since $b_m^{-1} \in C$, also $b_m^{-1} * \beta(r) = 0$. Thus the zeros of $\beta(x)$ are precisely those of its monic associates

$$\alpha(x) = b_m^{-1} * \beta(x) = a_0 + a_1 * x + a_2 * x^2 + \cdots + a_{m-1} * x^{m-1} + x^m.$$

When $m = 1$,

$$\alpha(x) = a_0 + x$$

has $-a_0$ as zero and when $m = 2$,

$$\alpha(x) = a_0 + a_1 x + x^2$$

has

$$\frac{1}{2}\left(-a_1 - \sqrt{a_1^2 - 4a_0}\right), \qquad \frac{1}{2}\left(-a_1 + \sqrt{a_1^2 - 4a_0}\right).$$

Every polynomial $x^n - a \in \mathbf{C}[x]$ has n zeros over $\mathbf{C}$. There exist formulae which yield the zeros of all polynomials of degrees 3 and 4. It is also known that no formulae can be devised for arbitrary polynomials of degree $m \geq 5$.

Any polynomial $\alpha(x)$ of degree $m \geq 1$ can have no more than m distinct zeros. The polynomial

$$\alpha(x) = a_0 + a_1 * x + x^2$$

will have two distinct zeros if and only if the discriminant $a_1^2 - 4 * a_0 \neq 0$. We then call each a simple zero of $\alpha(x)$. However, if $a_1^2 - 4 * a_0 = 0$, each formula yields $-\frac{1}{2} * a_1$ as a zero. We then call $-\frac{1}{2} * a_1$ a zero of multiplicity two of $\alpha(x)$ and exhibit the zeros as $-\frac{1}{2} * a_1, -\frac{1}{2} * a_1$.

Fundamental Theorem of Algebra. Every polynomial $\alpha(x) \in \mathbf{C}[x]$ of degree $m \geq 1$ has at least one zero in $\mathbf{C}$.

Theorem. Every polynomial $\alpha(x) \in \mathbf{C}[x]$ of degree $m \geq 1$ has precisely m zeros over $\mathbf{C}$, with the understanding that any zero of multiplicity n is to be counted as n of the m zeros.

Theorem. Any $\alpha(x) \in \mathbf{C}[x]$ of degree $m \geq 1$ is either of the first degree or may be written as a product of polynomials $\in \mathbf{C}[x]$, each of the first degree.

Next we study certain subsets of $\mathbf{C}[x]$ by restricting the ring of coefficients. First, let us suppose that

$$\alpha(x) = a_0 + a_1 * x + a_2 * x^2 + \cdots + a_m * x^m \in R[x]$$

of degree $m \geq 1$ has $r = a + b * i$ as zero, i.e.,

$$\alpha(r) = a_0 + a_1 * r + a_2 * r^2 + \cdots + a_m * r^m = s + t * i = 0.$$

We have

$$\alpha(\bar{r}) = a_0 + a_1 * \bar{r} + a_2 * \bar{r}^2 + \cdots + a_m * \bar{r}^m = \overline{s + t * i} = 0$$

so that

Theorem. If $r \in \mathbf{C}$ is a zero of any polynomial $\alpha(x)$ with real coefficients, then $\bar{r}$ is also a zero of $\alpha(x)$.

Let $r = a + b * i$, with $b \neq 0$, be a zero of $\alpha(x)$. Thus $\bar{r} = a - b * i$ is also a zero and we may write

$$\alpha(x) = [x - (a + b * i)][x - (a - b * i)] * \alpha_1(x) = \left[x^2 - 2 * a * x + a^2 + b^2\right] * \alpha_1(x)$$

where α_1 is a polynomial of degree two less than that of $\alpha(x)$ and has real coefficients. Since a quadratic polynomial with real coefficients will have imaginary zeros if and only if its discriminant is negative, we have

Theorem. The polynomials of the first degree and the quadratic polynomials with negative discriminant are the only polynomials $\in \mathbf{R}[x]$ which are primes over $\mathbf{R}$.

Theorem. A polynomial of odd degree $\in \mathbf{R}[x]$ necessarily has a real zero.

Suppose
$$\beta(x) = b_0 + b_1 * x + b_2 * x^2 + \cdots + b_m * x^m \in \mathbf{Q}[x].$$

Let c be the greatest common divisor of the numerators of the b_i's and d be the least common multiple of the denominators of the b_i's; then

$$\alpha(x) = \frac{d}{c} * \beta(x) = a_0 + a_1 * x + a_2 * x^2 + \cdots + a_m * x^m \in \mathbf{Q}[x]$$

has integral coefficients whose only common divisors are ± 1, the units of $\mathbf{Z}$. Moreover, $\beta(x)$ and $\alpha(x)$ have precisely the same zeros. If $r \in \mathbf{Q}$ is a zero of $\alpha(x)$, i.e. if

$$\alpha(r) = a_0 + a_1 * r + a_2 * r^2 + \cdots + a_m * r^m = 0$$

it follows that

(a) if $r \in \mathbf{Z}$, then $r|a_0$;
(b) if $r = s/t$, a common fraction in lowest terms, then

$$t^m * \alpha(s/t) = a_0 * t^m + a_1 * s * t^{m-1} + a_2 * s^2 * t^{m-2} + \cdots + a_{m-1} * s^{m-1} * t + a_m * s^m = 0$$

so that $s|a_0$ and $t|a_m$. We have proved

Theorem. Let $\alpha(x) = a_0 + a_1 * x + a_2 * x^2 + \cdots + a_m * x^m$ be a polynomial of degree $m \geq 1$ having integral coefficients. If $s/t \in \mathbf{Q}$ with $(s, t) = 1$, is a zero of $\alpha(x)$, then $s|a_0$ and $t|a_m$.

Let $\alpha(x)$ and $\beta(x)$ be non-zero polynomials in $F[x]$. A polynomial $d(x) \in F[x]$ having the properties

(a) $d(x)$ is monic;
(b) $d(x)|\alpha(x)$ and $d(x)|\beta(x)$;
(c) for every $c(x) \in F[x]$ such that $c(x)|\alpha(x)$ and $c(x)|\beta(x)$, we have $c(x)|d(x)$;

is called the *greatest common divisor* of $\alpha(x)$ and $\beta(x)$.

The greatest common divisor of two polynomials in $F[x]$ can be found in the same manner as the greatest common divisor of two integers.

Theorem. Let the non-zero polynomials $\alpha(x)$ and $\beta(x)$ be in $F[x]$. The monic polynomial

$$d(x) = s(x) * \alpha(x) + t(x) * \beta(x), \qquad s(x), t(x) \in F[x]$$

of least degree is the greatest common divisor of $\alpha(x)$ and $\beta(x)$.

Theorem. Let $\alpha(x)$ of degree $m \geq 2$ and $\beta(x)$ of degree $n \geq 2$ be in $F[x]$. Then non-zero polynomials $\mu(x)$ of degree at most $n-1$ and $v(x)$ of degree at most $m-1$ exist in $F[x]$ such that

$$\mu(x) * \alpha(x) + v(x) * \beta(x) = z, \qquad \text{where } z \text{ is the zero polynomial}$$

if and only if $\alpha(x)$ and $\beta(x)$ are not relatively prime.

Theorem. If $\alpha(x)$, $\beta(x)$, $p(x) \in F[x]$ with $\alpha(x)$ and $p(x)$ relatively prime, then

$$p(x) | \alpha(x) * \beta(x)$$

implies

$$p(x) | \beta(x).$$

Unique Factorization Theorem. Any polynomial $\alpha(x)$, of degree $m \geq 1$ and with leading coefficient a, in $F[x]$ can be written as

$$\alpha(x) = a * [p_1(x)]^{m_1} * [p_2(x)]^{m_2} * \ldots * [p_j(x)]^{m_j}$$

where the $p_i(x)$ are monic prime polynomials over F and the m_i's are positive integers. Moreover, except for the order of the factors, the factorization is unique.

2.9   Differentiation

Let $f : I \to \mathbf{R}$ be a function, where I is an open interval. We say that f is differentiable at $a \in I$ provided there exists a linear mapping $L : \mathbf{R} \to \mathbf{R}$ such that

$$\lim_{\epsilon \to 0} \frac{f(a + \epsilon) - f(a) - L(\epsilon)}{\epsilon} = 0.$$

The linear mapping L which, when it exists, is unique and is called the *differential* of f (or *derivative* of f at a) and is denoted by $d_a f$. It is customary in traditional texts to introduce the differentials df and dx and to obtain relations such as

$$df = \frac{df}{dx} dx.$$

Using the modern notation this relation would be written as

$$d_a f = f'(a) dz \mathcal{I}$$

where $\mathcal{I}(= id)$ denotes the identity function $x \to x$. If f and g are differentiable we find that

$$\frac{d}{dx}(f + g) = \frac{df}{dx} + \frac{dg}{dx} \qquad \text{summation rule}$$

$$\frac{d}{dx}(f - g) = \frac{df}{dx} - \frac{dg}{dx} \qquad \text{difference rule}$$

$$\frac{d}{dx}(f * g) = g * \frac{df}{dx} + f * \frac{dg}{dx} \qquad \text{product rule}$$

$$\frac{d}{dx}\left(\frac{f}{g}\right) = \frac{g * \frac{df}{dx} - f * \frac{dg}{dx}}{g^2}, \qquad g \neq 0 \text{ for } x \in I \qquad \text{quotient rule}$$

$$\frac{d}{dx} c = 0$$

where c is a constant.

2.10 Integration

A computer algebra system should be able to integrate formally elementary functions, for example

$$\int \frac{dx}{1-x^2} = \frac{1}{2}\ln\left(\frac{1+x}{1-x}\right).$$

In general it is assumed that the underlying field is **R**. Symbolic differentiation was undertaken quite early in the history of computer algebra, whereas symbolic integration (also called formal integration) was introduced much later. The reason is due to the big difference between formal integration and formal differentiation. Differentiation is an algorithmic procedure, and a knowledge of the derivatives of functions plus the sum rule, product rule, quotient rule and chain rule, enable us to differentiate any given function. The real problem in differentiation is the simplification of the result. On the other hand, integration seems to be a random collection of devices and special cases. There are only two general rules, i.e. the sum rule and the rule for integration by parts. If we integrate a sum of two functions, in general, we would integrate each summand separately, i.e.

$$\int (f_1(x) + f_2(x))dx = \int f_1(x)dx + \int f_2(x)dx.$$

This is the so-called *sum rule*. It can happen that the sum $f_1 + f_2$ could have an explicit form for the integral, but f_1 and f_2 do not have any integrals in finite form. For example

$$\int (x^x + (\ln x)x^x)dx = x^x.$$

However

$$\int x^x dx, \qquad \int (\ln x)x^x dx$$

do not have any integrals in finite form. The sum rule may only be used if it is known that two of the three integrals exist. For combinations other than addition (and subtraction) there are no general rules. For example, because we know how to integrate $\exp x$ and x^2 it does not follow that we can integrate $\exp(x^2)$. This function has no integral simpler than $\int \exp x^2 dx$. So we learn several "methods" such as: integration by parts, integration by substitution, integration by looking up in tables of integrals, etc. In addition we do not know which method or which combination of methods will work for a given integral. In the following presentation we follow closely Davenport et al. [12], MacCallum and Wright [34] and Risch [44].

Since differentiation is definitely simpler than integration, it is appropriate to rephrase the problem of integration as the "inverse problem" of differentiation, that is, given a function f, instead of looking for its integral g, we ask for a function g such that $dg/dx = f$.

Definition. Given two classes of functions A and B, the integration problem for A and B is to find an algorithm which, for every member f of A, either gives an element g of B such that $f = dg/dx$, or proves that there is no element g of B such that $f = dg/dx$.

For example, if $A = Q(x)$ and $B = Q(x)$, where $Q(x)$ denotes the rational functions, then the answer for $1/x^2$ must be $-1/x$, whilst for $1/x$ there is no solution in this set. On the other hand, if $B = Q(x, \ln x)$, then the answer for $1/x$ must be $\ln x$.

We consider now integration of rational functions. We deal with the case of $A = C(x)$, where C is a field of constants. Every rational function f can be written in the form $p + q/r$, where p, q and r are polynomials, q/r are relatively prime, and the degree of q is less than that of r. A polynomial p always has a finite integral, so the sum rule holds for $f_1(x) = p(x)$ and $f_2(x) = q(x)/r(x)$. Therefore the problem of integrating f reduces to the problem of the integration of p (which is very simple) and of the proper rational function q/r.

The *naive method* is as follows. If the polynomial r factorises into linear factors, such that

$$r(x) = \prod_{i=1}^{n} (x - a_i)^{n_i}$$

we can decompose q/r into partial fractions

$$\frac{q(x)}{r(x)} = \sum_{i=1}^{n} \frac{b_i(x)}{(x - a_i)^{n_i}}$$

where the b_i are polynomials of degree less than n_i. These polynomials can be divided by $x - a_i$, so as to give the following decomposition:

$$\frac{q(x)}{r(x)} = \sum_{i=1}^{n} \sum_{j=1}^{n_i} \frac{b_{i,j}}{(x - a_i)^j}$$

where the $b_{i,j}$ are constants. This decomposition can be integrated to give

$$\int \frac{q(x)}{r(x)} dx = \sum_{i=1}^{n} b_{i,1} \log(x - a_i) - \sum_{i=1}^{n} \sum_{j=2}^{n_i} \frac{b_{i,j}}{(j-1)(x - a_i)^{j-1}}.$$

Thus, we have proved that every rational function has an integral which can be expressed as a rational function plus a sum of logarithms of rational functions with constant coefficients – that is, the integral belongs to the field $C(x, \ln x)$. This algorithm requires us to factorise the polynomial r completely, which is not always possible without adding several algebraic quantities to C. Manipulating these algebraic extensions is often very difficult. Even if the algebraic extensions are not required, it is quite expensive to factorise a polynomial r of high degree. It also requires a complicated decomposition into partial fractions.

In the *Hermite's method* we determine the rational part of the integral of a rational function without bringing in any algebraic quantity. Similarly, it finds the derivative of the sum of logarithms, which is also a rational function with coefficients in the same field. We have seen that a factor of the denominator r which appears to the power n, appears to the power $n-1$ in the denominator of the integral. This suggests square-free decomposition. Let us suppose, then, that r has a square-free decomposition of the form $\prod_{i=1}^{n} r_i^i$. The r_i are then relatively prime, and we can construct a decompostion into partial fractions:

$$\frac{q(x)}{r(x)} = \frac{q(x)}{\prod_{i=1}^{n} r_i^i(x)} = \sum_{i=1}^{n} \frac{q_i(x)}{r_i^i(x)}.$$

Every element on the right-hand side has an integral, and therefore the sum rule holds, and it suffices to integrate each element in turn. Integration yields

$$\int \frac{q_i(x)}{r_i^i(x)} dx = -\left(\frac{q_i b/(i-1)}{r_i^{i-1}} \right) + \int \frac{q_i a + d(q_i b/(i-1))/dx}{r_i^{i-1}} dx$$

where a and b satisfy $ar_i + bdr_i/dx = 1$. Consequently, we have been able to reduce the exponent of r_i. We can continue in this way until the exponent becomes one, when the remaining integral is a sum of logarithms. Hermite's method is quite suitable for manual calculations. The disadvantage is that it needs several sub-algorithms and this involves some fairly complicated programming.

The *Horowitz method* is as follows. The aim is still to be able to write $\int (q(x)/r(x))dx$ in the form $q_1/r_1 + \int (q_2/r_2)dx$, where the integral remaining gives only a sum of logarithms when it is resolved. We know that r_1 has the same factors as r, but with the exponent reduced by one, that r_2 has no multiple factors, and that its factors are all factors of r. We have $r_1 = \gcd(r, dr/dx)$, and r_2 divides $r/\gcd(r, dr/dx)$. We may suppose that q_2/r_2 is written in reduced from, and therefore $r_2 = r/\gcd(r, dr/dx)$. Then

$$\frac{q(x)}{r(x)} = \frac{d}{dx}\left(\frac{q_1}{r_1}\right) + \frac{q_2}{r_2} = \frac{1}{r_1}\frac{dq_1}{dx} - \frac{q_1}{r_1^2}\frac{dr_1}{dx} + \frac{q_2}{r_2} = \frac{r_2 dq_1/dx - q_1 s + q_2 r_1}{r}$$

where $s = (r_2 dr_1/dx)/r_1$ (the division here is without remainder). Thus we arrive at

$$q = r_2\frac{dq_1}{dx} - q_1 s + q_2 r_1$$

where q, s, r_1 and r_2 are known, and q_1 and q_2 have to be determined. Since the degrees of q_1 and q_2 are less than the degrees m and n of r_1 and r_2 respectively we write

$$q_1(x) = \sum_{i=0}^{m-1} a_i x^i, \qquad q_2(x) = \sum_{i=0}^{n-1} b_i x^i.$$

Thus the equation for q can be rewritten as a system of $m + n$ linear equations in $n + m$ unknowns. Moreover, this system can be solved, and integration (at least this sub-problem) reduces to linear algebra.

Next we describe the *logarithmic part method*. The two methods described above can reduce the integration of any rational function to the integration of a rational function (say q/r) whose integral would be only a sum of logarithms. This integral can be resolved by completely factorising the denominator, but this is not always necessary for an expression of the results. The real problem is to find the integral without using any algebraic numbers other than those needed in the expression of the result. Let us suppose that

$$\int \frac{q(x)}{r(x)}dx = \sum_{i=1}^{n} c_i \log v_i(x)$$

is a solution to this integral where the right hand side uses the fewest possible algebraic extensions. The c_i are constants and, in general, the v_i are rational functions. Since $\ln(a/b) = \ln a - \ln b$, we can suppose, without loss of generality, that the v_i are polynomials. Furthermore, we can perform a square-free decomposition, which does not add any algebraic extensions, and we can apply the identity

$$\ln \prod_{i=1}^{n} p_i^i \equiv \sum_{i=1}^{n} i \ln p_i.$$

From the identity

$$c * \ln(p * q) + d * \ln(p * r) \equiv (c + d) * \ln p + c * \ln q + d * \ln r$$

we can suppose that the v_i are relatively prime, whilst still keeping the minimality of the number of algebraic extensions. Moreover, we can suppose that all the c_i are different. Differentiating the integral, we find

$$\frac{q(x)}{r(x)} = \sum_{i=1}^{n} \frac{c_i}{v_i} \frac{dv_i}{dx}.$$

The assumption that the v_i are square-free implies that no element of this summation can simplify, and the assumption that the v_i are relatively prime implies that no cancellation can take place in this summation. This implies that the v_i must be precisely the factors of r, i.e. that $r(x) = \prod_{i=1}^{n} v_i(x)$. Let us write $u_i = \prod_{j \neq i} v_j$. Then we can differentiate the product of the v_i, which shows that

$$r(x) = \sum_{i=1}^{n} u_i \frac{dv_i}{dx}.$$

We find that $q(x) = \sum_{i=1}^{n} c_i u_i dv_i/dx$. These two expressions for q and dr/dx permit the following deduction

$$v_k(x) = \gcd(0, v_k) = \gcd\left(q - \sum_{i=1}^{n} c_i u_i dv_i/dx, v_k\right) = \gcd(q - c_k u_k dv_k/dx, v_k)$$

since all the other u_i are divisible by v_k.

Next we consider algebraic solutions of the first order differential equation

$$\frac{dy}{dx} + f(x)y = g(x).$$

This leads to the *Risch algorithm*. We have introduced the problem of finding an algorithm which, given f and g belonging to a class A of functions, either finds a function y belonging to a given class B of functions, or proves that there is no element of B which satisfies the given equation. For the sake of simplicity, we shall consider the case when B is always the class of functions elementary over A. To solve this differential equation we substitute

$$y(x) = z(x) \exp\left(-\int\limits^{x} f(s)ds\right).$$

This leads to the solution

$$y(x) = \exp\left(-\int\limits^{x} f(s)ds\right) \int\limits^{x} \left(g(s)\left(\exp\int\limits^{s} f(t)dt\right)ds\right).$$

In general, this method is not algorithmically satisfactory for finding y, since the algorithm of integration described in the last section reformulates this integral as the differential equation we started with. Risch [44] found one method to solve these equations for the case when A is a field of rational functions, or an extension of a field over which this problem can be solved. The problem can be stated as follows: given two rational functions f and g, find the rational function y such that $dy/dx + f(x)y = g(x)$, or prove that there is none. f satisfies the condition that $\exp(\int^x f(s)ds)$ is not a rational function and its integral is not a sum of logarithms with rational coefficients. The problem is solved in two stages: reducing it to a purely polynomial problem, and solving that problem. The Risch algorithm is recursive. Before applying it one has (in principle) to check that the different extension variables are not algebraically related. For rational functions the Risch algorithm is the same as for the Horowitz method. For more details of the Risch algorithm and extensions of it we refer to the literature [12], [34], [44].

2.11 Commutativity and Noncommutativity

In computer algebra it is usually assumed that the symbols are commutative. Many mathematical structures are noncommutative. Here we discuss some of these structures.

We recall that an *associative algebra* is a vector space V over a field F which satisfies the ring axioms in such a way that addition in the ring is addition in the vector space and such that $c * (A * B) = (c * A) * B = A * (c * B)$ for all $A, B \in V$ and $c \in F$. Moreover the associative law holds, i.e. $A * (B * C) = (A * B) * C$. An example of an associative algebra is the set of the $n \times n$ matrices over the real or complex numbers with matrix multiplication as composition. There, in general, we have

$$A * B \neq B * A.$$

Another important example of a non-commutative structure is that of a *Lie algebra*. A Lie algebra is defined as follows. A vector space L over a field F, with an operation $L \times L \to L$ denoted by

$$(x, y) \to [x, y]$$

and called the commutator of x and y, is called a Lie algebra over F if the following axioms are satisfied.

(L1) The bracket operation is bilinear.

(L2) $[x, x] = 0$ for all $x \in L$.

(L3) $[x, [y, z]] + [y, [z, x]] + [z, [x, y]] = 0, \qquad x, y, z \in L.$

A simple example of a Lie algebra is the vector product in the vector space $\mathbf{R}^3$.

Remark. The connection between an associative algebra and a Lie algebra is as follows. Let A, B be elements of the associative algebra. We define the commutator as follows:

$$[A, B] := A * B - B * A.$$

It can be proved easily that the commutator defined in this way satisfies the axioms given above. Thus we have constructed a Lie algebra from an associative algebra.

Another example of a non-commutative structure are the quaternions. The quaternions have a matrix representation.

2.12   Tensor and Kronecker Product

Let V, W be vector spaces over a field F. We define the tensor product [30] between elements of V and W. The value of the product should be in a vector space. If we denote $v \otimes w$ as the tensor product of elements $v \in V$ and $w \in W$, then we have the following relations. If $v_1, v_2 \in V$ and $w \in W$, then

$$(v_1 + v_2) \otimes w = v_1 \otimes w + v_2 \otimes w.$$

If $w_1, w_2 \in W$ and $v \in V$, then

$$v \otimes (w_1 + w_2) = v \otimes w_1 + v \otimes w_2.$$

If $c \in F$, then

$$(c * v) \otimes w = c * (v \otimes w) = v \otimes (c * w).$$

We now construct such a product, and prove its various properties.

Let U, V, W be vector spaces over F. By a bilinear map

$$g : V \times W \to U$$

we mean a map which to each pair of elements (v, w) with $v \in V$ and $w \in W$ associates an element $g(v, w)$ of U, having the following property.

For each $v \in V$, the map $w \mapsto g(v, w)$ of W into U is linear, and for each $w \in W$, the map $v \mapsto g(v, w)$ of V into U is linear. For the proofs of the following theorems we refer to the literature [30].

Theorem. Let V, W be finite dimensional vector spaces over the field F. There exists a finite dimensional space T over F, and a bilinear map $V \times W \to T$ denoted by

$$(v, w) \mapsto v \otimes w,$$

satisfying the following properties.

1. If U is a vector space over F, and $g : V \times W \to U$ is a bilinear map, then there exists a unique linear map

$$g_* : T \to U$$

such that, for all pairs (v, w) with $v \in V$ and $w \in W$ we have

$$g(v, w) = g_*(v \otimes w).$$

2. If $\{ v_1, \ldots, v_n \}$ is a basis of V, and $\{ w_1, \ldots, w_m \}$ is a basis of W, then the elements

$$v_i \otimes w_j, \qquad i = 1, \ldots, n \quad \text{and} \quad j = 1, \ldots, m$$

form a basis of T.

The space T is called the *tensor product* of V and W, and is denoted by $V \otimes W$. Its dimension is given by

$$\dim(V \otimes W) = (\dim V)(\dim W).$$

The element $v \otimes w$ associated with the pair (v, w) is also called a tensor product of v and w.

Frequently we have to take a tensor product of more than two spaces. We have associativity for this product.

Theorem. Let U, V, W be finite dimensional vector spaces over F. Then there is a unique isomorphism

$$U \otimes (V \otimes W) \to (U \otimes V) \otimes W$$

such that

$$u \otimes (v \otimes w) \mapsto (u \otimes v) \otimes w$$

for all $u \in U$, $v \in V$ and $w \in W$.

This theorem allows us to omit the parentheses in the tensor product of several factors. Thus if $V_1, \ldots, V_r$ are vector spaces over F, we may form their tensor product

$$V_1 \otimes V_2 \otimes \cdots \otimes V_r$$

and the tensor product

$$v_1 \otimes v_2 \otimes \cdots \otimes v_r$$

of elements $v_i \in V_i$. The theorems described above give the general useful properties of the tensor product.

Next we introduce the *Kronecker product* of two matrices. It can be considered as a realization of the tensor product.

Definition. Let A be an $m \times n$ matrix and let B be a $p \times q$ matrix. Then the Kronecker product of A and B is an $(mp) \times (nq)$ matrix defined by

$$A \otimes B := \begin{pmatrix} a_{11}B & a_{12}B & \cdots & a_{1n}B \\ a_{21}B & a_{22}B & \cdots & a_{2n}B \\ \vdots & & & \\ a_{m1}B & a_{m2}B & \cdots & a_{mn}B \end{pmatrix}.$$

Sometimes the Kronecker product is also called direct product or tensor product. Obviously we have

$$(A + B) \otimes C = A \otimes C + B \otimes C$$

where A and B are matrices of the same size. Analogously

$$A \otimes (B + C) = A \otimes B + A \otimes C$$

where B and C are of the same size. Furthermore, we have

$$(A \otimes B) \otimes C = A \otimes (B \otimes C).$$

Example. Let

$$A = \begin{pmatrix} 2 & 3 \\ 0 & 1 \end{pmatrix}, \qquad B = \begin{pmatrix} 0 & -1 \\ -1 & 1 \end{pmatrix}.$$

Then

$$A \otimes B = \begin{pmatrix} 0 & -2 & 0 & -3 \\ -2 & 2 & -3 & 3 \\ 0 & 0 & 0 & -1 \\ 0 & 0 & -1 & 1 \end{pmatrix}, \qquad B \otimes A = \begin{pmatrix} 0 & 0 & -2 & -3 \\ 0 & 0 & 0 & -1 \\ -2 & -3 & 2 & 3 \\ 0 & -1 & 0 & 1 \end{pmatrix}.$$

We see that $A \otimes B \neq B \otimes A$.

Example. Let

$$\mathbf{e}_1 = \begin{pmatrix} 1 \\ 0 \end{pmatrix}, \qquad \mathbf{e}_2 = \begin{pmatrix} 0 \\ 1 \end{pmatrix}.$$

Then

$$\mathbf{e}_1 \otimes \mathbf{e}_1 = \begin{pmatrix} 1 \\ 0 \\ 0 \\ 0 \end{pmatrix}, \qquad \mathbf{e}_1 \otimes \mathbf{e}_2 = \begin{pmatrix} 0 \\ 1 \\ 0 \\ 0 \end{pmatrix}, \qquad \mathbf{e}_2 \otimes \mathbf{e}_1 = \begin{pmatrix} 0 \\ 0 \\ 1 \\ 0 \end{pmatrix}, \qquad \mathbf{e}_2 \otimes \mathbf{e}_2 = \begin{pmatrix} 0 \\ 0 \\ 0 \\ 1 \end{pmatrix}.$$

Obviously, $\{\, \mathbf{e}_1, \mathbf{e}_2 \,\}$ is the standard basis in $\mathbf{R}^2$. We see that

$$\{\, \mathbf{e}_1 \otimes \mathbf{e}_1, \quad \mathbf{e}_1 \otimes \mathbf{e}_2, \quad \mathbf{e}_2 \otimes \mathbf{e}_1, \quad \mathbf{e}_2 \otimes \mathbf{e}_2 \,\}$$

is the standard basis in $\mathbf{R}^4$.

2.13 Exterior Product

Next we introduce the *exterior product* (also called *alternating product* or *Grassmann product*).   Let V be a finite dimensional vector space over $\mathbf{R}$, r be an integer ≥ 1 and $V^{(r)}$ be the set of all r-tuples of elements of V, i.e. $V^{(r)} = V \times V \times \ldots \times V$. An element of $V^{(r)}$ is therefore an r-tuple $(\mathbf{v}_1, \ldots, \mathbf{v}_r)$ with each $\mathbf{v}_i \in V$. Let U be another finite dimensional vector space over $\mathbf{R}$. An r-multilinear map of V into U $f : V \times V \times \ldots \times V \to U$ is linear in each component. In other words, for each $i = 1, \ldots, r$ we have

$$f(\mathbf{v}_1, \ldots, \mathbf{v}_i + \mathbf{v}_i', \ldots, \mathbf{v}_r) = f(\mathbf{v}_1, \ldots, \mathbf{v}_i, \ldots, \mathbf{v}_r) + f(\mathbf{v}_1, \ldots, \mathbf{v}_i', \ldots, \mathbf{v}_r)$$

$$f(\mathbf{v}_1, \ldots, c * \mathbf{v}_i, \ldots, \mathbf{v}_r) = c * f(\mathbf{v}_1, \ldots, \mathbf{v}_r)$$

for all $\mathbf{v}_i, \mathbf{v}_i' \in V$ and $c \in \mathbf{R}$.

We say that a multilinear map f is alternating if it satisfies the condition

$$f(\mathbf{v}_1, \ldots, \mathbf{v}_r) = 0$$

whenever two adjacent components are equal, i.e. whenever there exists an index $j < r$ such that $\mathbf{v}_j = \mathbf{v}_{j+1}$. Note that the conditions satisfied by multilinear maps are similar to the properties of the determinants. The following theorem handles the general case of alternating products.

Theorem. Let V be a finite dimensional vector space over F, of dimension n. Let r be an integer $1 \leq r \leq n$. There exists a finite dimensional space over F, denoted by $\bigwedge^r V$, and an r-multilinear alternating map $V^{(r)} \to \bigwedge^r V$, denoted by

$$(\mathbf{u}_1, \ldots, \mathbf{u}_r) \mapsto \mathbf{u}_1 \wedge \cdots \wedge \mathbf{u}_r$$

satisfying the following properties.

1. If U is a vector space over F, and $g : V^{(r)} \to U$ is an r-multilinear alternating map, then there exists a unique linear map

$$g_* : \bigwedge^r V \to U$$

such that for all $\mathbf{u}_1, \ldots, \mathbf{u}_r \in V$ we have

$$g(\mathbf{u}_1, \ldots, \mathbf{u}_r) = g_*(\mathbf{u}_1 \wedge \cdots \wedge \mathbf{u}_r).$$

2. If $\{\mathbf{v}_1, \ldots, \mathbf{v}_n\}$ is a basis of V, then the set of elements

$$\{\mathbf{v}_{i_1} \wedge \cdots \wedge \mathbf{v}_{i_r}\}, \qquad 1 \leq i_1 < \cdots < i_r \leq n$$

is a basis of $\bigwedge^r V$.

Thus if $\{\mathbf{v}_1, \ldots, \mathbf{v}_n\}$ is a basis of V, then every element of $\bigwedge^r V$ has a unique expression as a linear combination

$$\sum_{i_1 < \cdots < i_r}^{n} c_{i_1 \ldots i_r} \mathbf{v}_{i_1} \wedge \cdots \wedge \mathbf{v}_{i_r}$$

the sum being taken over all r-tuples $(i_1, \ldots, i_r)$ of integers from 1 to n, satisfying

$$i_1 < \cdots < i_r.$$

One can also shorten this notation, by writing $(i) = (i_1, \ldots, i_r)$. Thus the above sum would be written

$$\sum_{(i)}^{n} c_{(i)} \mathbf{v}_{i_1} \wedge \cdots \wedge \mathbf{v}_{i_r}.$$

The structure introduced above is also called a *Grassmann algebra*.

The dimension of the vector space $\bigwedge^r V$ is given by

$$\binom{n}{r}.$$

Example. We consider the vector space $\mathbf{R}^4$ with the standard basis

$$\mathbf{e}_1, \quad \mathbf{e}_2, \quad \mathbf{e}_3, \quad \mathbf{e}_4.$$

Then a basis for the two forms is given by

$$\mathbf{e}_1 \wedge \mathbf{e}_2, \quad \mathbf{e}_1 \wedge \mathbf{e}_3, \quad \mathbf{e}_1 \wedge \mathbf{e}_4, \quad \mathbf{e}_2 \wedge \mathbf{e}_3, \quad \mathbf{e}_2 \wedge \mathbf{e}_4, \quad \mathbf{e}_3 \wedge \mathbf{e}_4.$$

A basis for the three forms is given by

$$\mathbf{e}_1 \wedge \mathbf{e}_2 \wedge \mathbf{e}_3, \quad \mathbf{e}_1 \wedge \mathbf{e}_2 \wedge \mathbf{e}_4, \quad \mathbf{e}_1 \wedge \mathbf{e}_3 \wedge \mathbf{e}_4, \quad \mathbf{e}_2 \wedge \mathbf{e}_3 \wedge \mathbf{e}_4.$$

The basis for the four forms consists of only one element, namely

$$\mathbf{e}_1 \wedge \mathbf{e}_2 \wedge \mathbf{e}_3 \wedge \mathbf{e}_4.$$

The determinant of a square matrix can be calculated using the exterior product. Consider the 4×4 matrix

$$\begin{pmatrix} 1 & 2 & 5 & 2 \\ 0 & 1 & 2 & 3 \\ 1 & 0 & 1 & 0 \\ 0 & 3 & 0 & 7 \end{pmatrix}$$

then

$$\begin{pmatrix} 1 \\ 0 \\ 1 \\ 0 \end{pmatrix} \wedge \begin{pmatrix} 2 \\ 1 \\ 0 \\ 3 \end{pmatrix} \wedge \begin{pmatrix} 5 \\ 2 \\ 1 \\ 0 \end{pmatrix} \wedge \begin{pmatrix} 2 \\ 3 \\ 0 \\ 7 \end{pmatrix} = 24\, e_1 \wedge e_2 \wedge e_3 \wedge e_4$$

where e_1, e_2, e_3, e_4 is the standard basis of $\mathbf{R}^4$. Thus 24 is the determinant of the matrix. In Chapter 8, we give an implementation of this (obviously slow) technique to find the determinant.

Now, we consider some practical methods to evaluate the determinant of a matrix. The method employed usually depends on the nature of the matrix — numeric or symbolic.

- **Numeric matrix**

  Let A be an $n \times n$ matrix. The determinant of A is the sum taken over all permutations of the columns of the matrix, of the products of elements appearing on the principal diagonal of the permutated matrix. The sign with which each of these terms is added to the sum is positive or negative according to whether the permutation of the column is even or odd.

  An obvious case is a matrix consisting of only a single element, for which we specify that

  $$\det(A) = a_{11}, \text{ when } n = 1.$$

  No one actually computes a determinant of a matrix larger that 4×4 by generating all permutations of the columns and evaluating the products of the diagonals. A more efficient way rests on the following facts:

  - Adding a numerical multiple of one row (or column) of matrix A to another leaves $\det A$ unchanged.

– If B is obtained from A by exchanging two rows (or two columns), then $\det B = -\det A$. If A has two rows or columns proportional to each other then $\det A = 0$.

The idea is to manipulate the matrix A with the help of these two operations in such a way that it becomes triangular, i.e. all matrix elements below the principle diagonal are equal to zero. It follows from the definition that the determinant of a triangular matrix is the product of the elements of the principal diagonal.

- **Symbolic matrix**
  For a symbolic matrix, the determinant is best evaluated using *Leverrier's method*.

  The *characteristic polynomial* of an $n \times n$ matrix A is a polynomial of degree n in terms of λ. It may be written as

  $$P(\lambda) = \det(\lambda I - A) = \lambda^n - c_1\lambda^{n-1} - c_2\lambda^{n-2} - \ldots - c_{n-1}\lambda - c_n.$$

  Now, we present Leverrier's method to find the coefficients, c_i, of the polynomial. It is fairly insensitive to the individual peculiarities of the matrix A. The method has an added advantage that the inverse of A, if it exists, is also obtained in the process of determining the coefficients, c_i. Obviously we also obtain the determinant. The coefficients, c_i, of $P(\lambda)$ are obtained by evaluating the trace of each of the matrices, $B_1, B_2, \ldots, B_n$, generated as follows. Set $B_1 = A$ and compute $c_1 = \text{tr}(B_1)$. Then compute

  $$B_k = A(B_{k-1} - c_{k-1}I), \quad c_k = \left(\frac{1}{k}\right)\text{tr}(B_k), \quad k = 2, 3, \ldots, n.$$

  The inverse of a non-singular matrix A can be obtained from the relationship

  $$A^{-1} = \left(\frac{1}{c_n}\right)(B_{n-1} - c_{n-1}I)$$

  and the determinant of the matrix is

  $$\begin{cases} c_n & \text{if } n \text{ is odd} \\ -c_n & \text{if } n \text{ is even.} \end{cases}$$

Chapter 3

Computer Algebra Systems

3.1 Introduction

In this chapter we survey some computer algebra systems and give some applications. There are a large number of computer algebra systems available, so we will concentrate on a few of them. Reduce and Axiom are based on Lisp. On the other hand Mathematica, Maple and MuPAD are based on C. Besides symbolic manipulations, all systems can also do numerical manipulations. The systems are not only an interactive environment operating in response to on-line commands, but all of them are also programming languages.

Reduce is one of the oldest computer algebra systems around. It is based on Portable Standard Lisp. The system was designed in the late 1960s by Anthony C. Hearn [18], [47], [49]. It allows one to mix Reduce and Lisp code.

Maple [10] is a product of Waterloo Maple Software. It is a system for mathematical computation — symbolic, numerical and graphical.

Axiom is a symbolic, numerical and graphical system developed at the IBM Thomas J. Watson Research Center. It gives the user all foundation and algebra instruments necessary to develop a computer realization of sophisticated mathematical objects in exactly the way a mathematician would do it. Axiom is distributed by The Numerical Algorithms Group Limited [24].

Mathematica [62] is a product of Wolfram Research, Inc. It is a general computer software system and language which handles symbolic, numerical and graphical computations.

MuPAD is a symbolic-numerical computer algebra system developed by the Department of Mathematics of the University of Paderborn [39]. MuPAD lets you define your own data types.

Besides the computer algebra systems described above there are a number of other excellent computer algebra systems available:

- Macsyma is a product of Macsyma Inc. It is based on Lisp. It implements the Aslaksen test for complex functions. A detailed description of Macsyma is given by Davenport et al. [12].

- Derive is a small but powerful computer algebra system of the Software Warehouse, Hawaii. It is also based on Lisp. Derive applies the rules of algebra, trigonometry, calculus, and matrix algebra to solve a wide range of mathematical problems. It also has graphical capabilities.

- Magma (Computational Algebra Group, School for Mathematics and Statistics, University of Sydney) is a system for computation in algebraic, geometric and combinatorical structures such as groups, rings, fields, algebras, modules, graphs and codes.

- MathCad is a product of MathSoft, Cambridge, Massachusetts. It provides a platform for engineers, scientists and academicians to perform, share and document symbolic and numerical calculations.

Further packages exist for special tasks such as number theory, graph theory and algebra. The web page of the International School for Scientific Computing (see preface) provides links to web pages where more information on other computer algebra packages can be found.

3.2 Reduce

3.2.1 Basic Operations

We give a summary of the most commonly used commands in Reduce. Reduce does not distinguish between capital and small letters. Thus the commands `sin(x)`, `Sin(x)` or `SIN(X)` are the same.

The commands we use most in this book are differentiation and integration. To differentiate $x^3 + 2x$ with respect to x we write

    df(x**3+2*x,x);

The output is `3*x^2 + 2`. To integrate $x^2 + 1$ we write

    int(x**2 + 1,x);

The output is `x*(x^2 + 3)/3`. Both `**` and `^` denote the power operator.

The command `solve()` solves a number of algebraic equations, systems of algebraic equations and transcendental equations. For example, the command

    solve(x**2 + (a+1)*x + a=0,x);

gives the solution `x = -1` and `x = -a`.

Another important command is the substitution command `sub()`. For example, the command

    sub(x=2,x*y+x**2);

yields `2*y + 4`.

Amongst others, Reduce includes the following mathematical functions: `sqrt(x)` (square root, $\sqrt{x}$), `exp(x)` (exponential function, $\exp(x)$), `log(x)` (natural logarithm, $\ln(x)$), and the trigonometric functions `sin(x)`, `cos(x)`, `tan(x)` with arguments in radians.

Reduce reserves `i` (or `I`) to represent $\sqrt{-1}$ and `pi` (or `PI`, `pI`, `Pi`) for the number π. Thus the input

    i*i;

gives `-1` and

    sin(pi);

results in `0`. Other predefined constants are `T`, `nil`, `E`, `Infinity` where `T` stands for true and `nil` stands for false.

For differentiation Reduce offers two options for implementation. In the first option we declare the function to be differentiated as operator. The following example shows how to use this option. The 2 in df() indicates that we differentiate twice.

```
operator f;
f(x) := x*x + sin(x);
result := df(f(x),x,2);
```

The output is

```
result := -sin(x) + 2
```

On the other hand we can also declare that f depends on x and then differentiate f with respect to x.

```
depend f, x;
f := x*x + sin(x);
result := df(f,x,2);
```

In Reduce the default data type for numbers is rational numbers. The command

```
2 + 0.1 + 1/3;
```

gives the output

```
73/30
```

The switch on rounded allows the calculation with real numbers. The commands

```
on rounded;
2 + 0.1 + 1/3;
```

gives the output

```
2.43333333333
```

Reduce only knows the most elementary identities such as

```
cos(-x)     = cos(x)
sin(pi)     = 0
log(e)      = 1
e^(i*pi/2) = i
```

The user can add further rules for the reduction of expressions by using the LET command, such as the trigonometry identity

```
for all x let sin(x)^2 + cos(x)^2 = 1;
```

then

```
sin(y)^2 + cos(y)^2 - 5;
```

gives the output

```
-4
```

For other commands we refer to the user's manual for Reduce [18].

3.2.2 Example

As an example we show how soliton equations can be derived from pseudospherical surfaces. We show how the sine-Gordon equation can be derived. Extensions to other soliton equations are straightforward. Soliton equations can be described by pseudospherical surfaces, i.e. surfaces of constant negative Gaussian curvature. An example is the *sine-Gordon equation*

$$\frac{\partial^2 u}{\partial x_1 \partial x_2} = \sin(u).$$

Here we show how Reduce can be used to find the sine-Gordon equation from the line element of the surface. The *metric tensor field* is given by

$$g = dx_1 \otimes dx_1 + \cos(u(x_1, x_2))dx_1 \otimes dx_2 + \cos(u(x_1, x_2))dx_2 \otimes dx_1 + dx_2 \otimes dx_2$$

i.e. the *line element* is

$$\left(\frac{ds}{d\lambda}\right)^2 = \left(\frac{dx_1}{d\lambda}\right)^2 + 2\cos(u(x_1, x_2))\frac{dx_1}{d\lambda}\frac{dx_2}{d\lambda} + \left(\frac{dx_2}{d\lambda}\right)^2.$$

Here u is a smooth function of x_1 and x_2. First we have to calculate the Riemann curvature scalar R from g. Then the sine-Gordon equation follows when we impose the condition

$$R = -2.$$

The calculation of the curvature scalar is well described in many textbooks (see for example [51]). For the sake of completeness we give the equations. We have

$$g_{11} = g_{22} = 1, \qquad g_{12} = g_{21} = \cos(u(x_1, x_2)).$$

The quantity g can be written in matrix form

$$g = \left(\begin{array}{cc} g_{11} & g_{12} \\ g_{21} & g_{22} \end{array} \right).$$

Then the inverse of g is given by

$$g^{-1} = \left(\begin{array}{cc} g^{11} & g^{12} \\ g^{21} & g^{22} \end{array} \right).$$

where

$$g^{11} = g^{22} = \frac{1}{\sin^2 u}, \qquad g^{12} = g^{21} = -\frac{\cos u}{\sin^2 u}.$$

Next we have to calculate the *Christoffel symbols*. They are defined as

$$\Gamma^a_{mn} := \frac{1}{2} g^{ab} \left(g_{bm,n} + g_{bn,m} - g_{mn,b} \right)$$

where the *sum convention* is used and

$$g_{bm,1} := \frac{\partial g_{bm}}{\partial x_1}, \qquad g_{bm,2} := \frac{\partial g_{bm}}{\partial x_2}.$$

Next we have to calculate the *Riemann curvature tensor* which is given by

$$R^r_{msq} := \Gamma^r_{mq,s} - \Gamma^r_{ms,q} + \Gamma^r_{ns} \Gamma^n_{mq} - \Gamma^r_{nq} \Gamma^n_{ms}.$$

The *Ricci tensor* follows as

$$R_{mq} := R^a_{maq} = -R^a_{mqa}$$

i.e. the Ricci tensor is constructed by contraction. From R_{nq} we obtain R^m_q via

$$R^m_q = g^{mn} R_{nq}.$$

Finally the *curvature scalar* R is given by

$$R := R^m_m.$$

With the metric tensor field given above we find that

$$R = -\frac{2}{\sin u} \frac{\partial^2 u}{\partial x_1 \partial x_2}.$$

If

$$R = -2$$

then we obtain the sine-Gordon equation.

We apply the concept of operators. Operators are the most general objects available in Reduce. They are usually parametrized, and can be parametrized in a completely general way. Only the operator identifier is declared in an operator declaration. The number of parameters is not declared. Operators represent mathematical operators or functions. We declare u as an operator and it depends on x. x itself is also declared as an operator and depends on 1 and 2. Since terms of the form $\cos^2(x)$ and $\sin^2(x)$ result from our calculation we have to include the identity

$$\sin^2(u) + \cos^2(u) \equiv 1$$

in order to simplify expressions.

```
% tensor.red

matrix g(2,2);
matrix g1(2,2);  % inverse of g;
array gamma(2,2,2); array R(2,2,2,2); array Ricci(2,2);

operator u, x;

g(1,1) := 1;  g(2,2) := 1;
g(1,2) := cos(u(x(1),x(2))); g(2,1) := cos(u(x(1),x(2)));

g1 := g^(-1);    % calculating the inverse
for a := 1:2 do
   for m := 1:2 do
      for n := 1:2 do
      gamma(a,m,n) := (1/2)*
                      (for b := 1:2 sum g1(a,b)*(df(g(b,m),x(n))
                       + df(g(b,n),x(m)) - df(g(m,n),x(b))));

for a := 1:2 do
   for m := 1:2 do
      for n := 1:2 do
      write "gamma(",a,",",m,",",n,") = ", gamma(a,m,n);

for b := 1:2 do
   for m := 1:2 do
      for s := 1:2 do
         for q := 1:2 do
         R(b,m,s,q) := df(gamma(b,m,q),x(s))-df(gamma(b,m,s),x(q))
                     + (for n := 1:2 sum gamma(b,n,s)*gamma(n,m,q))
                     - (for n := 1:2 sum gamma(b,n,q)*gamma(n,m,s));

cos(u(x(1),x(2)))**2 := 1 - sin(u(x(1),x(2)))**2;

for m := 1:2 do
```

```
for q := 1:2 do
Ricci(m,q) := for s := 1:2 sum R(s,m,s,q);

for m := 1:2 do
   for q := 1:2 do
   write "Ricci(",m,",",q,") = ", Ricci(m,q);

array Ricci1(2,2);
for m := 1:2 do
   for q := 1:2 do
   Ricci1(m,q) := (for b := 1:2 sum g1(m,b)*Ricci(q,b));

CS := for m := 1:2 sum Ricci1(m,m);
```

The calculation of the curvature scalar from a metric tensor field (for example Goedel metric, Schwarzschild metric) is one of the oldest applications of computer algebra. Here we show that it can be extended to find soliton equations. By modifying the metric tensor field we can obtain other soliton equations. For example

```
g(1,2) := cos(u(x(1),x(2)));  g(2,1) := cos(u(x(1),x(2)));
```

could be replaced by

```
g(1,2) := cosh(u(x(1),x(2)));  g(2,1) := cosh(u(x(1),x(2)));
```

Additionally, we have to include the identity rule

```
sinh(u(x(1),x(2)))**2 := cosh(u(x(1),x(2)))**2 - 1;
```

in order to simplify expressions.

A large number of application programs in Reduce can be found in [11], [19], [49], [50], [51]. In [49] applications in quantum mechanics are described. In [50] applications for differential equations are given. In [52] applications for nonlinear dynamical systems are provided.

3.3 Maple

3.3.1 Basic Operations

Maple distinguishes between small and capital letters. The command Sin(0.1) gives Sin(.1), whereas sin(0.1) gives the desired result 0.09983341665.

In Maple the differentiation command is diff(). The input

```
diff(x^3 + 2*x,x);
```

yields as output 3x^2 + 2. The integration command is int(). The input

```
int(x^2 + 1,x);
```

yields x^3/3 + x.

Maple has two different commands for solving equations. The command

```
solve(x^2 + (1+a)*x + a=0,x);
```

solves the equation x^2 + (1+a)x + a = 0 and gives the result x = -1 and x = -a. The command

```
fsolve(x^2 - x - 1=0,x)
```

solves x^2 - x - 1.0 = 0 and gives the output -0.6180339887 and 1.618033989.

The substitution command is given by subs(). For example, the command

```
subs(x=2,x*y + x^2);
```

gives 2y + 4.

Amongst others, Maple includes the following mathematical functions: sqrt(x) (square root, $\sqrt{x}$), exp(x) (exponential function, e^x), log(x) (natural logarithm, $\ln(x)$), and the trigonometric functions sin(x), cos(x), tan(x) with arguments in radians.

Predefined constants are

```
Catalan, E, Pi, false, gamma, infinity, true.
```

For other commands we refer to the user manual for Maple [10].

3.3.2   Example

As an example we consider a quantum mechanical problem. Given a trial function for a one-dimensional potential, we find an approximation for the ground state energy. The *eigenvalue equation* in one-space dimension is given by

$$-\frac{\hbar^2}{2m}\frac{d^2u}{dx^2} + V(x)u(x) = Eu(x).$$

We use the variational principle to estimate the ground state energy of a particle in the potential

$$V(x) := \begin{cases} cx & \text{for} \quad x > 0 \\ \infty & \text{for} \quad x < 0 \end{cases}$$

where $c > 0$. Owing to this potential the spectrum is discrete and bounded from below. We use

$$u(x) = \begin{cases} x\exp(-ax) & \text{for} \quad x > 0 \\ 0 & \text{for} \quad x < 0 \end{cases}$$

as a *trial function*, where $a > 0$. We have to keep in mind that the trial function is not yet normalized. From the eigenvalue equation we find that the *expectation value* for the energy is given by

$$\langle E \rangle := \frac{\langle u|\hat{H}|u\rangle}{\langle u|u\rangle} = \frac{\int\limits_0^\infty xe^{-ax}\left(-\frac{\hbar^2}{2m}\frac{d^2}{dx^2} + cx\right)xe^{-ax}dx}{\int\limits_0^\infty x^2\exp(-2ax)dx} = \frac{3c}{2a} + \frac{\hbar^2 a^2}{2m}$$

where $\langle \,|\, \rangle$ denotes the scalar product in the Hilbert space $L_2(0, \infty)$. The expectation value for the energy depends on the parameter a. The expectation value has a minimum for

$$a = \left(\frac{3mc}{2\hbar^2}\right)^{1/3}.$$

In the program we evaluate $\langle E \rangle$ and then determine the minimum of $\langle E \rangle$ with respect to a. Thus the ground-state energy is greater than or equal to

$$\frac{9}{4} \left(\frac{2\hbar^2 c^2}{3m} \right)^{1/3}.$$

```
# energy.map

# potential
V := c*x;
# trial ansatz
u := x*exp(-a*x);
# eigenvalue equation
Hu := -hb^2/(2*m)*diff(u,x,x) + V*u;

# integrating for finding expectation value
# not normalized yet
res1 := int(u*Hu,x);

# collect the exponential functions
res2 := collect(res1,exp);

# substitution of the boundary
res3 := 0 - subs(x=0,res2);

# finding the norm of u to normalize
res4 := int(u*u,x);
res5 := -subs(x=0,res4);

# normalized expectation value
expe := res3/res5;

# finding the minimum with respect to a
minim := diff(expe,a);
res6 := solve(minim=0,a);
a := res6[1];

# approximate ground state energy
appgse = subs(a=0,expe);
```

Remark. Only the real solution of res6, namely res6[1] is valid in our case.

3.4 Axiom

3.4.1 Basic Operations

Axiom emphasizes strict typechecking. Unlike other computer algebra systems, types in Axiom are dynamic objects. They are created at run-time in response to user commands. Types in Axiom range from algebraic type (e.g. polynomials, matrices and power series) to data structures (e.g. lists, dictionaries and input files). Types may be combined in meaningful ways. We may build polynomials of matrices, matrices of polynomials of power series, hash tables with symbolic keys and rational function entries and so on.

Categories in Axiom define the algebraic properties which ensure mathematical correctness. Through categories, programs may discover that polynomials of continued fractions have commutative multiplication whereas polynomials of matrices do not. Likewise, a greatest common divisor algorithm can compute the "gcd" of two elements for any Euclidean domain, but foil the attempts to compute meaningless "gcds" of two hash tables. Categories also enable algorithms to be compiled into machine code that can be ruled with arbitrary types.

Type declarations in Axiom can generally be omitted for common types in the interactive language. Basic types are called *domains of computation*, or simply, *domains*. Domains are defined in the form:

   `Name(...): Exports == Implementation`

Each domain has a capitalized `Name` that is used to refer to the class of its members. For example, `Integer` denotes "the class of integers", whereas `Float` denotes "the class of floating point numbers" and so on. The "..." part following `Name` lists the parameter(s) for the constructor. Some basic types like `Integer` take no parameters. Others, like `Matrix`, `Polynomial` and `List`, take a single parameter that again must be a domain. For example,

   `Matrix(Integer)`   denotes "matrices over the integers"
   `Polynomial(Float)`   denotes "polynomial with floating point coefficients"

There is no restriction on the number of types of parameters of a domain constructor.

The `Exports` part in Axiom specifies the operations for creating and manipulating objects of the domain. For example, the `Integer` type exports constants 0 and 1, and operations +, - and *. The `Implementation` part defines functions that implement the exported operations of the domain. These functions are frequently described in terms of another lower-level domain used to represent the objects of the domain.

Every Axiom object belongs to a unique domain. The domain of an object is also called its type. Thus the integer 7 has type `Integer` and the string "willi" has type `String`. The type of an object, however, is not unique. The type of the integer 7 is not only an `Integer` but also a `NonNegativeInteger`, a `PositiveInteger` and possibly any other "subdomain" of the domain `Integer`.

A subdomain is a domain with a "membership predicate". `PositiveInteger` is a subdomain of `Integer` with the predicate "is the integer > 0 ?". Subdomains with names are defined by the abstract data type programs similar to those for domains. The `Exports` part of a subdomain, however, must list a subset of the exports of the domain. The `Implementation` part optionally gives special definitions for subdomain objects. The following gives some examples in Axiom.

Axiom uses D to differentiate an expression

```
f := exp exp x
D(f,x)
```

An optional third argument n in D instructs Axiom for the n-th derivative of f, e.g. `D(f,x,3)`.

Axiom has extensive library facilities for integration. For example

```
integrate((x**2+2*x+1)/((x+1)**6+1),x)
```

yields

$$\frac{\arctan(x^3 + 3x^2 + 3x + 1)}{3}.$$

Axiom uses the `rule` command to describe the transformation rules one needs. For example

```
sinCosExpandRules := rule
    sin(x+y) == sin(x)*cos(y) + sin(y)*cos(x)
    cos(x+y) == cos(x)*cos(y) - sin(x)*sin(y)
    sin(2*x) == 2*sin(x)*cos(x)
    cos(2*x) == cos(x)**2 - sin(x)**2
```

Thus the command

```
sinCosExpandRules(sin(a+2*b+c))
```

applies the rules implemented above.

For more commands, we refer to the literature [24].

3.4.2   Example

In solving systems of polynomial equations, *Gröbner basis* theory plays a central role [12], [24], [34]. If the polynomials

$$\{ p_i \ : \ i = 1, \ldots, n \}$$

vanish totally, so does the combination

$$\sum_i c_i p_i$$

where the coefficients c_i are in general also polynomials. All possible such combinations generate a space called an *ideal*. The ideal generated by a family of polynomials consists of the set of all linear combinations of those polynomials with polynomial coefficients. A system of generators (or a basis) G for an ideal I is a Gröbner basis (with respect to an ordering $<$) if every complete reduction of an $f \in I$ (with respect to G) gives zero. The ordering could be lexicographic (meaning lexicographic variable ordering followed by variable degree ordering), inverse lexicographic (meaning inverse lexicographic then variable degree), total degree (meaning total degree then lexicographic) and inverse total degree (meaning total degree then inverse lexicographic). Most computer algebra systems can find the Gröbner basis for a given system of polynomials. Here we consider Axiom.

DMP stands for DistributedMultivariatePolynomial. Consider the commands:

```
(d1, d2, d3) : DMP([z,y,x],FRAC INT)
d1 := -4*z + 4*y**2*x + 16*x**2 + 1
d2 := 2*z*y**2 + 4*x + 1
d3 := 2*z*x**2 - 2*y**2 - x
groebner [d1,d2,d3]
```

This gives the output

$$z - \frac{1568}{2745}x^6 - \frac{1264}{305}x^5 + \frac{6}{305}x^4 + \frac{182}{549}x^3 - \frac{2047}{610}x^2 + \frac{103}{2745}x - \frac{2857}{10980}$$

$$y^2 + \frac{112}{2745}x^6 - \frac{84}{305}x^5 - \frac{1264}{305}x^4 - \frac{13}{549}x^3 + \frac{84}{305}x^2 + \frac{1772}{2745}x + \frac{2}{2745}$$

$$x^7 + \frac{29}{4}x^6 - \frac{17}{16}x^4 - \frac{11}{8}x^3 + \frac{1}{32}x^2 + \frac{15}{16}x + \frac{1}{4}$$

3.5 Mathematica

3.5.1 Basic Operations

Mathematica distinguishes between capital letters and small letters. The command sin[0.1] for the evaluation of the sine of 0.1 gives the error message, possible spelling error, whereas Sin[0.1] gives the right answer 0.0998334.

In Mathematica the differentiation command

    D[x^3 + 2*x,x]

gives the output 2 + 3 x^2. The command

    Integrate[x^2 + 1,x]

gives x + x^3/3.

The command Solve[ ] can solve a number of algebraic equations, systems of algebraic equations and transcendental equations. For example the command

    Solve[x^2 + (a+1)*x + a==0,x]

gives the solution x = -1 and x = -a.

The replacement operator /. applies rules to expressions. Consider the expression

    x*y + x*x

Then

    x*y + x*x /. x -> 2

yields as output 4 + 2*y.

Amongst others, Mathematica includes the following mathematical functions: Sqrt[x] (square root, $\sqrt{x}$), Exp[x] (exponential function, e^x), Log[x] (natural logarithm, $\ln(x)$), and the trigonometric functions Sin[x], Cos[x], Tan[x] with arguments in radians.

Predefined constants are

    I, Infinity, Pi, Degree, GoldenRatio, E, EulerGamma, Catalan

For other commands we refer to the user's manual for Mathematica [62].

3.5.2   Example

As an example we consider the *spin-1 matrices*

$$s_+ := \begin{pmatrix} 0 & \sqrt{2}\hbar & 0 \\ 0 & 0 & \sqrt{2}\hbar \\ 0 & 0 & 0 \end{pmatrix}$$

and

$$s_- := \begin{pmatrix} 0 & 0 & 0 \\ \sqrt{2}\hbar & 0 & 0 \\ 0 & \sqrt{2}\hbar & 0 \end{pmatrix}.$$

We calculate the *commutator* of the two matrices

$$[s_+, s_-] := s_+ s_- - s_- s_+$$

and then determine the eigenvalues of the commutator. The Mathematica program is as follows:

```
(* spin.m *)

sp = {{ 0, Sqrt[2]*hb, 0}, {0, 0, Sqrt[2]*hb}, {0, 0, 0}}
sm = {{0, 0, 0}, {Sqrt[2]*hb, 0, 0}, {0, Sqrt[2]*hb, 0}}
comm = sp . sm - sm . sp
Eigenvalues[comm]
```

The output is

```
            2       2
{0, -2 hb , 2 hb }
```

3.6 MuPAD

3.6.1 Basic Operations

MuPAD is a computer algebra system which has been developed mainly at the University of Paderborn. It is a symbolic, numeric and graphical system. MuPAD provides native parallel instructions to the user. MuPAD syntax is close to that of MAPLE and has object-orientated capabilities close to that of AXIOM. MuPAD distinguishes between small and capital letters. The command Sin(0.1) gives Sin(0.1), whereas sin(0.1) gives the desired result 0.09983341664. In MuPAD the differentiation command is diff(). The input

```
diff(x^3 + 2*x,x);
```

yields as output

$$3 x^2 + 2$$

The integration command is int(). The input

```
int(x^2 + 1,x);
```

yields

$$x + \frac{x^3}{3}$$

The command

```
solve(x^2 + (1+a)*x + a=0,x);
```

solves the equation x^2 + (1+a)*x + a=0 and gives the result {-a, -1}. The substitution command is given by subs(). For example, the command

```
subs(x*y + x^2,x=2);
```

gives 2 y + 4. Amongst others, MuPAD includes the following mathematical functions: sqrt(x) (square root, $\sqrt{x}$), exp(x) (exponential function, e^x), ln(x) (natural logarithm), and the trigonometric functions sin(x), cos(x), tan(x) with arguments in radians.

Predefined constants are

```
I, PI, E, EULER, TRUE, FALSE, gamma, infinity
```

3.6.2   Example

As an example we consider *Picard's method* to approximate a solution to the differential equation $dy/dx = f(x, y)$ with initial condition $y(x_0) = y_0$, where f is an analytic function of x and y. Integrating both sides yields

$$y(x) = y_0 + \int_{x_0}^{x} f(s, y(s))ds.$$

Now starting with y_0 this formula can be used to approach the exact solution iteratively if the procedure converges. The next approximation is given by

$$y_{n+1}(x) = y_0 + \int_{x_0}^{x} f(s, y_n(s))ds.$$

The example approximates the solution of $dy/dx = x + y$ using five steps of Picard's method. To input a file the command read(filename) is used. So in this case the command read("picard"); will give the output

```
                3     4     5     6
          2     x     x     x     x
    x + x  + -- + -- + -- + --- + 1
                3    12    60    720
```

```
/*picard*/


x0:=0:          /*initial x*/
y0:=1:          /*initial y*/
y:=y0:
y1:=subs(y,x=s):
f:=func(x+y,x,y):            /*declare function f(x,y)=x+y */


for i from 1 to 5 do
    y:=(y0+subs(int(f(s,y1),s),s=x)-subs(int(f(s,y1),s),s=x0)):
    y1:=subs(y,x=s):
end_for:


print(y);
```

Chapter 4

Object-Oriented Programming

In this chapter, we discuss the basic concepts of object-oriented programming. We begin by introducing objects, classes and abstract data types. Other concepts include message passing, inheritance, polymorphism, etc. Many examples have been used to illustrate these concepts. In Section 4.5 we describe the object-oriented languages C++ and Java. Finally, we give a brief introduction to the object-oriented languages Eiffel, Smalltalk and Oberon.

By the end of the chapter, the reader will be aware of the main concepts in object-oriented systems. In the later chapters, we will see how these concepts benefit the development of computer algebra systems.

4.1 Objects, Classes and Abstract Data Types

4.1.1 Objects

Anything in the world is an object. Flowers, cars, matrices are examples of objects. Cars have size, colour and other characteristics. They have operations like starting the engine, switch on the headlights, pressing the brake and so on. Similarly, matrices have numbers of rows, columns and so on. They have operations like addition, multiplication, Kronecker product, etc.

In this section, we investigate the use of object-oriented techniques (OOT) to model the world. As indicated in the name OOT, objects play an important role. An object in a computer has a unique identity (ID) that is independent of the values of its attributes. The ID is used to distinguish an object from others. It cannot be altered at any time during the lifetime of the object. Figure 4.1 shows the fundamental constituents of an object.

The internal structure of an object consists of two parts: attributes (data) and methods (operations).

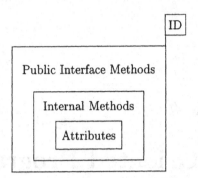

Figure 4.1: *Schematic diagram of an object*

Attributes describe the properties of an object at a specific moment in time. They tell something about the object but not how the object acts. They can be considered as the local data associated with the object. For example a cat has weight, species, sex and age. A polynomial has degree, number of terms, and coefficients of each term.

The attributes we specify in an object depend on the needs of the problem. Sometimes, we might not want to remember that a cat has four legs, or we might just simply assume it. Sometimes, the nature of a matrix might be important; is it a sparse or a dense matrix? In fact, deciding what to include as the set of attributes is a matter of design.

Methods describe the behaviour possessed by an object and what the object can do or will do when something happens. For example, employees have name, age, sex, salary and employee ID. However, the attributes alone could not fully describe the employees. Operations applied on employees are needed. They might include promotion, salary increment, car allowance etc. They represent the actions that can be performed by the object or on the object. The results of such actions might change the state of the object.

In fact, there are two kinds of methods. Firstly there are public interface methods, which provide a way to communicate with the object. Secondly these are internal methods. They specify the object behaviour but they are not accessible from outside the object. In other words, they only need to be known to the designer of that object. Users of the object only need to know about the public interface methods.

A method of an object is activated by a message sent by another object to the object containing the method. Alternatively, it could be invoked by another method in the same object by a local message. A *contract* is established between the caller and the called object. This mechanism is known as *message passing* between objects and will be discussed in greater detail in the next section.

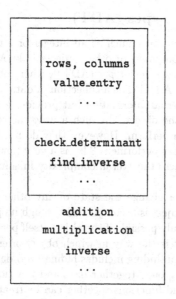

Figure 4.2: *Schematic diagram of the class* Matrix

As an example, consider an object representing a Matrix shown in Figure 4.2. It has a unique identity (ID) which distinguishes it from all other objects. This object contains information such as number of rows/columns, value of each entry and so on. It also contains some public interface methods used to manipulate the information such as inverse, determinant and so on. It is known that finding the inverse of a matrix requires us to ensure that the determinant of the matrix is non-zero. Therefore, the method inverse might contain the following internal methods:

check_determinant	check if the determinant of the matrix is zero
find_inverse	find the inverse of the matrix.

Note that check_determinant and find_inverse are hidden from outside. Only methods within the class have access to them. This concept of *data hiding* is important. It enhances the modularity and reduces the knock-on effects during the development of the class. Knock-on effects are caused when changes to one module require changes in other modules as well [63].

In object-oriented systems, the knock-on effects are caused by changes to an object's interface, not by changes to the internal details. In order to reduce the knock-on effects to minimum, an object's interface should be kept as simple as possible and it should hide as much detail as possible from the user.

4.1.2   Abstract Data Types (ADT)

An attribute can be a primitive type such as an integer or a character or it can be another object. An object composed of other objects is known as an *aggregation*. This is achieved through *abstract data types* (ADTs), where objects are constructed from existing data structures. An ADT is a collection of data and a set of operations on that data. It hides the internal information but provides a public set of methods for the users to manipulate the object. In such a way, the others "know" exactly what operations the ADT can perform. However, they do not know how the data is stored or how the operations are performed. This is a very powerful technique which supports modularity and reduces the overall complexity in software systems.

No object can directly read or change the state of any other object. Accessing or altering the attributes of an object is accomplished through its public interface methods. However, such access is only possible if the object itself permits it. This property is known as *encapsulation*. It is the way in which object-oriented systems perform *information hiding*. Information hiding includes hiding both data and operations implementation. Encapsulation groups together data and functions that work on that data. Thus, the set of data and functions together can be treated as a whole. These properties lead to weak coupling between objects.

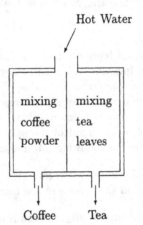

Figure 4.3: *Schematic diagram of a drinks dispenser*

As an example, consider a drinks dispenser in Figure 4.3. It accepts hot water as input and dispenses either tea or coffee according to which one of the two buttons is pressed. The machine is analogous to an abstract data type. The hot water, tea and coffee powder are analogous to the data collection and the operations are dispensing tea or coffee. The machine is treated as a black box. We are not concerned with the detail on how the operations are performed. What we expect is that the machine will dispense the right drink.

The interior mechanisms are surrounded by the metal casing. The openings are only for the input (hot water) to the machine and its output (tea or coffee). Thus, the internal mechanisms not only are hidden from the users but are also inaccessible. Furthermore, the mechanism of one operation is hidden from and inaccessible to the other operation. Thus, we can improve the operation of mixing coffee powder without affecting the operation of mixing tea leaves. We could also add another operation, such as dispensing hot chocolate milk, without affecting the original two operations. Thus, both abstraction and information hiding work here.

4.1.3 Classes

Many similar objects can be specified by the same general description. For example, an even number and a prime number. They can be described by a more general abstraction – an **Integer**. They share many common properties and behaviours. They can be arranged in a fixed ordering and they can be added, multiplied and so on. For the precise definition of integer, please refer to Chapter 2. The collection of instance objects that satisfy the properties could be classified as an **Integer**.

An object is tested against the general concepts to see if it can be considered as an **Integer**. These considerations lead to the notion of a *class*. The description of objects that have similar characteristics is called a *class*. Every object is an *instance* of a class. A class consists of data structure and operations applicable to the class. It is useful for classifying objects and it provides a basis for creation of a new instance object of the class.

In an object-oriented world, there are three different kinds of attributes in a class [25]:

1. The *class attributes* describe the characteristics of the class as a whole. These may be the totals or averages related to the class. For example, consider the **Card** class which is the superclass of all card games available in a departmental store, such as poker card, UNO, Magic, etc. The class attributes may be the total number of stock available, number of different types of card, or the average price of a deck of cards.

2. The *shared instance attributes* describe the common attributes possessed by all instances of a class. For example, consider a subclass of the **Card** called **Poker_Card**. A shared instance attribute may contain the number of cards in each deck. It may be the size of each card such as 6 cm × 9 cm. Sometimes, the shared instance attributes are referred to as class variables. They are different from the class attributes as defined above which describe the characteristics of the set of instance object as a whole.

3. The *default instance attributes* describe the attributes for each object. Even though the names of these attributes are specified in the class, their values are

only assigned in the individual objects. The `Poker_Card` class could have the following default instance attributes:

type of suit   e.g. diamonds, hearts, etc.
rank   e.g. Jack, King, Ace, etc.
colour   e.g. black or red.

4.2 Message Passing

Objects with attributes and methods communicate with each other by *message passing*. A message contains a name which is associated to a method within the object. It may also contain some arguments. Basically, message passing involves two parties: the *sender* and the *receiver*. When a message is received by an object, the associated method will be activated. Upon the completion of the execution of the method, it returns the result to the sender object.

Suppose an employee of a company, John, is organizing a birthday party for all the employees who are born in January. To find out the date of birth (DOB) of all the employees in the company, he has to request the information from the employees' record. This process involves an object called 'John', sending a message `Request_DOB` to each object of the class `Employee`. The DOB of each employee is then returned as the outcome. Figure 4.4 depicts the situation.

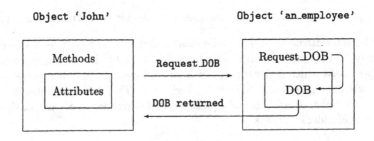

Figure 4.4: *Message passing between objects*

In this example, the message originated from the method in the object called 'John'. The destination of the message is another object called `an_employee`. The message name is `Request_DOB` and the action is to obtain the DOB of `an_employee`.

In fact, objects from different classes can respond to the same message in different ways. This is one form of polymorphism which will be discussed later in this chapter.

4.3 Inheritance

Inheritance is a powerful concept in object-oriented systems. It is most commonly used as a mechanism to create specialized versions of a more general class. This is known as *generalized-specialization*. It applies to families of classes that have similar but not identical attributes and methods. In such a situation, it is useful to inherit properties. This is achieved by multiple levels of abstraction where each level represents different versions or refinements of a class. Lower levels are built upon the earlier ones. In other words, a superclass represents a *generalization* of the subclasses. Similarly, a subclass of a given class represents a *specialization* of the class above. Inheritance also provides a mechanism for managing classes and sharing common codes.

As an example, consider the invertible matrices, lower triangular matrices, and symmetric matrices. They have certain common properties, such as number of rows, columns and so on. These general properties can be inherited from a more generic class called `Matrix`. Figure 4.5 shows the inheritance hierarchy. The subclasses inherited characteristics from the superclass `Matrix`. The subclasses represent the specialized versions of the superclass.

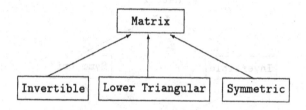

Figure 4.5: *Inheritance hierarchy of* `Matrix`

The superclass `Matrix` could also inherit from some superclasses like `Mathematical Structure`, which in turn could inherit from a class called `Structure`. Hence an inheritance hierarchy can be constructed.

Different object-oriented languages support different types of inheritance. Smalltalk only has single inheritance. This means a subclass can only inherit directly from one superclass. Other languages such as C++ support *multiple inheritance*. It allows a class to inherit from more than one immediate superclass. For example, if an invertible symmetric matrix is both an invertible matrix and a symmetric matrix, then multiple inheritance occurs, as shown in Figure 4.6. It is required that the rules for defining multiple inheritance are able to handle any conflict which may arise. For example, the naming conflicts which arise when the same name is used in two different superclasses, and is inherited into the same subclass. This illustrates that although multiple inheritance is a powerful mechanism, it increases the level of complexity.

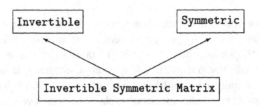

Figure 4.6: *An example of multiple inheritance*

The situation becomes more complicated when the Invertible and Symmetric classes inherit from the same super-superclass Matrix, as shown in Figure 4.7. The subclass Invertible Symmetric Matrix could potentially inherit properties from the super-superclass twice. This situation is known as *repeated inheritance*.

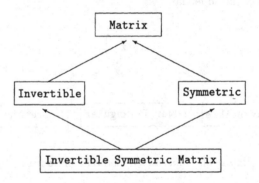

Figure 4.7: *An example of repeated inheritance*

There is another form of inheritance called *selective inheritance*. In this case, the subclass inherits only some of the properties from the superclass. This makes some information inaccessible to some subclasses. For example, the database of a company might consist of several parts, which is accessible to either manager, accountant or marketing person. A marketing person has access to the company's client records which are inaccessible to an accountant. On the other hand, an accountant has access to the company's sales records which are not accessible to a marketing person. The manager has access to all information in the database, including the employee records which are not accessible to both the accountant and the marketing person. This situation is depicted in Figure 4.8.

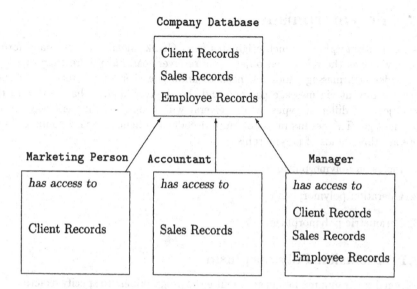

Figure 4.8: *An example of selective inheritance*

Subclasses inherit methods and attributes from superclasses. However, a subclass can redefine any of the methods or attributes inherited from the superclass as well as defining additional methods and attributes of its own. This is an extremely powerful feature when combined with polymorphism. This concept will be discussed in a later part of this chapter.

Sometimes, it is useful to introduce classes at a certain level which may not exist in reality but are useful conceptual constructs. These are known as *abstract classes*. An abstract class usually occupies a suitable position in the class hierarchy. It specifies the behaviours of its descendents, but not the actual implementation. The purpose of this abstract class is to define what must be done, but not how the tasks are carried out. It is up to its descendents to provide the implementation of the behaviours. These abstract classes do not have any instances created by themselves. In contrast, instances are created by the subclasses of these abstract classes which correspond to some real-world objects.

Class libraries with many classes are usually linked through an inheritance relationship. Inheritance reduces redundancy and code duplication. This is one of the major benefits derived from the concept. Classes are arranged in hierarchies with more generic classes towards the top of the hierarchy, and more specialized classes towards the bottom of the hierarchy.

4.4   Polymorphism

The word *polymorphism*, which originates from Greek, means 'having many forms'. It simply means the referenced object (the receiver) can have more than one type. The sender of a message does not need to know the class of the receiver. It only requests an event via message passing, and the receiver knows what to do. In this way, objects of different types (the receivers) will respond in different ways to the same message. This results in more compact code, and hence is easier to understand. There are three kinds of polymorphism

1. Inclusion polymorphism

2. Operation polymorphism

3. Parametric polymorphism

4.4.1   Inclusion Polymorphism

Traditional programming languages require the programmers to specify exactly which method to use during compilation time. This is known as a *static binding*. Inclusion polymorphism allows decisions to be made only at run-time, known as *dynamic binding* (or *late binding*). Using dynamic binding, the system ensures that the right method for the right class will be invoked during run-time. Although dynamic binding is flexible, it reduces performance. This is because the look-and-match algorithm is now carried out during run-time. Fortunately, this overhead is minimal in most languages. Static binding, on the other hand, is more secure and efficient.

This form of polymorphism applies on strongly type programming languages, such as Eiffel, Simula and C++, and it is related to inheritance. It is a means that provides common interfaces to different classes in the same inheritance tree, using the same name for the method. However, each subclass may implement the method differently.

Consider the inheritance tree in Figure 4.9. `Circle` and `Triangle` are subclasses of the abstract superclass `Shape`. Suppose the method `Calculate_Area` returns the area of a geometric shape. The method `Calculate_Area` is defined in the class `Shape`. However, no method can be used to calculate the area of a shape without explicitly specifying which kind of shape it is. The subclasses inherit this abstract method and override its definition by using appropriate algorithms to calculate the area of a circle or a triangle. In other words, when the same message is sent to each class, the response of each may be different. This behaviour provides a mechanism to handle heterogeneous objects easily. If we have many different shapes, such as parallelogram, rhombus, rectangle and so on, we could issue the same message `Calculate_Area` and expect each of them to return the right area. We do not need to know ahead of time which kind of shapes they are.

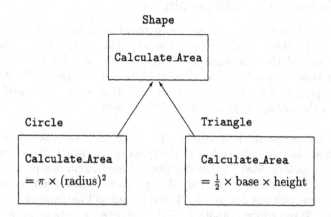

Figure 4.9: *An example of inclusion polymorphism*

This form of polymorphism allows us to avoid some type checking with strongly typed languages. It also makes development simpler and more dynamic.

4.4.2 Operation Polymorphism

This form of polymorphism refers to the same method with several definitions. The actual method is decided via the type of the parameters in the message. As an example, consider the following C++ class definition called Document:

```
class Document
{
    ....
    void print(char* s);
    void print(char c);
    void print(int n);
};
```

In the class, the same method print is defined three times. Each version differs by the type of the input data. Therefore, when a message is sent to an object of type Document, the appropriate print method will be invoked depending on the data type of the message parameters. This is known as *function overloading*.

In fact, this type of polymorphism has been used for the arithmetic operators in most languages. The operators +, -, * and / can be used to add, subtract, multiply and divide integers or floating-point numbers. The compiler automatically generates the appropriate code based on the type of the operands (integer or floating point).

4.4.3   Parametric Polymorphism

This form of polymorphism uses *types* as parameters in the class declarations. For example, use of a stack to store integer numbers. For strongly typed languages such as Eiffel, Java and C++, a stack of strings requires rewritten code for the strings as the data type. Generally, it requires replicate code for each different data type, making the programming task too tedious and error prone. This problem can be overcome by the concept of parametric polymorphism.

Parametric polymorphism is usually used to implement data structures and algorithms that are largely independent of the types of the object they operate on. For example, a parametric `Stack` might describe the implementation of `Stack` that contain objects with arbitrary types. By doing so, the `Stack` could be used to store strings, integers, pointers and so on. The compiler will automatically generate the code of these `Stack` classes. Another good candidate for parametric polymorphism is the `Vector` structure. In principle, the data content in a vector is largely independent of the operations available for the structure. A sorting algorithm is another ideal candidate for parametric polymorphism. We usually only care about whether the data items are sorted properly, but not the data types of the data items, provided that they can be arranged in a proper ordering. In short, code sharing is the most important advantage of parametric polymorphism.

The Standard Template Library (STL) implements these concepts. The Standard Template Library includes the classes: `deque`, `list`, `queue`, `set` and `vector`. We describe the Standard Template Library in detail in Chapter 5.

So far, we have introduced the essence of object-oriented programming, such as object, class, abstract data type, message passing, inheritance and polymorphism. In the next section, we discuss the most popular object-oriented programming languages available today. We describe how the concepts of object-oriented programming are implemented in these languages.

4.5 Object-Oriented Languages

There are many object-oriented languages available in the market. Here we discuss some of them and compare them with C++. The commonly used object-oriented languages are: C++, Java, Smalltalk, Eiffel, Oberon and Delphi.

In the early 1960s in Europe, a group led by Ole-Johan Dahl and Kristan Nygaard in the Norwegian Computer Centre constructed a simulation language known as Simula 67 [7]. It was designed for system description and simulation. The system here was a collection of independent objects with a common objective, which examined the life cycle of the elements of the system. Simula was based on Algol 60 with one very important addition – the *class* concept. Using this concept, it is possible to declare a class, create objects of that class, name these objects and form a hierarchical structure of class declaration. Although this concept was only introduced to describe the life cycles of the elements in the discrete simulation, it was later recognized as a general programming tool ideal for describing and designing programs in an abstract way. The basic idea was that the data (or data structure) and the operations performed on it can be considered as one, and this forms the basis for the implementation of abstract data types. The class concept has been adapted and used in many modern languages, such as Concurrent Pascal, Modula-2, CLU and Ada. It has also been proved useful in concurrent programming. Classes in Simula are based on procedure declarations and the block structure of Algol 60, but free the latter concept from its inherently nested structure by allowing several block instances to co-exist. Simula 67 had a considerable influence on programming languages as diverse as Smalltalk, C++ and Ada.

Smalltalk was developed in the 1970s at Xerox's Palo Alto Research Center. Smalltalk-80 was the first commercially available version, released by Xerox.

Delphi can be considered as a successor of Turbo Pascal whereas Oberon can be considered as a successor of Modula-2.

Eiffel is a programming language designed to encourage the correct construction of software components using the object-oriented approach. An early specification of the Eiffel language can be found in the book *Object-Oriented Software Construction* [36], in which Bertrand Meyer outlined how good software engineering techniques could be incorporated into the language based on the object-oriented paradigm.

The object-oriented language Java, developed by Sun Microsystems, was designed from the ground up to allow for secure execution of code across a network. It shares many similarities with C++, but all constructions considered unsafe by the developers (such as pointers and references for basic data types) are eliminated.

4.5.1   C++

Since C++ was first invented, it has undergone three major revisions, with each revision adding to and altering the language. The first revision was in 1985 and the second occurred in 1990. The third revision occurred during the C++ standardization process. Several years ago, work began on a standard for C++. Towards that end, a joint ANSI and ISO (International Standards Organization) standardization committee was formed. The first draft of the proposed standard was created on 25 January, 1994. In that draft, the ANSI/ISO C++ committee kept the features first defined by Stroustrup and added some new ones. This initial draft reflected the state of C++ at the time. After the completion of the first draft of the C++ standard, an event occurred that caused the standard to expand greatly: the creation of the Standard Template Library (STL). The STL is a set of generic routines such as sorting and searching that we can use to manipulate data. Subsequent to the first draft, the committee voted to include the STL in the specification for C++. The addition of the STL expanded the scope of C++ beyond its original definition. The standardization of C++ took longer than anyone had expected when it began. In the process, many new features were added to the language and many small changes were made. The version of C++ defined by the C++ committee is larger and more complex than Stroustrup's original design. However, the standard is now complete. The final draft was passed out of committee on 14 November, 1997. The material in this book uses Standard C++ to implement SymbolicC++. Standard C++ is the version that is currently accepted by all major compilers.

Since object-oriented programming was fundamental to the development of C++, it is important to define precisely what object-oriented programming is. Object-oriented programming has taken the best ideas of structured programming and has combined them with several powerful concepts that allow us to organize our programs more effectively. In general, when programming in an object-oriented fashion, we decompose a problem into its constituent parts. Each component becomes a self-contained object that contains its own instructions and data related to the object. Through this process, complexity is reduced. All object-oriented programming languages have three things in common: encapsulation, polymorphism and inheritance.

All programs are composed of two fundamental elements: program statements (code) and data. Code is that part of a program that performs actions, and data is the information affected by those actions. Encapsulation is a programming mechanism that binds together code and the data it manipulates, and that keeps both safe from outside interference and misuse.

In an object-oriented language code and data may be bound together in such a way that a self-contained *black box* is created. Within the box are all necessary data and code. When code and data are linked together in this fashion, an *object* is created. In other words, and object is the device that supports encapsulaton. Within an object, the code, data, or both may be *private* to that object or *public*. Private code or data is known to, and accessible only by, another part of the object. This means that private code or data may not be accessed by a piece of the program that exists outside the object. When code or data is public, other parts of our program may access it even though it is defined within an object. Typically the public parts of an object are used to provide a controlled interface to the private elements of the object.

Polymorphism is the quality that allows one interface to be used for a general class of actions. The specific action is determined by the exact nature of the situation. For example, consider a stack (which is a first-in, last-out list). We might have a program that requires three different types of stacks. One stack is used for integer values, one for floating-point values, and one for characters. In this case, the algorithm that implements each stack is the same, even though the data being stored differs. In a non-object-oriented language, we would be required to create three different sets of stack routines, calling each set by a different name, with each set having its own interface. However, because of polymorphism, in C++ we can create one general set of stack routines (one interface) that works for all three specific situations. This way, once we know how to use one stack, we can use them all. The concept of poly-morphism is often expressed by the phrase "one interface, multiple methods." This means that it is possible to design a generic interface to a group of related activities. Polymorphism helps reduce complexity by allowing the same interface to be used to specify a general class of action. It is the compiler's job to select the specification (i.e., method) as it applies to each situation. The first object-oriented program-ming languages were interpreters, so polymorphism was, of course, supported at run time. However, C++ is a compiled language. Therefore, in C++, both run-time and compile-time polymorphism are supported.

Inheritance is the process by which one object can acquire the properties of another object. The reason this is important is that it supports the concept of hierarchical classification. Most knowledge is made manageable by hierarchical (i.e., top-down) classifications. For example, a Student is a Person. Without the use of hierarchies, each object would have to explicitly define all of its characteristics. However, using inheritance, an object needs to define only those qualities that make it unique within its class. It can inherit its general attributes from its parent. Thus, it is the inheri-tance mechanism that makes it possible for one object to be a specific instance of a more general case.

Many of the features of C++ exist to provide support for encapsultion, polymorphism, and inheritance. However we can use C++ to write any type of program, using any type of approach. The fact that C++ supports object-oriented programming does not mean that we can only write object-oriented programs. As with its predecessor, C, one of C++'s strongest advantages is its flexibility.

4.5.2   Java

Java is an object-oriented programming language developed by Sun Microsystems. It shares many similarities with C, C++, and Objective C (for instance the for-loops have the same syntax in all four languages); but it is not based on any of those languages, nor have efforts been made to make it compatible with them. Java is the language of the Internet. It was strongly influenced by C++. Java and C++ both use the same basic syntax, and Java's object-oriented features are similar to C++'s. In fact, at first glance it is possible to mistake a Java program for a C++ program. Because of their surface similarities, it is a common misconception that Java is simply an alternative to C++. Although related, Java and C++ were designed to solve different sets of problems. C++ is optimized for the creation of high-performance programs. Towards this end, C++ compiles to highly efficient, executable code. Java is optimized for the creation of portable programs. To obtain portability, Java compiles to pseudo-code (called Java bytecode), which is usually interpreted. This makes Java code very portable, but not very efficient. Thus, Java is excellent for Internet applications, which must work on a wide variety of computers. But for high-performance programs, C++ will remain the language of choice.

Because of the similarities between Java and C++, most C++ programmers can readily learn Java. The skills and knowledge we gain in C++ will translate easily. However some important differences do exist, so be careful not to jump to false conclusions.

Originally, Java was created because C++ is inadequate for certain tasks. Since the designers were not burdened with compatibility with existing languages, they were able to learn from the experience and mistakes of previous object-oriented languages. A few new features which C++ does not have, like garbage collection and multithreading, have been added. A few C++ features that had been proved to be better in theory than in practice, like multiple inheritance, operator overloading and templates, have been thrown away. There is still argument over whether the designers have made the right choice.

More importantly, Java was designed from the ground up to allow for secure execution of code across the network, even when the source of that code was untrusted and possibly malicious. This requires the elimination of more features in C and C++. Most notably there are no pointers in Java. The programs cannot (at least in theory) access arbitrary addresses in memory.

Furthermore Java was designed not only to be cross-platform like C, but also in compiled binary form. Since this is impossible across different processor architectures, Java is compiled to an intermediate byte-code which is interpreted on the fly by the Java interpreter. Thus to port a Java program to a new platform, all that is needed is to port the interpreter.

Moreover, Java was designed to write bug-free code. Shipping C code has, on average, one bug per 55 lines of code. About half of these bugs are related to memory allocation and deallocation. On the other hand, Java has a number of features that make these bugs less common:

- strong typing;

- no unsafe constructs;

- the language is small so it is easy to become fluent;

- the language is easy to read and write;

- there are no undefined or architecture-dependent constructs;

- Java is object-oriented so reuse is easy;

- Java has concurrency;

- Java has a large number of built-in classes.

The syntax of Java is deliberately similar to C and C++. It is also case-sensitive. Here we list the programming syntax of Java.

- **Data Types**

  The primitive data types of Java are very similar to those in C. Boolean, String and true arrays have been added. However the implementation of the data types has been substantially cleaned up in several ways:

  1. C and C++ leave a number of issues to be machine and compiler dependent (for instance the size of an int), Java specifies everything.
  2. Java prevents casting between arbitrary variables. Only cast between numeric variables and cast between subclasses and superclasses of the same object are allowed.
  3. All numeric variables in Java are signed.

  Below, we listed the built-in data types available in Java:

  - boolean
    1-bit. May take on the values true and false only.
    true and false are defined constants of the language. They are not the same as True and False, TRUE and FALSE, zero and nonzero, 1 and 0 or any other numeric value. The data type boolean may not be cast into any other type of variable nor may any other variable be cast into a boolean.
  - byte
    1 byte, signed (two's complement). Covers values from -128 to 127.

- `short`
  2 bytes, signed (two's complement). Covers from $-32,768$ to $32,767$.

- `int`
  4 bytes, signed (two's complement).
  From $-2,147,483,648$ to $2,147,483,647$. Like all numeric types, `int`s may be cast into other numeric types (`byte`, `short`, `long`, `float`, `double`). When lossy casts are done (e.g. `int` to `byte`) the conversion is done on modulo the length of the smaller type.

- `long`
  8 bytes, signed (two's complement).
  From $-9,223,372,036,854,775,808$ to $9,223,372,036,854,775,807$.

- `float`
  4 bytes, IEEE 754. Covers a range from
  $1.40129846432481707e-45$ to $3.40282346638528860e+38$ (positive or negative).

  Like all numeric types, `float`s may be cast into other numeric types (`byte`, `short`, `int`, `long`, `double`). When lossy casts to integer types are done (e.g. `float` to `short`) the fractional part is truncated and the conversion is done on modulo the length of the smaller type.

- `double`
  8 bytes, IEEE 754.
  Covers from $4.94065645841246544e-324d$ to $1.79769313486231570e+308d$ (positive or negative).

- `char`
  2 unsigned bytes, Unicode.
  `char`s are not the same as `byte`s, `int`s, `short`s or `String`s. `char`s may not be cast into any other type nor may those types be cast to `char`s.

- `String`
  String is an object. It contains zero or more characters enclosed in double quotes.

- `array`
  Arrays are objects. Multi-dimensional arrays are created via arrays of arrays.

`sizeof` is not necessary in Java because all sizes are precisely defined. For example, an `int` always contains 4 bytes.

Java contains `if`, `else`, `for`, `while`, `do while` and `switch` statements, which are identical to C. However all condition tests must return boolean values. Since assignment and arithmetic statements do not return a boolean value, some of the more obfuscated condition tests in C are prohibited.

- **Command Line Arguments**

  Command line arguments are similar to C except that `argv` has become a string array called `args` and `args[0]` has become the first command line argument, not the name of the program. The other arguments are all shifted one to the left accordingly.

- **Comments**

  Java supports both forms of comment adopted in C and C++:

  ```
  /* This is a C and Java comment */
  // This is a C++ and Java comment
  ```

  However comments that begin with `/**` are treated specially by the compiler. These comments should only be used before a declaration. They indicate that the comment should be included in automatically generated documentation for that declaration.

- **Classes**

  Java does not support multiple inheritance. Superclasses of a class are indicated with the keyword `extends`.

- **Methods**

  Methods must be defined within the block that defines the class, which is different from C++.

- **Concurrency**

  Java is internally multi-threaded. The model includes threads, synchronization, and monitors.

Many language constructs have been removed from C++, because the developers of Java believe that they make C++ unsafe and hard to read. Features removed include `#define`, `typedef`, `operator` overloading, `enums`, `union` and `struct`. Another important feature that has been removed is the pointer arithmetic. Other features that have been removed include global variables, standalone functions, friend functions and virtual functions.

A number of features have been added to Java such as true arrays with bounds checking, garbage collection, concurrency, interfaces (from Objective C) and packages. There is no need to explicitly free memory in Java.

We mentioned that there is no pointer in Java. Does that mean that there are no linked lists in Java? How do we make a linked list without pointers? Java uses the `Vector` class in `java.util`. It can do anything a linked list can. After all, code reuse is one of the main features of OOP. Object variables in Java are all references,

where they may be pointers in many other languages. The main difference is that we
cannot do pointer arithmetic on references. Therefore whenever we need a pointer to
an object in C++, we have to use the object itself in Java. On the other hand the
primitive data types

```
byte, short, int, long, float, double, char, boolean
```

are not references. If we want to get a reference to one of these, we need to wrap it
in a class first. Java provides ready-made type-wrapper classes for

```
Byte, Short, Integer, Long, Float, Double, Character, Boolean
```

Java also has a built-in String class, Vector class and Bitset class. Furthermore,
Java has the built-in classes BigInteger and BigDecimal to deal with large inte-
gers and large decimal numbers. The following program shows an application of the
BigInteger class.

```java
// Verylong.java

import java.math.*;

public class Verylong
{
    public static void main(String[] argv)
    {
    BigInteger b1 = new BigInteger("12345");
    BigInteger b2 = new BigInteger("56789");
    BigInteger b3 = new BigInteger("-56712");

    BigInteger b4 = b1.add(b2);
    System.out.println("b4 = " + b4);
    BigInteger b5 = b1.multiply(b3);
    System.out.println("b5 = " + b5);
    BigInteger b6 = b3.abs();
    System.out.println("b6 = " + b6);
    BigInteger b7 = b1.and(b2);
    System.out.println("b7 = " + b7);
    double x = b1.doubleValue();
    System.out.println("x = " + x);
    }
}
```

Java defines one special class, the Object class, which is the ancestor of every other class. It can also be considered as a container class. It declares twelve members: a constructor and eleven methods. Four of them, clone(), hashCode(), equals(), toString() are intended to be overriden. Thereby they provide clean and consistent facilities for duplicating objects and determining when they are the same. The following program shows a small application of the Object class.

```java
// MyObject.java

import java.math.*;       // for BigInteger, BigDecimal

public class MyObject
{
   public static void main(String args [])
   {
   Object[] a = new Object[5];
   a[0] = new String("Good Morning");
   a[1] = new Integer(4567);
   a[2] = new Character('X');
   a[3] = new Double(3.14);
   a[4] = new BigInteger("234567890");

   for(int i=0; i<a.length; i++)
   {
   System.out.println(a[i]);
   }

   a[3] = new BigDecimal("11.5678903452111112345");

   System.out.println(a[3]);

   boolean b = a[4].equals(a[3]);
   System.out.println("b = " + b);  // => false

   Object [] x = new Object[1];
   x[0] = new Float(2.89);
   boolean r = x.equals(a);
   System.out.println("r = " + r);  // => false
   }
}
```

Java passes everything by value. When we are passing primitive data types into a method we get a distinct copy of the primitive data type. When we are passing a handle into a method we get a copy of the handle. Thus everything is passed by value. It seems to allow us to think of the handle as "the object", since it implicitly dereferences it whenever we make a method call. The following program illustrates this.

```java
// Pass.java

import java.math.*;

public class Pass
{
    static void inverse(int x)
    {
    x = -x;
    System.out.println("x = " + x);   // => -7
    }

    static void twice(BigInteger b)
    {
    BigInteger b2 = new BigInteger("2");
    b = b2.multiply(b);
    System.out.println(b);   // => 666666
    }

    public static void main(String[] argv)
    {
    int j = 7;
    inverse(j);
    System.out.println("j = " + j);   // => 7

    BigInteger b1 = new BigInteger("333333");
    twice(b1);
    System.out.println(b1);   // => 333333
    }
}
```

Arrays are passed by reference. The following program illustrates this for basic data types and abstract data types.

```java
// Pass1.java

import java.math.*;

public class Pass1
{
   static void inverse(int x[])
   {
   x[0] = -x[0];
   }

   static void twice(BigInteger b[])
   {
   BigInteger b1 = new BigInteger("2");
   b[0] = b1.multiply(b[0]);
   }

   public static void main(String[] argv)
   {
   // allocating memory for a one-dimensional array
   // with one element
   int [] j = new int[1];
   j[0] = 7;

   inverse(j);
   System.out.println("j[0] = " + j[0]);   // => -7

   BigInteger [] b = new BigInteger[1];
   b[0] = new BigInteger("333333");
   twice(b);
   System.out.println("b[0] = " + b[0]);   // => 666666
   }
}
```

In the following program we show how rational numbers can be implemented using Java. The class `Rational.java` implements rational numbers and their arithmetic functions: addition (`add`), subtraction (`subtract`), multiplication (`multiply`), division (`divide`). This class extends `java.lang.Number`, implementing that class's abstract methods. The methods `equals`, `toString`, `clone` from the `Object` class are overriden.

```java
// Rational.java

import java.lang.*;

class Rational extends Number
{
  private long num;
  private long den;

public Rational(long num,long den)
{
   this.num = num;  this.den = den;
}

private void normalize()
{
   long num = this.num;
   long den = this.den;
   if(den < 0)
   {
   num = (-1)*num;  den = (-1)*den;
   }
}

private void reduce()
{
   this.normalize();
   long g = gcd(this.num,this.den);
   this.num /= g;
   this.den /= g;
}

private long gcd(long a,long b)
{
   long g;
   if(b == 0)
   {
```

```
    return a;
    }
    else
    {
    g = gcd(b,(a%b));
    if(g < 0) return -g;
    else return g;
    }
}

public long num() { return this.num; }

public long den() { return this.den; }

public void add(long num,long den)
{
    this.num = (this.num*den) + (num*this.den);
    this.den = this.den*den;
    this.normalize();
}

public void add(Rational r)
{
    this.num = (this.num*r.den()) + (r.num()*this.den);
    this.den = this.den * r.den();
    this.normalize();
}

public void subtract(long num,long den)
{
    this.num = (this.num*den) - (num*this.den);
    this.den = this.den * den;
    this.normalize();
}

public void subtract(Rational r)
{
    this.num = (this.num*r.den()) - (r.num()*this.den);
    this.den = this.den*r.den();
    this.normalize();
}

public void multiply(long num,long den)
{
```

```
    this.num = (this.num*num);
    this.den = (this.den*den);
    this.normalize();
}

public void multiply(Rational r)
{
    this.num = (this.num*r.num());
    this.den = (this.den*r.den());
    this.normalize();
}

public void divide(long num,long den)
{
    this.num = (this.num*den);
    this.den = (this.den*num);
    this.normalize();
}

public void divide(Rational r)
{
    this.num = (this.num*r.den());
    this.den = (this.den*r.num());
    this.normalize();
}

public static boolean equals(Rational a,Rational b)
{
    if((a.num()*b.den()) == (b.num()*a.den()))
    {  return true;  }
    else
    {  return false; }
}

public boolean equals(Object a)
{
    if(!(a instanceof Rational))
    { return false; }
    return equals(this,(Rational) a);
}

public Object clone() { return new Rational(num,den); }

public String toString()
```

```
{
    StringBuffer buf = new StringBuffer(32);
    long num, den, rem;
    this.reduce();
    num = this.num;
    den = this.den;
    if(num == 0) return "0";
    if(num == den) return "1";
    if(num < 0)
    {
    buf.append("-");
    num = -num;
    }
    rem = num%den;
    if(num > den)
    {
    buf.append(String.valueOf(num/den));
    if(rem == 0)
    {
    return buf.toString();
    } else
    {
    buf.append(" ");
    }
    }
    buf.append(String.valueOf(rem));
    buf.append("/");
    buf.append(String.valueOf(den));
    return buf.toString();
}

public float floatValue()
{
    return (float) ((float)this.num/(float)this.den);
}

public double doubleValue()
{
    return (double) ((double)this.num/(double)this.den);
}

public int intValue()
{
    return (int) ((int)this.num/(int)this.den);
```

```
}

public long longValue()
{
   return (long) ((long)this.num/(long)this.den);
}

public void print()
{
   System.out.print(this.toString());
}

public void println()
{
   System.out.println(this.toString());
}

// main() method used for testing other methods.
public static void main(String args[])
{

   Rational r1 = new Rational(-4,6);
   Rational r2 = new Rational(13,6);
   r1.add(r2);
   System.out.println(r1.toString());

   Rational r3 = new Rational(123,236);
   Rational r4 = new Rational(-2345,123);
   r3.multiply(r4);
   System.out.println(r3.toString());

   Rational r5 = new Rational(3,6);
   Rational r6 = new Rational(1,2);
   boolean b1 = equals(r5,r6);
   System.out.println("b1 = " +b1);
   boolean b2 = equals(r4,r5);
   System.out.println("b2 = " +b2);

   Rational r7 = new Rational(3,17);
   Rational r8 = (Rational) r7.clone();
   System.out.println("r8 = " + r8.toString());
} // end main
} // end class Rational
```

The following programs shows how a `Matrix` class could be implemented. Two constructors are implemented and the methods `equals` and `toString` from the `Object` class are overriden.

```java
// Matrix.java

class Matrix
{
    private int rows, columns;
    public double entries[][];

    Matrix(int m,int n)
    {
    rows = m;
    columns = n;
    entries = new double[m][n];
    int i, j;
    for(i=0; i< rows; i++)
    for(j=0; j< columns; j++)
    entries[i][j] = 0.0;
    }

    Matrix(int m,int n,double[][] A)
    {
    int i, j;
    rows = m;
    columns = n;
    entries = A;
    }

    public void add(Matrix M)
    {
    if((this.rows != M.rows) || (this.columns != M.columns))
    {
    System.out.println("matrices cannot be added");
    System.exit(0);
    }
    int i, j;
    for(i=0; i<columns; i++)
    for(j=0; j<rows; j++)
    this.entries[i][j] = this.entries[i][j] + M.entries[i][j];
    }
```

```
public Matrix multiply(Matrix M)
{
int i, j, t;
if(columns != M.rows)
{
System.out.println("matrices cannot be multiplied");
System.exit(0);
}
Matrix product = new Matrix(rows,M.columns);
for(i=0; i<rows; i++)
{
for(j=0; j<M.columns; j++)
{
double tmp = 0.0;
for(t =0; t<columns; t++)
tmp = tmp + entries[i][t]*M.entries[t][j];
product.entries[i][j] = tmp;
}
}
return product;
}


public void randomize()
{
int i, j;
for(i=0; i<rows; i++)
for(j=0; j<columns; j++)
entries[i][j] = Math.random();
}


public boolean equals(Matrix A,Matrix B)
{
int i, j;
for(i=0; i<rows; i++)
{
for(j=0; j<columns; j++)
{
if(A.entries[i][j] != B.entries[i][j])
return false;
}
}
return true;
}
```

```
public boolean equals(Object ob)
{
if(!(ob instanceof Matrix))
{
return false;
}
return equals(this,(Matrix) ob);
}

public Object clone()
{
return new Matrix(rows,columns,entries);
}

public String toString()
{
int i, j;
String result = new String();
for(i=0; i<rows; i++)
{
for(j=0; j<columns; j++)
{
result = result + String.valueOf(entries[i][j])+"   ";
}
result = result + "\n";
}
result = result + "\n";
return result;
}

public void onStdout()
{
System.out.println(toString());
}

public static void main(String args[])
{
Matrix M = new Matrix(2,2);
M.entries[0][0] = 3.4; M.entries[0][1] = 1.2;
M.entries[1][0] = 4.5; M.entries[1][1] = 5.8;

Matrix N = new Matrix(2,2);
N.entries[0][0] = 6.4; N.entries[0][1] = -1.2;
N.entries[1][0] = 8.5; N.entries[1][1] = 6.8;
```

```
M.add(N);
System.out.println("M = \n" + M.toString());

System.out.println("\n");

Matrix X = new Matrix(2,2);
X = M.multiply(N);
System.out.println("X = \n" + X.toString());

Matrix Y = new Matrix(2,2);
Matrix Z = new Matrix(2,2);
boolean b1 = Y.equals(Z);
System.out.println("b1 = " +b1);
Z.randomize();
System.out.println("Z = \n" +Z);
boolean b2 = Y.equals(Z);
System.out.println("b2 = " +b2);

System.out.println("\n");

double d[][] = new double[2][2];
d[0][0] = 2.1; d[0][1] = -3.4;
d[1][0] = 0.9; d[1][1] = 5.6;
Matrix B = new Matrix(2,2,d);
B.add(B);
System.out.println("B = \n" + B.toString());

System.out.println("\n");

Matrix U = (Matrix) X.clone();
System.out.println("U = \n" + U.toString());
    }
}
```

To illustrate the differences between C++ and Java, we present programs on the binary tree. The program constructs a binary search tree of 100,000 nodes. The content of the tree is traversed using inorder traversal. Note that the elements are arranged in ascending order with inorder traversal. In general, C++ is considerably faster than Java.

The C++ binary tree:

```
// tree.cpp

   #include <iostream.h>
   #include <stdlib.h>

   class Tree
   {
    private:
      int  data;
      Tree *left, *right;
    public:
      Tree(int);
      void insert(int);
      void inorder();
   };

   Tree::Tree(int n) : data(n), left(NULL), right(NULL) { }

   void Tree::insert(int n)
   {
      if(n < data)
      {
      if(left != NULL)
      left->insert(n);
      else
      left = new Tree(n);
      }
      else
      if(n > data)
      {
      if(right != NULL)
      right->insert(n);
      else
      right = new Tree(n);
      }
   }
```

```
void Tree::inorder()
{
   if(this != NULL)
   {
   left->inorder();
   // print data here ...
   right->inorder();
   }
}

void main()
{
Tree *root = NULL;
int i, m, n = 100000;

for(i = 0; i < n; i++)
{
m = rand();
if(root == NULL)
root = new Tree(m);
else
root -> insert(m);
}
root -> inorder();
}
```

The Java binary tree:

```
// tree.java

import java.util.Random;

class Tree
{
   private int data;
   private Tree left = null;
   private Tree right = null;
   public Tree(int n)
   { data=n; }

   public void insert(int n)
   {
```

```
        if(n<data)
        {
        if(left != null)
        left.insert(n);
        else
        left = new Tree(n);
        }
        else
        if(n>data)
        {
            if(right != null)
            right.insert(n);
            else
            right = new Tree(n);
        }
    }

    public void inorder()
    {
        if(left != null)
        left.inorder();
        // print data here ...
        if(right != null)
        right.inorder();
    }

    public static void main(String args[])
    {
        Random r = new Random(start);
        Tree t = null;
        int n = 100000;
        int i;
        for(i=0; i<n; i++)
        {
        int m = Math.abs(r.nextInt()%1000);
        if(t == null)
        t = new Tree(m);
        else
        t.insert(m);
        }
        t.inorder();
    }
}
```

4.5.3   Other Object-Oriented Languages

Besides C++ and Java there are a number of other object-oriented languages. In this section we discuss briefly Eiffel, Smalltalk and Oberon.

Eiffel [57] is a pure object-oriented programming language with Pascal-like control structures. It satisfies the following requirements for implementing abstract data types (ADTs) for a programming language:

- The language must be able to define new user-defined data types, that is, provide facilities for building ADTs.

- It must be able to create one or more instances of a user-defined data type, which corresponds to instances of an ADT.

- There must be facilities to support procedural abstractions, i.e. it must provide procedures and functions with appropriate parameter passing mechanisms for implementing the operations of ADTs.

- The language must allow the implementation of generic ADTs.

- It must fully support encapsulation.

- It must support software component reuse (some of the previous requirements also contribute to this requirement).

Abstract data types can be implemented easily in Eiffel as classes (equivalent to types). Programming in Eiffel involves constructing new classes from existing classes using either the client-supplier relationship or inheritance, or both. The idea of a contract between software components leads to a programming form that is more likely to produce correct code. Eiffel can be described as a language that:

- is based on the manipulation of objects;

- is strongly typed;

- supports the construction of correct programs;

- enables easy reuse of code;

- is based on the theory of abstract data types;

- enables new types to be built from existing types;

- supports a programming method known as programming by *contract*.

An object in Eiffel is an instance of a class, and a class is the description of a data type. A typical class contains the description of the operations which apply to instances of the type, together with descriptions of the data stored. Therefore, we have to examine:

1. How to create individual objects which are instances of a data type.

2. How to write the descriptions of classes which implement the data types.

3. How to construct and execute an Eiffel program.

We begin by showing how to declare objects (using declarations) and how to apply the data type operations to them (using instructions). In Eiffel, data type operations are implemented as **features** of a **class**. Therefore, we examine, in outline, how new classes are constructed. In the final section, we explain how to build an Eiffel system (a program that can be executed) which will enable us to tackle practical programming problems.

A typical Eiffel program manipulates on objects. To facilitate the construction of the *instructions* which define the manipulation to be carried out, *identifiers* are used to name objects. For example, to declare an identifier as PERSON we write:

    p: PERSON

where p refers to an object and PERSON is the name of a previously defined data type. We say that p is of type PERSON. To declare an identifier for a queue of PERSON we write:

    q: QUEUE[PERSON]

where QUEUE is a *generic* data type – that is, the data type QUEUE has been implemented in such a way that the type of the items to be held in a particular queue can be specified at the time when the queue object is declared. Thus, in the construct QUEUE[PERSON], the square brackets [PERSON] is known as a *generic parameter*, which specifies that the queue contains items of type PERSON. Hence, QUEUE[PERSON] is said to be the specialization of a previously defined generic data type QUEUE.

Eiffel has a number of basic data types: INTEGER, REAL, BOOLEAN, CHARACTER and DOUBLE, which can be used to declare identifiers. For example, in the following i and r refer to objects of type INTEGER and REAL, respectively:

    i: INTEGER
    r: REAL

In the examples, single letter identifiers have been used to denote objects but it is usually better to have more meaningful names so that the resulting code is easier to read. In Eiffel, an identifier may consist of one or more letters and/or digits provided that the first character is a letter. We may also include the underscore character (_) in an identifier. There is no limit on the number of characters in an identifier, nor is the case of the letters significant. Here are some examples of legitimate Eiffel identifiers:

```
address, make, phone_number, size, C3Po, the_Date,
customer1, customer2, put_char, addToQueue, PERSON
```

Eiffel distinguishes the language used for defining data types from the data types themselves. It is quite common for vendors to provide a comprehensive library of commonly used classes, such as lists, trees and so on, giving the users a set of reusable components from which they start building new applications. Despite the separation of the language from the class library, Eiffel does assume the existence of the basic data types: `INTEGER`, `REAL`, `BOOLEAN`, `CHARACTER` and `DOUBLE`. Each of them has a set of standard operations and is used exactly the same way as the types we have introduced so far. However, there are some differences which we now explore.

In Eiffel, the mechanism for declaring objects, in which the creation and initialization activities are separated, is somewhat cumbersome and inefficient for the basic data types. Therefore, a short-cut is provided. For example, the declaration:

```
i: INTEGER
```

specifies a memory location at which an integer i is stored. Moreover, i is initialized to a default value *zero*. Thus, i is not a reference to an integer, but is directly associated with the storage location of the integer. This is equivalent to a normal (non-pointer) variable in other languages. Such a data type is said to be an *expanded type*. Eiffel provides the capability to associate an identifier directly with the storage for all data types, thereby avoiding the reference mechanism. However, it is usual for objects to be viewed as dynamic, in the sense that each of them have a limited lifetime because they are created, manipulated and destroyed during the execution of the program. The reference mechanism supports this view. Thus, Eiffel has both the reference types (the normal situation) and expanded types (used mainly for the basic types). The following table gives the default initial values for the basic types:

Class	Default value
INTEGER	0
REAL	0.0
CHARACTER	'%U' (the null character)
DOUBLE	0.0 (double precision)
BOOLEAN	false

Next, let us consider the construct of a class:

```
class <class_name>
    creation
        -- the name(s) of the procedure(s) used in the creation of new instances
    feature
        -- declarations of the features of the class
    end -- CLASS_NAME
```

A class declaration may contain different sections, each specified by keywords such as
creation or feature. Although not essential, it is recommended that the name of
the class be included as a comment following the keyword end.

In the case of reference types, the initialization process means that we can build
several creation procedures for one data type to provide initializations appropriate
to different situations. Note that the existence of a creation procedure implies that
the default initialization does not create a valid object. Therefore, Eiffel insists that
we invoke one of them or generate an error otherwise. For expanded types, however,
since the creation procedure is automatically invoked by the Eiffel system when the
object is created, only one creation procedure is allowed to avoid ambiguity.

Smalltalk can be described as a combination of a programming language, an operating system, a programming environment and a design and programming method-
ology [17], [35]. It can be considered as a tool for knowledge representation, thus
making it a serious competitor to Lisp and Prolog. The application areas in which it
excels are application prototyping, interactive graphics systems and simulation. Although there are still some problems with the speed of execution (mainly stemming
from the interpretive nature of the system), spectacular results have been achieved in
cost savings with certain types of software, especially in applications containing high
graphics or interactive components. The principal reason for this is that the Smalltalk
environment takes the maximum advantages of reuse of software components.

All the entities used in Smalltalk-80 are objects. An object may represent a number,
a character, a drawing, a list, a program or an editor, etc. An object has two main
characteristics:

- It allows data to be stored and accessed.

- It can reply to messages.

For example, the object 0.0 will respond to the message sin by returning the value
0, which is written as

```
0.0 sin
```

The only means to manipulate an object is via *message passing*. Messages therefore constitute the interface between objects and the external world. The way in which an object responds to a message is private and need not be known externally. Writing an application in Smalltalk consists of determining the objects that describe the problem and defining the operations for each object. When the basic objects of an application and the operations on these objects are defined, the programmer can use them without knowing the detail of their construction; he can also reuse them in contexts different from where they were constructed.

In the Smalltalk environment, several types of scope variable are defined:

1. *Instance variables*: They are variables that belong to an object and there are two types of instance variables — *named instance variables* and *indexed instance variables*. Named instance variables are instance variables identified by a name. For example, the class Point has two named instance variables, x and y. When the name of an instance variable appears in an expression, it refers to the value of the corresponding instance. When a new instance is created, it contains instance variables specified by its class; the default value of these instance variables is an object called nil. Indexed instance variables are instance variables that are not accessed by name but by a numeric index. In contrast to objects that possess only named instance variables, those possessing indexed instance variables may have a varying number of instance variables.

2. *Class variables*: They are variables that are shared by all the instances of a class. For example, in the case of a computer-aided drawing application, the outline of a circle is made up of several line segments and the number of these segments is the same for all circles; one can therefore define it as a class variable.

3. *Global variables*: They are variables that are shared by all objects. They are stored in a dictionary. For example, all the classes are referred to by global variables whose name is the access key to the dictionary.

4. *Pool variables*: They are variables that are accessible by instances of several classes. In order to define a set of variables shared by several classes, it is necessary to define a dictionary that is common to these classes, where this dictionary itself is a global variable.

Now, let us consider the classes and subclasses in Smalltalk. Each object in Smalltalk belongs to one and only one class. It can, however, be useful to share some elements of the description between several classes or to describe one class by means of another. This is implemented in Smalltalk, where each class is described in terms of another class called its *superclass*. The instances of the new class are identical to those of its superclass, except for additions made explicit in the new class. The new class is called a *subclass* of its superclass.

A subclass itself can also have one or more subclasses. The set of classes thus takes on a tree structure whose root is the class called `Object`. `Object` is the only class that does not possess a superclass. The instances of each subclass inherit all the instance variables, class variables as well as the methods from the superclass.

To give a complete definition of a class, we need to specify:

- Name: The name of the class is obligatory and must be different from the name of any other existing class.

- Superclass: The superclass is also obligatory and must correspond to an existing class.

- New class variables: Variables are optional, but if they exist their name must be different from any name already defined in the set of its superclasses. New methods may also be added, and methods already defined in the set of its superclasses may be redefined.

- Its new pool variables.

- Its new instance variables.

- The list of its new methods.

Every entity in Smalltalk is an object. Consequently, a class itself is an object. We have also seen how each object belongs to one and only one class. A class itself is therefore an instance of another class which we call the *metaclass*. In earlier versions of Smalltalk, all the classes were instances of one single metaclass. However, the current version of Smalltalk makes each class the sole instance of a metaclass for the reason of flexibility (especially when creating new instances of a class).

The metaclasses are classes and therefore contain methods that allow their instances (that is, the classes) to respond to the messages they receive. As a result, these methods are called the *class methods*, while the other methods are called *instance methods*. When a class is created, a new metaclass is automatically created. In contrast to other classes, metaclasses have no name; neither do they have metaclass instances. They are all instances of the same class called `Metaclass`.

One can access the metaclass of a class by sending the message `class`. For example, consider a class of points named `Point`, we can access its metaclass by the expression:

```
Point class
```

The metaclasses are classes, therefore they have a superclass called `Class`. To summarize, we can regard metaclasses in two ways:

1. They are instances of the class `Metaclass`.

2. They are subclasses of the class `Class`.

`Metaclass` and `Class` are two subclasses of the class `ClassDescription`.

A variable name is a simple identifier consisting of an initial letter followed by a sequence of letters or digits. All characters are significant. The normal convention in Smalltalk is to concatenate words into one identifier, with each of the words in the variable name beginning with a capital letter, except perhaps the first word. Simple variables begin with lower case letters (e.g. `hydeParkCorner` or `redPen`); class names, global and class variables conventionally begin with an upper case letter (e.g. `LargePositiveInteger, DisplayScreen` and `BigBen`). This distinction is not rigid, but it is advisable to adhere to it, otherwise the code written may not be comprehensible to others.

Smalltalk supports five kinds of literals. They are characters, numbers, strings, symbols and arrays of literals. Smalltalk literals always refer to the same object and can be thought as literal constants.

- *Character Literals*: Character literals are instances of the class `Character`. They are represented by an ASCII character preceded by a dollar symbol, e.g.

    `$4   $t   ${   $$   $|`

- *Numeric Literals*: Numeric literals represent numbers. They are a sequence of digits preceded by an optional minus sign. The sequence of digits may contain a decimal point with at least one digit on either side. The following are examples of numeric literals:

    `123    -12    3.14    -3.14    0.0    0`

    Numeric literals in scientific notation are also acceptable, e.g.

    `1.0e1    -3e3    -3.13e-3`

    Numeric literals can also be expressed in a non-decimal base by preceding the number with '<base>r'. For bases greater than 10, the extra digits are represented by capital letters, starting with 'A'.

- *String Literals*: String literals are objects that refers to sequences of characters. They are all instances of class `String` and they are represented by a sequence of characters delimited by apostrophes, e.g.

    `'There'   'isn"t'   'any'   'more!'`

Note that a double single quote (") is used here to represent the character literal
$' within the string 'isn"t'. The individual characters in a string are in fact
represented by their corresponding character literals.

- *Symbol Literals*: Symbol literals are strings of characters used to name objects
  in the Smalltalk system. They are all instances of the class Symbol. They
  represent identifiers and message selectors, and are denoted by a *unique* sequence
  of characters preceded by a # symbol. The symbol must begin with a letter and
  may contain digits. Some examples of symbols are given below.

  #a   #alpha2   #intersectsWith   #max

- *Array Literals*: Array literals are a numerically indexed data structure with
  all the elements represented by literals. The literal array elements must be
  separated by spaces and all enclosed in parentheses with a single leading #
  character. Several kinds of literal arrays are given below:

#(1 2 3)	an array of numeric literals
#($a $B)	an array of character literals
#('to' 'be' 'or' 'not' 'to' 'be')	an array of string literals
#(jack frost)	an array of symbols
#(jddf 8r45 'crazy' $b (1 2))	an array of literals

  Notice that in the last two examples, the symbols #jack, #frost, #jddf and the
  array #(1 2) do not require a preceding # character if they are within another
  array literal.

Objects can be assigned to variables. This is accomplished by the use of the assignment prefix. The assignment prefix consists of a variable name followed by a left
arrow. It can also be concatenated. Smalltalk variables are much like pointers in C,
in the way they always point to an object and they can point to any kind of object (or
memory location in C) at any time throughout their life. A variable can be assigned
to a *pseudo-variable*. A pseudo-variable is a particular defined object in Smalltalk
that can only be accessed through a variable name. Since pseudo-variables represent
predefined objects, their use is reserved. The pseudo-variables we may encounter
includes nil, which represents an uninitialized object of unspecified type, true for
logical truth and false for logical falsehood. Variable names must be declared before they are used. However, it is legal for a variable to share the same name as a
message. The syntax of the message expression will determine when the name refers
to a variable or a message.

Oberon [43], [61] is the latest descendant in the family of languages whose root is
Algol 60 (1960). Other members include Pascal (1970) and Modula-2 (1979). Pascal has some deficiencies. They are not significant in introductory courses, but they

become relevant in the realm of larger systems. While Pascal encourages structured design, *modular design* becomes more and more important in software engineering. This notion has at least two aspects. The first one is known as *information hiding*. Any large system is composed of modules that are designed in relative isolation. This implies that an interface exists and it specifies all the properties accessible to the other modules and hides all others. The second aspect is of a technical nature, the *separate compilation* of modules, where the compiler checks the correct use of interfaces. When a module is compiled, a description of its interface is written to a *symbol file*. During the compilation of a client module, the compiler gets the symbol files of the imported modules and thus obtains all the necessary information for type checking.

The principal innovation of the language Modula-2 with respect to Pascal was indeed the module concept, incorporating information hiding and separate compilation. In contrast to independent compilation known from assemblers and other language compilers, separate compilation enables a compiler to perform the same type-consistency checks across module boundaries as within a module. The explicit definition of interfaces and the retention of full type safety turned out to be a tremendous benefit.

Modules exporting one or several data types, together with a set of procedures operating on variables of these types, represent the notion of *abstract data type*. In these cases, only the names of the types appear in the module interface, whereas the structure of the record's fields is accessible via exported procedures only, which can therefore rely on the validity of certain invariants governing the abstract types.

Furthermore, Modula-2 removed one of the most aggravating handicaps found in strong typing: it introduced dynamic arrays as parameters of procedures. Also noteworthy was the introduction of procedure types for variables, and facilities for concurrent processes and low-level programming. The latter allow a programmer to refer directly to specific machine facilities, such as interface registers for controlling input/output operations. Once again, these features contributed to the widening of the language's applications, particularly in the areas of system design and process control. Last but not least, certain syntactic properties of Pascal were remedied, notably the open-ended if, while and for statements. These were precisely the structures that were adopted from Algol 60 and left unchanged in order to maintain tradition and to avoid alienating the Algol community.

Several years of experience in practising modular design with Modula-2 and other system programming languages revealed that the ultimate goal was *extensible design*. Structured programming and modular programming were merely intermediate steps towards the goal. The introduction of abstractions represented by modules and the use of procedures calling procedures declared at lower levels of the abstraction hierarchy embodies the extensibility in the procedural domain. Equally important for a successful design, however, is the extensibility in the domain of data definition. In this respect, Modula-2 is inadequate, because types cannot be extended and remain compatible at the same time.

In this respect, *object-oriented languages* provided a viable solution, and became the wave of the 1980s. They offer a facility to define subtypes $T1$, called subclasses, of a given type (class) $T0$ with the property that all operations applicable to instances of $T0$ are also applicable to instances of $T1$. We recognize at this point that the ultimate innovation was data type extensibility, which unfortunately remained obscure behind the term object-oriented. This term was also accompanied by a whole new nomenclature for the many familiar concepts with the aim of perpetrating a new view or metaphor of programming at large. Thus types became classes, variables instances, procedures methods and procedure activations messages, etc.

The primary merit of the language Oberon, developed in 1986, lies in the provision of data type extensibility on the basis of the established, well-understood notions of data type and procedure. The consequence is that no break with traditional programming technique is necessary and no familiarization with a whole new class of concepts and notions is required. The only new facility is the extension of a record type. Oberon thereby unifies the traditional concepts of procedural programming with the techniques required to obtain data extensibility.

This single new facility might well have been added to Modula-2. Why was yet another new language created? The reason was the desire to have a language available that upholds the principle of processors. Oberon is easily an order of magnitude smaller than systems of comparable (or even lesser) functionality.

The Oberon system is a hierarchy of modules, most of which export one or a few abstract data types. Each user is encouraged to extend the system – extensions are created by simply adding new modules. There is no boundary between the system and the application program. Except for modules *Kernel* and *Display*, the entire Oberon system is expressed in the programming language Oberon [42].

4.6   Summary

This chapter introduced the fundamental concepts of object-oriented programming, which include the followings:

- *Object*: It represents both attributes and methods through the process of encapsulation.

- *Abstract data type*: It is a collection of data and a set of operations on that data.

- *Class*: It is a collection of objects that satisfy some common properties.

- *Message passing*: It provides a means for objects to communicate with one another.

- *Inheritance*: It provides a mechanism for managing classes and sharing common codes.

- *Polymorphism*: It gives extra meanings to methods, making them more flexible and extensible.

The concepts presented here are general features of object-oriented programming. They are not restricted to any programming language. Last but not least, five object-oriented programming languages were described, namely C++, Java, Eiffel, Smalltalk, and Oberon.

Each object-oriented language has its own merits. The question arises which languages to use to write a computer algebra system ? We have chosen C++ as our programming language for the development of our computer algebra system. The reasons for our choice include

1. C++ is faster than the other languages.

2. We need the flexibility of pointers.

3. We need the template feature provided by C++.

4. A large amount of C/C++ code already exists for problems in physics and mathematics. This code can be linked with the computer algebra system.

5. We need to link our system to Parallel Virtual Machine (PVM) [16].

6. We can use the Standard Template Library with SymbolicC++.

The following chapter describes the basic tools in C++, which is one of the most popular object-oriented languages. Subsequent chapters apply the concepts we learned here to mathematics and physics, using the programming language C++.

Chapter 5

Basic Tools in C++

C++ not only corrects most of the deficiencies found in C, it also introduces many completely new features that were designed for the language to provide data abstraction and object-oriented programming. Here are some of the prominent new features:

- *Classes*, the basic language construct that consists of data structure and operations applicable to the class.

- *Member variables*, which describe the attributes of the class.

- *Member functions*, which define the permissible operations of the class.

- *Operator overloading*, which gives additional meaning to operators so that they can be used with user-defined data types.

- *Function overloading*, which is similar to operator overloading. It allows the same function to have several definitions whereby reducing the need for unusual function names, making code easier to read.

- *Programmer-controlled automatic type conversion*, which allows us to blend user-defined types with the fundamental data types provided by C++.

- *Derived classes*, also known as subclasses, inherit member variables and member functions from their base classes (also known as superclasses). They can be differentiated from their base classes by adding new member variables, member functions or overriding existing functions.

- *Virtual functions*, which allow a derived class to redefine member functions inherited from a base class. Through *dynamic binding*, the run-time system will choose an appropriate function for the particular class.

In the next few sections, we demonstrate how to apply these new features to create data types for particular applications, and combine these abstract data types into object-oriented programs.

121

5.1   Pointers and References

A *pointer* is a reference to data or code in a program. It is literally the address in memory of the item pointed at. Pointers enables us to write more flexible programs, especially for object-oriented programs. We use pointers when the following situations occur:

- *If the program handles large amounts of data (more than 64K).*

  This problem arises only for personal computers using DOS as the operating system. For a large, complex program the 64K area C++ sets aside for data might not be large enough to hold all the data. Pointers let us get around this.

  When we declare global variables in C++, the compiler allocates space for them in an area called the *data segment.* The data segment has a maximum size of 64K where all the global variables must be accommodated. For small programs, this limit does not matter, but there are times when we might need more. For example, suppose a program requires an array of 400 strings, each of them holding up to 100 characters. The array would take up roughly 40K bytes, which is less than the maximum 64K. An array of this size is not a problem, assuming the other variables fit in the remaining 24K.

  But what if we need two such arrays at the same time? That would require about 80K, which cannot be fitted in the 64K data segment. To handle larger amounts of data, we need to use a *heap.* The program can allocate the 80K on the heap, keeping a pointer as a reference to the location of the data. This pointer takes up only 4 bytes in the data segment.

- *If the program uses data of unknown size at compile time.*

  A heap is all the memory the operating system makes available that is not being used by the program code, its data segment and its stack. Using C++ we can reserve space on the heap, access it through a pointer, and then release the space again. Some C++ data items (for example arrays of characters) need to have their sizes specified at compile-time, even though they might not need all the allocated space when the program runs. A simple example would be a program that reads a string from the user, such as the user's name. To store that name in a regular string variable, we would have to reserve enough space to handle the largest possible string, even if the name is only a few letters. If the allocation of variables on the heap is carried out during run-time, we can allocate just the right size to hold the actual string data.

  This is a trivial example, but in applications with hundreds of such data items (such as multiple windows or lists read from files), allocating only as much space as needed can mean a difference between running successfully or running out of memory.

- *If the program uses temporary data buffers.*

  Pointers and the heap are extremely handy for situations when we need memory allocated temporarily, but do not want to commit that memory for the entire duration of the program. For example, a file editor usually needs a data buffer for every file being edited. Rather than declaring at compilation time that we will have a certain number of buffers of a certain size allocated for files, we can allocate just as much as we need at a given time, making memory available for other purposes.

  Another common usage of temporary memory is sorting. Usually when we sort a large array of data, we make a copy of the array, sort the copy, and then copy the sorted data back into the original array. This protects the integrity of the data. However, it also requires to have two copies of the data while we are sorting. If we allocate the sorting array on the heap, we can sort it and copy it back into the original, then dispose of the sorting array, freeing that memory for other uses.

- *If the program handles multiple data types.*

  One less common use of pointers is to point at variable data structures. For instance, a block of memory is set aside to hold a "history list" of different-length string items typed into a data-entry field. To read the history list, a routine would scan through the block looking for individual strings. We could use a simple pointer to indicate where the blocks begins. This simply tells us where something is, without specifying what it is.

- *If the program uses linked lists.*

  Another common use of pointers is to tie together *linked lists*. In many simple database-type applications, we can hold data records in arrays or typed files. However sometimes we need something more flexible than a fixed-size array. By allocating dynamic records so that each record has a pointer that points to the next record, we can construct a list that contains as many elements as we need.

A pointer is a memory address in a computer. It could be the address of a variable, a data record or a function. Normally, we do not care where the variable resides in memory. All we need is to refer to it by name. The compiler knows where to look for it. That is exactly what happens when we declare a variable. For example, if the program includes the following code,

```
int number;
```

The compiler will set aside an area in memory which is referred to as number. We can find out the memory address of number by using the & operator. This memory address can be assigned to a *pointer variable*, which holds the address of the data or code in memory.

```
int number = 10;
int *p;            // p is a pointer variable
p = &number;       // the memory address of number is assigned to p
```

So far we have seen how values are assigned to pointers, but that is not much use if we cannot get the values back. We could treat a typed pointer as if it were a variable of the type by *dereferencing* it. To dereference a pointer, we use the *(prefix) operator. Suppose p contains the memory address of number, *p gets the value of number. Consider the following:

```
// pointer.cxx

#include <iostream.h>

void main()
{
    double *p1, *p2;        // declares two pointers to double
    double a = 2.5, b = 5.1;
    p1 = &a;               // p1 holds the address of a
    p2 = p1;
    cout << "*p1 = " << *p1 << " and *p2 = " << *p2 << endl;
    *p2 = b;
    cout << "*p1 = " << *p1 << " and *p2 = " << *p2 << endl;
}
```
Result
======
*p1 = 2.5 and *p2 = 2.5
*p1 = 5.1 and *p2 = 5.1

In the first cout statement, both p1 and p2 pointed at a. Therefore, *p1 and *p2 have values of 2.5. In the second cout statement, the content of p2 is assigned as b using the dereferencing operator. As a result, both p1 and p2 point at a value of 5.1.

Now we describe how pointers can be used dynamically:

- *Allocating dynamic variables*: C++ uses the keywords new and delete for allocating and deallocating dynamic variables respectively. The new operator allocates dynamics variables on the heap. For example,

    ```
    new int;
    ```

    allocates one object of type int whereas

    ```
    new int[100];
    ```

allocates an array of 100 objects of type int. The new operator returns a pointer to the type specified. It can be assigned to a pointer variable, such as

```
int *p = new int[500];
```

The dynamically allocated int array here is uninitialized. To initialize class objects allocated with new, we use their constructor. It has the following form:

```
Thing *p = new Thing(argument list);
```

The new operator allocates an object of the class Thing, which is initialized by its constructor that matches the argument list.

- *Deallocating dynamic variables*: Variables allocated with new must be deallocated when we are finished with them to make the heap space available for other dynamic variables. To achieve this purpose, we use the delete operator. For example,

```
delete p;
```

It returns the memory, previously allocated by the new operator, back to free store. To release an array of objects, we use

```
delete [] p;
```

The brackets [] are used here to free an array of objects.

As an example, suppose we are reading in some strings from a file and storing them in the memory. We are not sure how long any of the strings will be, therefore we need to declare memory spaces that are long enough to accommodate the longest possible string. Assuming that not all the strings take up the maximum length, then many spaces would be wasted. To get around this, we read each string into a buffer, then allocate just the right space to store the actual information of the string, as demonstrated in the following program:

```
// buffer.cxx

#include <iostream.h>
#include <fstream.h>    // for ifstream (input file stream)
#include <string.h>     // for strlen (length of a string)

const int dataSize  = 1000;
const int maxLength = 256;
```

```
void main()
{
   ifstream fin("LongString.dat");
   char buffer[maxLength];
   char *Long[dataSize];
   int i;
   for(i=0; i<dataSize; i++)
   {
      fin >> buffer;
      Long[i] = new char[strlen(buffer)+1];
      strcpy(Long[i],buffer);
   }
}
```

In the program, we read in strings from the data file "LongString.dat". Suppose this file consists of 1000 lines with each line not exceeding 256 characters. Our purpose is to store all the text into an array of strings called Long. In each step, we read in a string and store it in the buffer. We then measure the correct size of the string and allocate the right memory space for it. This process carefully avoids wasting the memory resource in a computer. Instead of allocating 256K for all the strings (256 characters per string times 1000), we use only the right amount of memory for the text plus a buffer which is 256 bytes large.

There is a special pointer in C++ called the this pointer. It is a constant pointer to an object of the class containing the member function, i.e. it denotes an implicitly declared *self-referential pointer* of the object. Let us illustrate the idea with a simple program:

```
// this.cpp
// (1) The member function increment uses the implicitly declared
//     pointer this to return the newly incremented value of
//     both c1 and c2.
// (2) The member function where_am_I displays the address
//     of the given object.
// (3) The this keyword provides a built-in self-referential pointer.

#include <iostream.h>

// declaration of class Cpair
class Cpair
{
  private:
    char c1, c2;
```

```
  public:
    Cpair(char);
    Cpair increment();
    void print();
    void where_am_I();
};

// Definition of class Cpair
Cpair::Cpair(char c) : c2(c), c1(1+c) { }

Cpair& Cpair::increment() { c1++; c2++; return *this; }

void Cpair::print() { cout << c1 << " and " << c2 << endl; }

void Cpair::where_am_I() { cout << this << endl; }

void main()
{
    Cpair x('A');
    x.print();              // output : B and A
    x.where_am_I();         // output : 0x7fffaee0

    Cpair z('X');
    z.where_am_I();         // output : 0x7fffaedc
    z.increment();
    z.print();              // output : Z and Y
    z.where_am_I();         // output : 0x7fffaedc

    Cpair n('1');
    n.increment().print();  // output : 3 and 2
    n.where_am_I();         // output : 0x7fffaed8
}
```

Pointers are powerful in C++. However a couple of common problems need to be avoided:

- *Dereferencing uninitialized pointers*: One common source of errors with pointers is to dereference a pointer that has not been initialized. Like all other C++ variables, a pointer's value remains undefined until we assign it. In principle, it could point at anywhere in the memory. If we dereference such a pointer, we will get some random bits. This becomes disastrous if we assign some values to the item pointed to, because we may overwrite some important data segment, such as the program code or even the operating system. This sounds a little ominous, but with a little discipline it is easy to manage.

To avoid dereferencing uninitialized pointers, which is potentially dangerous, C++ provides the keyword NULL that can be used for pointers that point at *nothing*. A NULL pointer is valid, but unattached. One of the common uses of a NULL pointer is to terminate a linked list. Suppose L_1, L_2, L_3 are elements of a linked list. To indicate that the linked list has ended, we usually make the last element point at the NULL pointer, as shown below:

$$L_1 \rightarrow L_2 \rightarrow L_3 \rightarrow \text{NULL}$$

- *Losing heap memory ("heap leaks"):* Another common problem when using dynamic variables is known as the *heap leak*. A heap leak is a situation where space is allocated on the heap and then lost – for some reason the pointer no longer points at the allocated memory area, so that it cannot be referred or deallocated.

A common cause of heap leaks is by reassigning dynamic variables without disposing of the previous ones. For instance,

```
#include <iostream.h>

void main()
{
    int* ptr;
    ptr = new int[500];
    ptr = NULL;
}
```

The pointer variable ptr is first allocated to an int array of 500 elements. It is then reassigned to NULL straight away. This action has made the memory area (the int array) initially pointed to by ptr lost, and it is not recoverable. In such a situation, memory has leaked. Serious memory leakage during run-time may cause a program to halt abnormally due to memory exhaustion.

C++ introduces a concept called *reference*. Basically, it defines an alias or alternative name for any instance of data. The syntax is to append an ampersand (&) to the name of the data type. For example,

```
int  ii =    5;    // An automatic variable ii
int *pi = &ii;     // A pointer to int which points at ii
int &ri =  ii;     // A reference to the variable ii
```

Now, we can use `ri` anywhere just as we would use `ii` or `*pi`. Suppose we write

```
ri *= 10;          // Multiply ri by 10
```

The values of `ii`, `*pi` and `ri` become 50. As an example, consider the following program:

```
// ref.cxx

#include <iostream.h>

void main()
{
    int i = 23;
    int *ip;
    ip = &i;

    cout << "i  = " << i << endl;      // i  = 23
    cout << "ip = " << ip << endl;     // ip = 0x7fffaed4
    cout << "&i = " << &i << endl;     // &i = 0x7fffaed4

    int &j = i;    // alias j with i
    int *jp;

    j = 45;        // the value of i has been changed too!
    jp = &j;

    cout << "jp = " << jp << endl;     // jp = 0x7fffaed4
    cout << "&j = " << &j << endl;     // &j = 0x7fffaed4

    cout << "i = " << i << endl;       // i = 45
    cout << "j = " << j << endl;       // j = 45
}
```

Notice that the values of the addresses for `&i`, `ip`, `jp` and `&j` are all identical. This indicates that references in C++ work by referring to the same memory location in the computer. This also explains why the value of `i` is also altered when we perform the statement `j = 45`.

By far the most important use for references is in the passing of arguments to functions (see Section 5.8 for an example on reference as an argument).

5.2   Classes

Class is a primary C++ construct used to create abstract data types (ADTs). It describes the behaviours of an object, such as attributes, operations, argument types of the operations and so on. The general syntax for declaring a `class` is:

```
class class-name
{
    private:
        <private data members>
        <private member functions>
    protected:
        <protected data members>
        <protected member functions>
    public:
        <public data members>
        <public member functions>
};
```

Classes in C++ offer three levels of visibility for the data members and member functions — `public`, `protected` and `private`. They are called the *access specifier*. Each access has its own merits, as described below:

- `private`: Only member functions of the class have access to the private members. Class instances are denied from accessing them.

- `protected`: Only member functions of the class and its descendant classes have access to the protected members. Class instances are denied from accessing them.

- `public`: Members are visible to the member functions of the class, class instances, member functions of descendant classes and their instances.

In the following, we list some simple guide-lines for the proper use of classes:

- The access specifier may appear in any order.

- The same access specifier may appear more than once.

- If an access specifier is omitted, the compiler treats the members as `protected`.

- Avoid placing data members in the `public` region, unless such a declaration significantly simplifies the program design.

- Data members are usually placed in the `protected` region so that the member functions of descendant classes can access them.

- Use member functions to alter or query the values of data members.

After a class has been declared, the class name can be used to declare new class instances. The syntax resembles declaring variables. In the rest of the discussion, we will construct a simplified version of the String class step by step. Along with the construction, we explain the concepts and the C++ language constructs. An improved version of the String class will be presented in Chapter 6.

In traditional C, a string is represented by a pointer to char. This means a string can be considered as an array of type char. In this representation, the end-of-string is denoted by '\0'. This convention has a major drawback. The '\0' requires one byte to store and it is part of the string. This means "abc" requires 4 bytes of memory to hold. The inconsistency between the length of the string and the memory required for the string has caused much confusion to many users. It is also a pitfall. Therefore, to avoid possible error and improve efficiency of the operations, it is best to implement the string as an abstract data type.

The listing shows a typical class description and implementation. As we see here, the class String has two data fields. The first field datalength maintains the size of the largest possible String. The second field data is a pointer to the type char. When it is allocated to some appropriate memory space, it will be used to store the characters of the String. Most of the time, data fields are declared as private for the reason of data hiding. However, it is perfectly alright to put them in the publicly accessible region.

```
// estring.h

#include <iostream.h>
#include <string.h>
#include <assert.h>

class String
{
 private:
   // Data fields
   int datalength;
   char *data;
 public:
   // Constructors and Destructor
   String();
   String(char);
   String(const char*);
   String(const String&);
   ~String();
   // Assignment operator
   const String& operator = (const String&);
```

```cpp
    // Member function
    int length() const;
    // Conversion operator
    operator const char *() const;
    // Friends : concatenation and output stream
    friend String operator + (const String&,const String&);
    friend ostream& operator << (ostream&,const String&);
};

// Class implementation

// Constructors and Destructor
String::String() : datalength(1), data(new char[1])
{
    assert(data != NULL);
    data[0] = '\0';
}

String::String(char c) : datalength(2), data(new char[2])
{
    assert(data != NULL);
    data[0] = c;
    data[1] = '\0';
}

String::String(const char *s)
    : datalength(strlen(s) + 1), data(new char[datalength])
{
    assert(data != NULL);
    strcpy(data,s);
}

String::String(const String &s)
    : datalength(strlen(s.data) + 1), data(new char[datalength])
{
    assert(data != NULL);
    strcpy(data,s.data);
}

String::~String()
{
    delete [] data;
}
```

```cpp
// assignment operator
const String & String::operator = (const String& s)
{
    if(&s != this)
    {
        delete [] data;
        datalength = strlen(s.data) + 1;
        data = new char[datalength];
        assert(data != NULL);
        strcpy(data,s.data);
    }
    return *this;
}

int String::length() const
{
    return strlen(data);
}

// conversion operator
String::operator const char *() const
{
    return data;
}

// concatenation operator
String operator + (const String& s1,const String& s2)
{
    String S(s1.length() + s2.length());
    strcpy(S.data, s1.data);
    strcat(S.data, s2.data);
    return S;
}

// friendship operators
ostream& operator << (ostream& out,const String& s)
{
    out << s.data;
    return out;
}
```

5.3   Constructors and Destructor

The two most important class member functions are the *constructor* and *destructor*. There could be many constructors in a class but it can have only one destructor. A constructor is responsible for the creation of class instances or objects. It can be used to handle initialization and allocation of dynamic memory for an object. Note that the constructors always have the same name as the class. A destructor, on the other hand, is used to clean-up after an object is destroyed. The clean up process usually releases the unused memory back to the system. It has the same name as the class except for a tilde (˜) prefix.

For our `String` class here, there are three overloaded constructors. Each of them is responsible for a different type of construction:

1. `String();`

   e.g. `String x; String y=0; String z="";`

   set x, y and z to an empty `String`. Note that 0 and "" may be used as NULL string.

2. `String(char);`

   e.g. `String x('A'), y('#');`

   set x to the character A and y to the character # respectively.

3. `String(const char*);`

   e.g. `String x("abc");`

   set x to the `String` "abc" .

Note that in each constructor, the `data` is allocated to a certain memory space. This is always required as memory spaces have to be reserved from the system to store the string. However, dynamically allocated memory is not freed automatically by the system, it must be freed by the users. Freeing unused memory is important for it to be "recycled" for subsequent allocations. In C++, a destructor is used to free memory when variables are out of scope. This is done implicitly by the system. A destructor for our string class is defined as follows:

```
String::~String()
{
    delete [] data;
}
```

The square brackets [] are used here to delete the entire array.

5.4   Copy Constructor and Assignment Operator

As the name suggested, the *copy constructor* is used to duplicate the content of an instance during a declaration statement. On top of that, it will also be invoked to generate temporary values when class arguments are passed as value parameters. If the data fields do not include any pointers that have to be initialized dynamically, the copy constructor generated by the compiler which performs memberwise copy, works correctly. However, a proper copy constructor is needed when the data fields involve pointer variables.

Let us take our `String` class for example. Suppose a string S1 = "Computer" is copied using a memberwise copy constructor to another string S2. The pointer address of S1 is copied over instead of its content "Computer". Both S1 and S2 point at the same memory location. If S2 is no longer needed, the `String` destructor will be invoked and the memory storage pointed to by S2 will be released. This leaves S1 with a *dangling* pointer — a pointer that does not point to any valid block of memory. When the remaining string S1 is destroyed, the `delete` operator will attempt to free the same memory block that has already been freed. This is disastrous if that memory block has already been allocated to other objects. Therefore, it is always good practice to provide a copy constructor for every user-defined class.

In our `String` class, the copy constructor takes a `String` as argument, generates another memory buffer and duplicates the content of the argument, as shown below:

```
String::String(const String &s)
    : datalength(strlen(s.data)+1), data(new char[datalength])
{
    assert(data != NULL);
    strcpy(data,s.data);
}
```

where `strlen()`, `strcpy()` are routines from `<string.h>` and `assert()` is from the library `<assert.h>`.

- `char *strcpy(char *s1,const char *s2);`
  Copies the string `s2` into the string `s1`, including the terminating null character. The value `s1` is returned.

- `size_t strlen(const char *s);`
  Returns the length of the string `s`, where the length is defined as the number of characters in the string, not counting the terminating null character.

- `void assert(int expr);`
  If `expr` is zero (false), then diagnostics are printed and the program is aborted. The diagnostics include the expression, the file name, and the line number in the file.

An *assignment operator* is needed whenever the data fields require dynamic memory allocation, or whenever the copy constructor of the class needs to be rewritten. For our `String` class, its definition is as follows:

```
const String &String::operator = (const String& s)
{
    if(&s != this)
    {
        delete [] data;
        datalength = strlen(s.data) + 1;
        data = new char[datalength];
        assert(data != NULL);
        strcpy(data,s.data);
    }
    return *this;
}
```

From the above listing, we notice that this function is similar to the copy constructor but there are two major differences:

- The assignment operator needs to handle the special case when an object is assigned to itself. This explains why the function started with an `if` statement. The conditional statement ensures that the operator works properly when a string is assigned to itself, e.g. `x = x;` because it is not only redundant to assign something to itself, more importantly the pointers `data` and `s.data` refer to the same address. It is impossible for `delete [] data;` and `strcpy(data,s.data);` to work correctly.

- Notice that the operator returns a reference `String&`. This allows statements like

  ```
  String x, y, z;
  x = y = z = "Space Shuttle";
  ```

  to work properly. The return reference type has made the left-hand side assignment possible.

5.5 Type Conversion

C++ has many type conversion rules which the compiler obeys when converting the value of an object from one fundamental type to another. These rules make mixed type operations possible. They are convenient but potentially dangerous if they are not used properly. Consider the following example,

```
// convert.cxx

void f(long);
void f(int);

void g(long);
void g(unsigned int);

void main()
{
    f('A');     // The compiler chooses f(int)
    g('P');     // Ambiguous !!!
}
```

Note that the ASCII table is used to convert the characters 'A' and 'P' into integers. However, the code generates a compilation error for the ambiguity between g(unsigned int) and g(long). Furthermore, implicit conversion may induce subtle run-time bugs which are hard to detect. Therefore, it is useful to know how the conversion rules work in C++. The general rules for implicit type conversion involve two steps.

1. First, a char, short, enum is promoted to int whenever an int is expected. It is converted to unsigned int if it is not representable as int.

2. If the expression still consists of mixed data types, then lower types are promoted to higher types, according to the following hierarchy:

$$\text{int} \rightarrow \text{unsigned int} \rightarrow \text{long} \rightarrow \text{unsigned long}$$
$$\rightarrow \text{float} \rightarrow \text{double} \rightarrow \text{long double}$$

The resultant value of the mixed expression has the higher type.
Suppose double a = 2; and long b = 3; then a+b does the following:

- b is promoted to double.

- Perform double addition a+b, and the type of the result is also double.

Explicit conversion uses *cast*. In traditional C, cast has the following form:

(*type*) expression

whereas in C++, there is another alternative (functional notation)

> *type* (expression)

For example, the following two equivalent expressions convert the `int n` to a `double`

```
int n = 10;
double x, y;
x = (double) n;
y = double(n);
```

However, the functional notation is preferred.

A construction with one argument serves as a type conversion operator from the argument's type to the constructor's class type. As an example, consider our string constructor:

```
String::String(const char *s)
    : datalength(strlen(s)+1), data(new char[datalength])
{
    assert(data != NULL);
    strcpy(data,s);
}
```

The constructor converts variables of `char*` to `String` when necessary. This kind of conversion only works for the conversion from an already defined type to a user-defined type. It is not possible to add a constructor to a built-in type (e.g. `int`, `double`). Sometimes, it is desirable to convert a user-defined type to a built-in data type. For example, we want the `String` class to work with the traditional C string manipulation functions:

```
// strlen.cxx

#include <iostream.h>
#include <string.h>
#include "MString.h"

void main()
{
    String s("Microscope");

    int i = strlen(s), j = s.length();

    if(i == j)
        cout << "The string " << s << " consists of " << i
```

```
              << " characters " << endl;
   }
```
Result
======

The string Microscope consists of 10 characters

In the code, the compiler automatically converts a String into a char * in the argument of strlen(). C++ provides a special conversion function to achieve this. The general form of such a conversion function is

```
operator type() { ... }
```

For our String class, the conversion function is

```
operator const char *() const { return data; }
```

The first const states that the memory pointed to by the return value should not be modified by code outside the class, and the second const says that this member function does not alter the contents of the String object operated on.

Type conversions also apply on pointers. Any pointer type can be converted to a generic pointer of type void*. However, a generic pointer needs to be cast to an explicit type when it is assigned to a non-generic pointer variable.

```
// generic.cxx

#include <iostream.h>

void main()
{
   int *n = NULL;
   void *generic_ptr;
   generic_ptr = n;            // OK
   n = (int*)generic_ptr;      // OK
   n = generic_ptr;            // Error !!!
}
```

The code generates a compilation error message:

```
a value of type "void *" cannot be assigned to an entity of type "int *
   n = generic_ptr;
     ^
```

The null pointer can be converted to any pointer type

```
char *ptr1 = 0;
int  *ptr2 = ptr1;  // Not OK: need (int *)ptr1;
int  *y = 0;        // OK
```

A pointer to a class can be converted to a pointer to a publicly derived base class. This also applies to references.

Let us consider a simple example on an abstract data type — the `Verylong` class.
The class we discuss here considers only positive long integer numbers which may
exceed the limit of the built-in **unsigned long** type. This is a simplified version of
the class which will be discussed in Chapter 6.

In this section, we focus on the type conversion in the class. For an abstract data
type, there are two types of conversion. The first is the conversion from the built-in
data types to the abstract data type. This is accomplished by the constructors of the
class. The second is the conversion from the abstract data type to a built-in type.
This type of conversion uses the conversion operator provided by C++.

```cpp
// cast.cxx

#include <iostream.h>
#include <string.h>
#include <math.h>

class Verylong
{
 private:
   char *data;
 public:
   Verylong(const char* = NULL);
   Verylong(unsigned long);
   Verylong(const Verylong&);
   Verylong& operator = (const Verylong&);
   ~Verylong();
   operator int() const;
   friend ostream & operator << (ostream&,const Verylong&);
};

Verylong::Verylong(const char *value)
{
   if(value)
   {
      data = new char[strlen(value)+1];
      strcpy(data,value);
   }
   else
   {
      data = new char[1];
      *data = '\0';
   }
}
```

```
Verylong::Verylong(unsigned long n)
{
   int digits;
   digits = (int)log10(n) + 1;
   data = new char[digits + 1];
   data[digits--] = '\0';

   while(n >= 1) // extract the number digit by digit
   {
      data[digits--] = n%10 + '0';
      n /= 10;
   }
}

Verylong::Verylong(const Verylong& x)
{
   data = new char[strlen(x.data) + 1];
   strcpy(data, x.data);
}

Verylong& Verylong::operator = (const Verylong& rhs)
{
   if(this == &rhs) return *this;
   else
   {
      delete [] data;
      data = new char [strlen(rhs.data) + 1];
      strcpy(data, rhs.data);
      return *this;
   }
}

Verylong::~Verylong() {  delete [] data; }

Verylong::operator int() const
{
   int a, number = 0;
   a = strlen(data);
   if(a > 5)
   {
      cerr << "Conversion not possible" << endl;
      return 0;
   }
```

```
    else
    {
       for(int j=a; j>=1; j--)
          number += (data[a-j] - 48)*pow(10,j-1);
    }
    return number;
}

ostream& operator << (ostream& os,const Verylong& x)
{  return os << x.data; }

void main()
{
    // string => Verylong
    Verylong s1("1234");
    cout << "s1 = " << s1 << endl; cout << endl;

    // Verylong => int
    int n = s1;
    cout << "n = " << n << endl; cout << endl;

    // Verylong number exceeded range of int
    Verylong s2("7777777777");
    cout << "s2 = " << s2 << endl; cout << endl;

    // Verylong => int
    int m = s2;
    cout << "m = " << m << endl; cout << endl;

    // unsigned long => Verylong
    Verylong n1(234567890);
    cout << "n1 = " << n1 << endl;
}
```

```
Result
======
s1 = 1234
n = 1234
s2 = 7777777777
Conversion not possible
m = 0
n1 = 234567890
```

5.6 Operator Overloading

In C++, a built-in operator may be given more than one meaning, depending on its arguments. Consider the binary addition operator a + b. The data type of a and b could be int, float, double or even a user-defined type such as Matrix. The operation for matrix addition is certainly very different from the floating point addition. We say that the operator + possesses more than one meaning. In fact, there are many more C++ operators that can be overloaded, as listed in the following:

Type	Operator Notation
Unary	++(prefix/postfix) --(prefix/postfix) &(address of)
	*(dereference) + - ~ ! (*type*)
Arithmetic	* / % + -
Shift	<< >>
Relational	> >= < <= == !=
Bitwise	& ^ |
Logical	&& ||
Assignment	= *= /= %= += -= <<= >>= &= ^= |=
Data Access	[] -> ->*
Others	() , new delete

Almost all C++ operators can be overloaded except

Member access operator	a.b
Dereferencing pointer to member	a.*b
Scope resolution operator	a::b
Conditional operator	a?b:c

Only predefined operators can be overloaded in C++. We cannot introduce any new operator notations, such as the exponentiation ** used in Fortran. Although the operators can be given extra meaning, the order of precedence cannot be changed. Suppose we wish to overload the XOR operator ^ as the exponentiation operator. We have to be very careful about the order of precedence. The expression a+b*c^d will be evaluated as (a+(b*c))^d, not as a+(b*(c^d)). Extra parentheses are needed for the expression to be correct.

In fact, there are two kinds of overloading in C++: *member overloading* and *global overloading*. Let us consider again the addition operator a + b where a and b are objects of class C. The general form for member overloading is as follows:

```
C C::operator + (C) { ... }
```

In the program, the expression a + b means a.operator+(b). We could also overload the operator using global overloading:

```
friend C operator + (C,C) { ... }
```

where **a** and **b** are the first and second parameters of the function, respectively. Note that a global overloaded operator needs a **friend** declaration in the class. The friendship between functions and classes will be discusses in Section 5.9.

Let us consider an example of operator overloading. Suppose we wish to overload the addition operator + to perform the concatenation of two **String** objects.

```
// concat.cxx

#include <iostream.h>
#include "MString.h"

void main()
{
    String S1 = "father", S2 = "mother", S3;
    S3 = "My " + S1;
    S3 = S3 + " and my ";
    S3 = S3 + S2;
    cout << "The concatenated string is " << S3 << endl;
}
```

```
Result
======
The concatenated string is My father and my mother
```

If we use the member overloading for the operator, "My " + S1 will be interpreted as "My ".operator+(S1). This is an error, because "My " is not an instance of the class. No member operator function can be applied to "My ". This happens because the C++ compiler does not automatically convert the left-hand operand of any member functions. It is therefore impossible for the code to work without global overloading.

The global overloading of +, on the other hand will convert the expression "My " + S1 to operator+(String("My "),S1). The compiler automatically converts the left-hand side of the + operator to a **String**. To understand how the concatenation works, let us look at the code:

```
String operator + (const String &s1, const String &s2)
{
    String S(s1.length() + s2.length());
    strcpy(S.data, s1.data);
    strcat(S.data, s2.data);
    return S;
}
```

The code generates a temporary buffer S with an appropriate size, copies S1 to S and appends S2 behind S using strcat().

As another example, consider a family of linear operators $\{ b, b^\dagger \}$ on an inner product space V. The commutation relations for these operators (so-called *Bose operators*) are given by

$$[b, b^\dagger] = I \tag{5.1}$$

where I is the identity operator and

$$[b, b] = [b^\dagger, b^\dagger] = 0$$

where 0 is the zero operator and $[\,,\,]$ denotes the commutator. The operator b is called an *annihilation operator* and the operator $b^\dagger$ is called a *creation operator*. The vector space V must be infinite-dimensional for (5.1) to hold. Let $|0\rangle$ be the vacuum state, i.e.

$$b|0\rangle = 0, \quad \langle 0|0\rangle = 1. \tag{5.2}$$

A state is given by

$$(b^\dagger)^n|0\rangle \quad \text{where } n = 0, 1, 2, \ldots$$

As an example, consider the operator $\hat{H} = b^\dagger b b^\dagger b$ and the state $|\phi\rangle = b^\dagger|0\rangle$, then

$$
\begin{aligned}
\hat{H}|\phi\rangle &= b^\dagger b b^\dagger b b^\dagger|0\rangle \\
[\text{by } (5.1)] &= b^\dagger b b^\dagger (I + b^\dagger b)|0\rangle \\
[\text{by } (5.2)] &= b^\dagger b b^\dagger|0\rangle \\
[\text{by } (5.1)] &= b^\dagger (I + b^\dagger b)|0\rangle \\
[\text{by } (5.2)] &= b^\dagger|0\rangle
\end{aligned}
$$

In the following program, we overload the addition operator as the creation operator and the subtraction operator as the annihilation operator. These operators apply on the vacuum state $|0\rangle$ and the outcome is another state. Several operators have been used for demonstrations:

$$b^\dagger|0\rangle, \quad b|0\rangle, \quad bb^\dagger|0\rangle, \quad bb^\dagger b^\dagger b^\dagger|0\rangle, \quad b^\dagger b b^\dagger b b^\dagger|0\rangle.$$

```
// bose.cxx

#include <iostream.h>

class State
{
 private:
   int m;       // number of creators acting on vacuum
   int factor;  // number of identical state
 public:
```

```
    // Constructor
    State();                // create a state
    // methods
    void display() const;   // display the state
    void bose_state(int);   // operation on a state
    void reset();           // reset the state
};

// Bose operator
class Bose
{
 public:
    // Constructor
    Bose();
    // Operators
    State operator + (State&); // creation operator
    State operator - (State&); // annihilation operator
};

State::State() : m(0), factor(1) { }

void State::bose_state(int bstate)
{ if(bstate == -1) factor *= m--; else m++; }

void State::display() const
{
    if(!factor) cout << "0";
    else
    {
        cout << factor << "*";
        if(m)
        {
            cout << "(";
            for(int i=0; i<m; i++) cout << "b+";
            cout << ")";
        }
        cout << "|0>";
    }
}

void State::reset() {  m = 0; factor = 1; }

Bose::Bose() { }
```

```
State Bose::operator + (State &s2)
{
    State s1(s2);
    s1.bose_state(1);
    return s1;
}

State Bose::operator - (State &s2)
{
    State s1(s2);
    s1.bose_state(-1);
    return s1;
}

void main()
{
    State g;
    Bose b;
    g = b+ g;
    cout << "g = "; g.display(); cout << endl;
    g.reset();
    g = b- g;
    cout << "g = "; g.display(); cout << endl;
    g.reset();
    g = b- (b+ g);
    cout << "g = "; g.display(); cout << endl;
    g.reset();
    g = b- (b+ (b+ (b+ g)));
    cout << "g = "; g.display(); cout << endl;
     g.reset();
    g = b+ (b- (b+ (b- (b+ g))));
    cout << "g = "; g.display(); cout << endl;
}
```

Result
======
```
g = 1*(b+)|0>
g = 0
g = 1*|0>
g = 3*(b+b+)|0>
g = 1*(b+)|0>
```

Next, let us consider the *Fermi operator*. Consider a family of linear operators $c_j, c_j^\dagger$ for $j = 1, \ldots, n$ defined on a finite dimensional vector space V satisfying the *anti-commutation relations*

$$[c_j, c_k]_+ = [c_j^\dagger, c_k^\dagger]_+ = 0 \qquad (5.3)$$

$$[c_j, c_k^\dagger]_+ = \delta_{jk} I \qquad (5.4)$$

where 0 is the zero operator and I is the unit operator with $j, k = 1, 2, \ldots, n$. Operators satisfying (5.3) and (5.4) are called the annihilation and creation operators for fermions.

We define a state $|0\rangle$ (the so-called *vacuum state*) with the properties

$$c_j |0\rangle = 0 \quad \text{for } j = 1, 2, \ldots, n \qquad (5.5)$$
$$\text{and} \quad \langle 0|0\rangle = 1 \qquad (5.6)$$

The state $|0\rangle$ is normalized. Other states can now be constructed from $|0\rangle$ and the creation operators $c_j^\dagger$.

Applying the rules given above, we find

$$
\begin{aligned}
& c_1 c_4 c_1^\dagger c_4^\dagger |0\rangle \\
[(\text{by } 5.4)] \quad &= -c_1 c_1^\dagger c_4 c_4^\dagger |0\rangle \\
[(\text{by } 5.4)] \quad &= -c_1 c_1^\dagger (I - c_4^\dagger c_4) |0\rangle \\
[(\text{by } 5.5)] \quad &= -c_1 c_1^\dagger |0\rangle \\
[(\text{by } 5.4)] \quad &= -(I - c_1^\dagger c_1) |0\rangle \\
[(\text{by } 5.5)] \quad &= -|0\rangle
\end{aligned}
$$

In the following program, we implement the Fermi operators and several examples are given:

```
// fermi.cxx

#include <iostream.h>
#include <assert.h>
#include "MString.h"

// It maintains the information of the operators still in active
class Power
{
public:
    String c; // the operator name, e.g. c1, c4+
    int n;    // the degree of the operator, e.g. (c4+)^2
```

```
};

class State
{
private:
    int factor; // the multiplication of the state
    int m;      // number of distinctive operators
    Power *p;   // a pointer to Power

public:
    // Constructor
    State();

    void operator = (State&);        // assignment operator
    void Fermi_creator(String,int);  // operation on the state
    void display() const;            // display the state
    void reset();                    // reset the state
};

// Fermi operator
class Fermi
{
private:
    String f;   // store the name of the operator, e.g. c1, c4

public:
    // Constructor
    Fermi(String);
    State operator + (State&);   // creation operator
    State operator - (State&);   // annihilation operator
};

State::State() : m(0), factor(1), p(NULL) {}

void State::operator = (State& s2)
{
    m = s2.m;
    p = new Power[m]; assert(p);
    for(int i=0; i<m; i++) p[i] = s2.p[i];
    factor = s2.factor;
}

void State::Fermi_creator(String ch,int s)
{
```

```
if(factor)
{
   // [c1,c2]_+ = [c1+,c2+]_+ = [c1,c2+]_+ = 0
   for(int i=0; i<m && (ch != (p+i)->c); i++)
      if((p+i)->n % 2) factor *= (-1);

   // if there is a new operator
   if(i==m)
   {
      if(s==1) // creation operator
      {
         Power *p2;
         m++;
         p2 = new Power[m]; assert(p2);
         for(int j=0; j<m-1; j++) p2[j] = p[j];
         (p2+m-1)->c = ch; (p2+m-1)->n = 1;

         delete [] p;
         p = p2;
      }
      else  // annihilation operator
      { m=0; factor=0; delete [] p;}
   }
   else // operator appears before
   {
      // creation operator, c1+ (c1+)^n = (c1+)^(n+1)
      if(s==1) (p+i)->n++;
      else // annihilation operator, [c1,c2+]_+ = I
      {
         if((p+i)->n % 2) // if power of operator is odd
         {
            (p+i)->n--;
            if(!(p+i)->n)
            {
               for(int j=i+1; j<m; j++) p[j-1] = p[j];
               m--;
            }
         }
         else // if power of operator is even
         { m=0; factor=0; delete [] p;}
      }
   }
}
```

```
void State::display() const
{
   if(!factor) cout << "0";
   else
   {
      if(factor != 1) cout << "(" << factor << ")*";

      for(int i=0; i<m; i++)
      {
         cout << " " << (p+i)->c << "+";
         if ((p+i)->n != 1) cout << "^" << (p+i)->n;
      }
      cout << "|0>";
   }
}

void State::reset()
{
   m=0; factor=1;
   delete [] p; p=NULL;
}

Fermi::Fermi(String st) : f(st) {}

State Fermi::operator + (State &s2)
{
   State s1(s2);
   s1.Fermi_creator(f,1);
   return s1;
}

State Fermi::operator - (State &s2)
{
   State s1(s2);
   s1.Fermi_creator(f,-1);
   return s1;
}

void main()
{
   State g;
   Fermi c1("c1"), c4("c4");
```

```
      g = c1-  g;
      cout << "g = "; g.display(); cout << endl;

      g.reset();
      g = c1+  g;
      cout << "g = "; g.display(); cout << endl;

      g.reset();
      g = c1- (c1+ g);
      cout << "g = "; g.display(); cout << endl;

      g.reset();
      g = c4- (c4+ (c1+ (c4+ g)));
      cout << "g = "; g.display(); cout << endl;

      g.reset();
      g = (c4- (c4+ (c4+ (c1+ (c4+ g)))));
      cout << "g = "; g.display(); cout << endl;

      g.reset();
      g = c4- (c4+ (c4- (c4+ (c4- (c4+ g)))));
      cout << "g = "; g.display(); cout << endl;

      g.reset();
      g = (c4+ (c4+ (c4+ (c4+ (c1+ g)))));
      cout << "g = "; g.display(); cout << endl;
   }
```

Result
======
```
g = 0
g =   c1+|0>
g = |0>
g = 0
g = (-1)* c4+^2 c1+|0>
g = |0>
g =   c1+ c4+^4|0>
```

5.7 Class Templates

Templates are usually used to implement data structures and algorithms that are independent of the data types of the objects they operate on. It allows the same code to be used with respect to different types, where the type acts as a parameter of the data structures or algorithms. They are also known as *parametrized types*. One important use for templates is the implementation of *container classes*, such as stacks, queues, lists, etc. A stack template may contain objects of type int, double or even user-defined types. The compiler will automatically generate the implementations of the Stack classes. The syntax of the class declaration is

```
template <class T>
class Stack
{
 public:
   Stack(int);
   ...
};
```

where the symbol T serves as the argument for the template which stands for arbitrary type. To declare a Stack, we use

```
Stack<char> s1(100); // 100 elements Stack of char
Stack<int> s2(50);  //  50 elements Stack of int
```

The keyword **template** always preceeds every forward declaration and definition of a template class. It is followed by a formal template list surrounded by angle brackets (< >). The formal parameter list cannot be empty. Multiple parameters are separated by commas, e.g.

```
template <class T1, class T2, class T3, class T4> class Myclass1
```

Declarations with specific type such as int, e.g.

```
template <class T, int size> class Myclass2
```

are also allowed. When the compiler encounters the object definitions, it substitutes the template type names by the actual data types. If the class template uses more than one template, the compiler will perform the type substitutions one at a time, beginning with the first class referred. As an example, consider a simplified version of a Vector template class (the full version will be presented in Chapter 6):

```
// svector.h

#include <iostream.h>
#include <string.h>
```

```
#include <assert.h>

// definition of class Vector
template <class T> class Vector
{
    private:
        // Data Fields
        int size;
        T *data;
    public:
        // Constructors
        Vector();
        Vector(int);
        Vector(const Vector<T>&);
        ~Vector();
        // Member Functions
        T &operator [] (int) const;
        void resize(int);
        // Assignment Operator
        const Vector<T> &operator = (const Vector<T>&);
};

// implementation of class Vector
template <class T> Vector<T>::Vector() : size(0), data(NULL) {}

template <class T> Vector<T>::Vector(int n) : size(n),data(new T[n])
{ assert(data != NULL); }

template <class T> Vector<T>::Vector(const Vector<T>& v)
    : size(v.size), data(new T[v.size])
{
    assert(data != NULL);
    for(int i=0; i<v.size; i++) data[i] = v.data[i];
}

template <class T> Vector<T>::~Vector()
{ delete [] data; }

template <class T> T &Vector<T>::operator[] (int i) const
{
    assert(i >= 0 && i < size);
    return data[i];
}
```

```
template <class T> void Vector<T>::resize(int length)
{
   int i;
   T *newData = new T[length]; assert(newData != NULL);
   if(length <= size)
      for(i=0; i<length; i++) newData[i] = data[i];
   else
      for(i=0; i<size; i++)   newData[i] = data[i];
   delete [] data;
   size = length;
   data = newData;
}

template <class T>
const Vector<T> &Vector<T>::operator = (const Vector<T> &v)
{
   if(this == &v) return *this;
   if(size != v.size)
   {
      delete [] data;
      data = new T[v.size]; assert(data != NULL);
      size = v.size;
   }
   for(int i=0; i<v.size; i++) data[i] = v.data[i];
   return *this;
}

void main()
{
   // The symbol Vector must always be accompanied by a
   // data type in angle brackets
   Vector<int> x(5);       // generates a vector of integers
   int i,j;

   for(i=0; i<5; ++i) x[i] = i*i;
   for(i=0; i<5; ++i) cout << x[i] << " ";
   cout << endl; cout << endl;

   Vector<char> ch(3);     // generates a vector of characters

   ch[0] = 'a'; ch[1] = 'b'; ch[2] = 'c';
   cout << ch[1] << endl;
   cout << ch[1]-ch[2] << endl;
   cout << endl;
```

```
    Vector<char *> c(2);    // generates a vector of strings

    c[0] = "aba"; c[1] = "bab";

    for(i=0; i<2; i++) cout << c[i] << endl;
    strcpy(c[0],c[1]);
    cout << c[0] << endl; cout << endl;

    Vector<Vector<int> > vec(3);
    for(i=0; i<3; i++) vec[i].resize(4);
    vec[0][0] = 7;   vec[0][1] = 4;  vec[0][2] = -3;   vec[0][3] = -17;
    vec[1][0] = 4;   vec[1][1] = 9;  vec[1][2] = 0;    vec[1][3] = 5;
    vec[2][0] = 7;   vec[2][1] = 11; vec[2][2] = 8;    vec[2][3] = 77;

    for(i=0; i<3; i++)
    {
        for(j=0; j<4; j++) cout << vec[i][j] << "   ";
        cout << endl;
    }
    cout << endl;

    cout << "vec[0][0] = " << vec[0][0] << endl;
    cout << "vec[0][0]*vec[2][3] = " << vec[0][0]*vec[2][3] << endl;
}
Result
======
0 1 4 9 16

b
-1

aba
bab
bab

7   4   -3   -17
4   9   0   5
7   11   8   77

vec[0][0] = 7
vec[0][0]*vec[2][3] = 539
```

5.8   Function Templates

In the last section, we discussed the class templates. In fact, functions can also be parameterized. In function template, all the type arguments must be mentioned in the arguments of the function. Therefore, we do not need to specify any template argument when calling a template function. The compiler will figure out the template type arguments from the actual arguments.

By using function templates, we can define a family of related functions by making the data type itself a parameter. It enhances the compactness of the code without forfeiting any of the benefits of a strongly-typed language. Traditionally, a function has to be defined for a specific data type to make it work. If we wish to write a mathematical library containing Bessel functions, then we have to define the Bessel functions for int, float, double, long double and complex, etc. If we want to improve the algorithm at a later stage, all the Bessel functions with different types have to be changed. This is cumbersome and prone to errors. By defining the functions on templates, the compiler will automatically generate the code for each data type.

The keyword **template** always preceeds the definition and forward declaration of a template function. It is followed by a comma-separated list of parameters, which are enclosed by a pair of angle brackets (< >). This list is known as the formal parameter list of the template. It cannot be empty. Each formal parameter consists of the keyword **class** followed by an identifier, which could be a built-in or user-defined type. The name of a formal parameter can occur only once within the parameter list. The function definition follows the formal parameter list. An example is

```
template <class T> T min(T x,T y) { ... }
```

We illustrate the function templates using a function that swaps two pointers. In the first version of the program we use pointers, whereas in the second version we use references. The schematic diagram is shown in Figure 5.1.

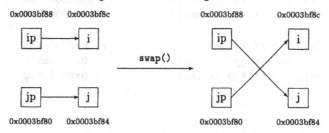

Figure 5.1: *Swapping two pointers*

```
// First version
// swap1.cxx

#include <iostream.h>

template <class T> void swap(T *a,T *b)
{
    T temp;
    temp = *a;
    *a = *b;
    *b = temp;
}

void main()
{
    int i, *ip;
    ip = &i;
    cout << "&i = " << &i   << endl;           // 0x0003bf8c
    cout << "ip = " << ip   << endl;           // 0x0003bf8c
    cout << "&ip = " << &ip << endl;           // 0x0003bf88

    int j, *jp;

    jp = &j;
    cout << "&j = " << &j << endl;             // 0x0003bf84
    cout << "jp = " << jp << endl;             // 0x0003bf84
    cout << "&jp = " << &jp << endl;           // 0x0003bf80

    swap(&ip,&jp);

    cout << "&i  = " << &i  << endl;           // 0x0003bf8c
    cout << "ip  = " << ip  << endl;           // 0x0003bf84
    cout << "&ip = " << &ip << endl;           // 0x0003bf88

    cout << "&j  = " << &j  << endl;           // 0x0003bf84
    cout << "jp  = " << jp  << endl;           // 0x0003bf8c
    cout << "&jp = " << &jp << endl;           // 0x0003bf80
}
```

```
// Second version
// swap2.cxx

#include <iostream.h>

template <class T> void swap(T& a,T& b)
{
   T temp;

   temp = a;
   a = b;
   b = temp;
}

void main()
{
   int i, *ip;

   ip = &i;
   cout << "&i =  " << &i   << endl;        // 0x0003bf8c
   cout << "ip =  " << ip   << endl;        // 0x0003bf8c
   cout << "&ip = " << &ip  << endl;        // 0x0003bf88

   int j, *jp;

   jp = &j;
   cout << "&j =  " << &j   << endl;        // 0x0003bf84
   cout << "jp =  " << jp   << endl;        // 0x0003bf84
   cout << "&jp = " << &jp  << endl;        // 0x0003bf80

   swap(ip,jp);

   cout << "&i  = " << &i   << endl;        // 0x0003bf8c
   cout << "ip  = " << ip   << endl;        // 0x0003bf84
   cout << "&ip = " << &ip  << endl;        // 0x0003bf88

   cout << "&j  = " << &j   << endl;        // 0x0003bf84
   cout << "jp  = " << jp   << endl;        // 0x0003bf8c
   cout << "&jp = " << &jp  << endl;        // 0x0003bf80
}
```

5.9   Friendship

From the previous discussions, we know that only member functions can access the private data of a class. Sometimes this rule is too restrictive and inefficient. In such cases, we may want to allow a nonmember function to access the private data directly. This can be achieved by declaring that nonmember function as a `friend` of the class concerned.

A friend **F** of class **X** is a function (or class) that, although not a member function of **X**, has full access rights to the private and protected members of **X**. In all other aspects, **X** is a normal function (or class) with respect to scope, declarations and definitions. Class friendship is not transitive: **X** is a friend of **Y** and **Y** is a friend of **Z** does not imply **X** is a friend of **Z**. However, friendship can be inherited. Like a member function, a friend function is explicitly specified in the declaration part of a class. The keyword `friend` is the function declaration modifier. It takes the general form

friend *return-type function-name* (*parameter list*)

As an example, consider the output stream operator in our `String` class. Note that it is declared as a `friend` in the class declaration. This means that the function has access to the private data of the class.

```
class String
{
   ...
   friend ostream& operator << (ostream&,const String&);
};

ostream& operator << (ostream& out,const String& s)
{
   out << s.data;
   return out;
}
```

The function puts the content of the string on the output stream. Note that the data field `s.data` is actually a private data member of the class. It is the friendship between the function and the class which has made the access possible. In fact, we have already made use of the friendship in the global overloading of the concatenation operator in the `String` class, as described in Section 5.6.

Let us consider another example. Suppose we want to construct a simplified `Complex` number class (the complete version will be discussed in Chapter 6). Instead of defining the `add()` and `display()` functions as members, we declare them as `friend` functions to the `Complex` class.

```
// sration.cxx

#include <iostream.h>

class Complex
{
 private:
    double real, imag;
 public:
    Complex();
    Complex(double);
    Complex(double,double);
    friend Complex add(Complex,Complex);
    friend void display(Complex);
};

Complex::Complex() : real(0.0), imag(0.0) { }
Complex::Complex(double r) : real(r), imag(0.0) { }
Complex::Complex(double r, double i) : real(r), imag(i) { }

Complex add(Complex c1,Complex c2)
{
    Complex result;
    result.real = c1.real + c2.real;
    result.imag = c1.imag + c2.imag;
    return result;
}

void display(Complex c)
{
    cout << "(" << c.real << " + i*" << c.imag << ")";
}

void main()
{
    Complex a(1.2, 3.5), b(3.1, 2.7), c;
    c = add(a, b);
    display(a); cout << " + "; display(b); cout << " = "; display(c);
}
```

```
Result
======
(1.2 + i*3.5) + (3.1 + i*2.7) = (4.3 + i*6.2)
```

5.10   Inheritance

Inheritance is a powerful reuse mechanism. It enables us to categorize related classes for sharing common properties. It is a mechanism for deriving new classes from old ones. For example Circle, Triangle and Square can be grouped under a more general abstraction — Shape. Instances of Shape possess some common properties such as area, perimeter, etc. Circle, Triangle and Square are subclasses (or derived classes) of Shape, whereas Shape is the superclass (or base class). Derived classes inherit properties and functionalities from the base class. Through inheritance, a class hierarchy can be created for sharing code and interface. The general form of a subclass that is derived from an existing base class is

```
class class-name : [ public/protected/private ] base-class-name
{
    ...
};
```

The keywords public, protected and private describe how the derived class inherits information from the base class. They allow us to control how the members of the superclass appear in the subclass.

- public inheritance: If the superclass is inherited publicly, the access of all the members of the superclass remain unchanged in the subclass.

- protected inheritance: In this type of inheritance, all public members of the superclass become protected in the subclass.

- private inheritance: Whenever the superclass is inherited privately, all the public and protected members of the superclass become private in the subclass. Therefore, the data members and member functions of the superclass are not accessible to the subclass.

The following table summarizes the behaviours of the three different kinds of inheritance:

Type of inheritance	Access in superclass $\Longrightarrow$ Access in subclass
Public inheritance	public members $\Longrightarrow$ public members protected members $\Longrightarrow$ protected members private members $\Longrightarrow$ private members
Protected inheritance	public members $\Longrightarrow$ protected members protected members $\Longrightarrow$ protected members private members $\Longrightarrow$ private members
Private inheritance	public members $\Longrightarrow$ private members protected members $\Longrightarrow$ private members private members $\Longrightarrow$ private members

Public inheritance corresponds to the *is-a* relationship, i.e. **X** *is-a* **Y** (**X** is derived from **Y**). It is used when the inheritance is part of the interface. For example, the inheritance described above pertaining to the geometric shape is a public inheritance because `Circle` is a `Shape`, `Triangle` is also a `Shape` and so on.

Protected and private inheritance have other meanings. Protected inheritance is used when the inheritance is part of the interface to the derived classes, but is not part of the interface to the users. On the other hand, private inheritance is used when the inheritance is not part of the interface but is an implementation detail.

In the following program, the base class `Shape` has a protected data member `area` and a public member function `print_Area()`. The classes `Triangle` and `Circle` are subclasses of `Shape` (Figure 5.2). Since `Triangle` is publicly inherited, the public member function `print_Area()` remain public. Therefore, it can be invoked from the main program. However, `Circle` is privately inherited, so the function `print_Area()` becomes private in the class. Hence, an attempt to call the function from the main program generates a compilation error.

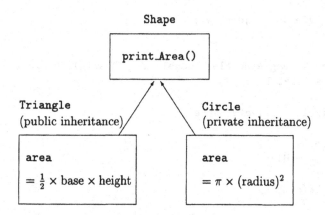

Figure 5.2: *Schematic diagram of the relationships among* **Shape**, **Triangle** *and* **Circle**

```
// shape1.cxx

#include <iostream.h>
const double PI = 3.14159265;

class Shape
{
  protected:
    double area;
  public:
    void print_Area() const;
};

void Shape::print_Area() const {   cout << area << endl; }

class Triangle : public Shape
{
  private:
    double base, height;
  public:
    Triangle(double,double);
};

Triangle::Triangle(double b,double h) : base(b), height(h)
{   area = 0.5*b*h; }

class Circle : private Shape
{
  private:
    double radius;
  public:
    Circle(double);
};

Circle::Circle(double r) : radius(r) { area = PI*radius*radius; }

void main()
{
    Triangle t(2.5,6.0);
    t.print_Area();   // 7.5
    Circle r(3.5);
    r.print_Area(); // function Shape::print_Area is inaccessible
                    // r.print_Area();
}
```

5.11 Virtual Functions

Dynamic binding in C++ is implemented by the *virtual function*. It allows the system to select the right method for a particular class during run-time. Virtual functions are defined in the base class using the keyword `virtual`. The definition of these functions can be deferred or overriden in any subclass. If a derived class does not supply its own implementation, the base class version will be used.

Let us consider the same example as in Section 4.4. An inheritance hierarchy is constructed with `Shape` as the abstract base class and `Circle`, `Triangle` as the derived classes (Figure 5.3). A virtual function named `Calculate_Area()` is declared in the abstract base class `Shape`. However, its implementation is deferred and redefined in its derived classes, where well-defined formulae can be applied to calculate the area of the geometric shapes.

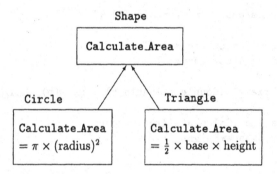

Figure 5.3: *Schematic diagram of the inheritance hierarchy*

```
// shape2.cxx

#include <iostream.h>
const double PI = 3.14159265;

class Shape
{
 public:
    virtual double Calculate_Area() const = 0;
};

class Circle : public Shape
{
 private:
```

```
      double radius;
   public:
      Circle(double);
      double Calculate_Area() const;
   };

   class Triangle : public Shape
   {
    private:
      double base, height;
    public:
      Triangle(double,double);
      double Calculate_Area() const;
   };

   Circle::Circle(double r) : radius(r) {}

   double Circle::Calculate_Area() const
   {   return PI*radius*radius; }

   Triangle::Triangle(double b,double h) : base(b), height(h) {}

   double Triangle::Calculate_Area() const
   {   return 0.5*base*height; }

   void main()
   {
       Shape *s;
       Circle c(5);
       Triangle t(3,8);
       s = &c;    // s points at Circle c
       cout << "s->Calculate_Area() = " << s->Calculate_Area() << endl;
       s = &t;    // s points at Triangle t
       cout << "s->Calculate_Area() = " << s->Calculate_Area() << endl;
   }
Result
======
s -> Calculate_Area() = 78.5398
s -> Calculate_Area() = 12
```

From the result, we observe that the appropriate `Calculate_Area()` function is in-voked depending on the object type pointed to by s during run-time. This form of polymorphism is called inclusion polymorphism.

5.12 Wrapper Class

In C++, data types are classified as basic and abstract data types. In some applications it would be useful to have the basic data type as an abstract data type. This leads to a concept called *wrapper class*. For example, Java has wrapper classes for all basic data types. In the following program, we show how the basic data type `double` is converted into an abstract data type `Double`.

```
// doublec.cxx

#include <iostream.h>
#include <stddef.h>     // for size_t

class Double
{
   private:
      double value;
   public:
      Double(double = 0.0);
      void* operator new (size_t);
      void operator delete(void*,size_t);
      operator double () const;      // type conversion operator
};

Double::Double(double f) : value(f) { }

void* Double::operator new(size_t s)
{  return (void*) new char[s]; }

void Double::operator delete(void *p, size_t)
{  delete [] p; }

Double::operator double () const { return value; }

void main()
{
   Double *s1, s2(5.23);
   s1 = new Double(3.14);
   cout << "Content of s1 = " << *s1 << endl;   // 3.14
   cout << "Content of s2 = " <<  s2 << endl;   // 5.23
   // conversion operator to 'double' is used !
   cout << "*s1 + s2 = " << *s1 + s2 << endl;   // 8.37
   delete s1;
}
```

5.13   Standard Template Library

5.13.1   Introduction

The *Standard Template Library*, or STL, is a C++ library of container classes, algorithms and iterators [1]. It provides many of the basic algorithms and data structures
of computer science. The intent of the STL is to provide a set of container classes
that are both efficient and functional. The presence of such a library will simplify
the creation of complex programs, and because the library is standard the resulting
programs ultimately will have a high degree of portability. This section is about the
new standard version of STL. The use of STL is likely to make software more reli-
able, more portable and more general and to reduce the cost of producing it. One of
the most interesting aspects of the STL is the way it radically departs in structure
from almost all earlier libraries. Since C++ is an object-oriented language, these
earlier libraries have tended to rely heavily on object-oriented techniques, such as
inheritance. The STL uses almost no inheritance. To see this non-object-oriented
perspective, consider that object-oriented programming holds encapsulation as a primary ideal. A well-designed object will try to encapsulate all the state and behaviour
necessary to perform the task for which it is designed, and at the same time, hide
as many of the internal implementation details as possible. In almost all previous
object-oriented container class libraries this philosophical approach was manifested
by collection classes with exceedingly rich functionality, and consequently with large
interfaces and complex implementations. The designers of STL moved in an entirely
different direction. Each component is designed to operate in conjunction with a rich
collection of generic algorithms. These generic algorithms are independent of the containers and can therefore operate with many different containers types. By separating
the functionality of the generic algorithms from the container classes themselves, the
STL realizes a great saving in size, in both the library and the generated code. Instead
of duplication of algorithms in each of the dozen or so different container classes, a
single definition of a library function can be used with any container. Furthermore,
the definition of these functions is so general that they can be used with ordinary
C-style arrays and pointers as well as with other data types. The boolean data type
bool is also implemented in the STL. At the core of the STL are three foundational
items:

1. containers

2. algorithms

3. iterators.

These items work in conjunction with one another. Containers are objects that hold
other objects. Algorithms act on the contents of containers. Iterators are objects
that are pointers. They provide us with the ability to cycle through the contents of

a container in the same way that we apply pointers to cycle through an array. The classes provided by the STL are:

algorithm deque functional iterator list map
.memory numeric queue set stack utility vector

5.13.2 The Namespace Concept

Namespaces are a recent addition to C++. A namespace creates a declarative region in which various program elements can be placed. Elements declared in one namespace are separated from elements declared in another. The line using namespace std; tells the compiler to use the STL namespace.

```
// namesp.cpp

#include <iostream>
using namespace std;

namespace A { int i = 10; }

namespace B { int i = 5; }

void fA()
{
   using namespace A;
   cout << "In fA: " << A::i << " " << B::i << " " << i << endl;
   // => 10 5 10
}

void fB()
{
  using namespace B;
  cout << "In fB: " << A::i << " " << B::i << " " << i << endl;
  // => 10 5 5
}

void main()
{
   fA();  fB();
   cout << A::i << " " << B::i << endl; // => 10 5
   using A::i;
   cout << i << endl; // => 10
}
```

5.13.3   The Vector Class

The vector class is a container class. The vector class supports a dynamic array. This is an array that can grow as needed. We can use the standard array subscript notation [] to access its elements.

The constructors and methods (member functions) in the class vector are summarized in the following.

```
Constructors
vector<T> v;            default constructor
vector<T> v(int);       initialized with explicit size
vector<T> v(int,T);     size and initial value
vector<T> v(aVector);   copy constructor
```

```
Element Access
v[i]                    subscript access, can be assignment target
v.front()               first value in collection
v.back()                last value in collection
```

```
Insertion
v.push_back(T)          push element on to back of vector
v.insert(iterator,value) insert new element after iterator
v.swap(vector<T>)       swap values with another vector
```

```
Removal
v.pop_back()            pop element from back of vector
v.erase(iterator)       remove single element
v.erase(iterator,iterator) remove range of values
```

```
Size
v.capacity()            maximum number of elements buffer can hold
v.size()                number of elements currently held
v.resize(unsigned,T)    change to size, padding with value
v.reserve(unsigned)     set physical buffer size
v.empty()               true if vector is empty
```

```
Iterators
vector<T>::iterator itr declare a new iterator
v.begin ()              starting iterator
v.end ()                ending iterator
```

```cpp
// Mvector.cpp

#include <iostream>
#include <vector>
using namespace std;

int main()
{
    vector<char> w(4);
    w[0] = 'X';    w[1] = 'Y';    w[2] = '+';    w[3] = '-';
    cout << w[2] << endl;  // => +
    int j;
    for(j=0; j<4; j++)
    cout << "w[" << j << "] = " << w[j] << endl;

    vector<double> v;
    double x;
    cout << "enter double number, followed by 0.0:\n";
    while(cin >> x, x != 0.0)
    v.push_back(x);    // adds the double value x at the end
                       // of the vector v
    vector<double>::iterator i;
    for(i=v.begin(); i != v.end(); ++i)
    cout << *i << endl; // dereferencing

    bool b = v.empty();
    cout << "b = " << b << endl;

    // insert element in the vector
    double y;
    cout << "enter value to be inserted: ";
    cin >> y;
    vector<double>::iterator p = v.begin();
    p += 2;
    v.insert(p,4,y);  // point to 3rd element, insert 4 elements

    for(i=v.begin(); i != v.end(); ++i)
    cout << *i << endl;

    int length = v.size();
    cout << "length = " << length << endl;

    return 0;
}
```

We read a textfile line by line into a vector. We read the file, line by line, until **eof**
is encountered. The only catch is that with the istream's operator **>>** string cannot
be used. The function only reads a word, and not a whole line. The global

```
getline(istream&,string&)
```

is the solution. Let the text file (ASCII file) be **william.txt** with the contents

```
to be or
not to be
```

After compiling and linking to obtain the execute file at the command line we enter:

```
readin1 william.txt
```

```
// readin1.cpp

#include <iostream>
#include <fstream>
#include <string>
#include <vector>

using namespace std;

int main(int argc,char *argv[])
{
   ifstream ifs(argv[1]);
   vector<string> v;
   string s;

   while(getline(ifs,s))
   {
   v.push_back(s);
   cout << s <<endl;
   }

   cout << "v[1] = " << v[1] << endl;   // => not to be

   return 0;
}
```

```
// bubblest.cpp

#include <algorithm>
#include <vector>
#include <stdlib.h>
#include <iostream.h>
using namespace std;

template <class RandomAccessIterator>
void bubbleSort(RandomAccessIterator first,RandomAccessIterator last)
{
    int exchanged;
    do
    {
    exchanged = 0; last--;
    for(RandomAccessIterator iter=first; iter < last; iter++)
    {
    if(*iter > *(iter+1))
    {
    iter_swap(iter,iter+1);
    exchanged++;
    }
    }
    } while(exchanged);
}

void main(void)
{
    const int nData = 6;
    vector<unsigned long> data(nData);

    for(int n=0; n < nData; n++)
    {
    data[n] = rand();
    cout << data[n] << " ";
    }
    cout << endl;

    bubbleSort(data.begin(),data.end());

    // after sorting
    for(n=0; n<nData; n++)
    cout << data[n] << " ";
}
```

5.13.4   The List Class

The list class is also a container class. A linked list is the data structure we choose
when the number of elements in a collection cannot be bounded, or varies widely during the course of execution. Like the vector class, the linked list class maintains values
of uniform type. Lists are not indexed. The following table gives the constructors
and the member functions of the class.

```
Constructors and Assignment
list<T> l;                       default constructor
list<T> l(aList);                copy constructor
l = aList                        assignment
l.swap (aList)                   swap values with another list

Element Access
l.front()                        first element in list
l.back()                         last element in list

Insertion and Removal
l.push_front(value)              add value to front of list
l.push_back(value)               add value to end of list
l.insert(iterator, value)        insert value at specified location
l.pop_front()                    remove value from front of list
l.pop_back()                     remove value from end of list
l.erase(iterator)                remove referenced element
l.erase(iterator,iterator)       remove range of elements
l.remove(value)                  remove all occurrences of value
l.remove_if(predicate)           removal all values that match condition

Size
l.empty()                        true if collection is empty
l.size()                         return number of elements in collection

Iterators
list<T>::iterator itr            declare a new iterator
l.begin()                        starting iterator
l.end()                          ending iterator
l.rbegin()                       starting backwards moving iterator
l.rend()                         ending backwards moving iterator

Miscellaneous
l.reverse()                      reverse order of elements
l.sort()                         place elements into ascending order
l.sort(comparison)               order elements using comparison function
l.merge(list)                    merge with another ordered list
```

The following program gives an application of the list class.

```cpp
// mlist.cpp

#include <iostream>
#include <list>
#include <string>
using namespace std;

void main(void)
{
    list<char> lst;
    int i;
    for(i=0; i<10; i++)
    lst.push_back('X'+i);
    cout << "size of list = " << lst.size() << endl;
    cout << "contents of list: ";
    list<char>::iterator p = lst.begin();

    while(p != lst.end()) { cout << *p << " ";  p++; }
    lst.push_back('A');
    cout << endl;

    lst.sort();
    cout << "sorted contents:\n";
    p = lst.begin();
    while(p != lst.end()) { cout << *p << " ";  p++; }
    cout << endl;

    list<string> str;
    str.push_back("Good");
    str.push_back(" ");
    str.push_back("Night");
    str.push_back(" ");
    str.push_back("Egoli");
    list<string>::iterator q = str.begin();

    while(q != str.end()) { cout << *q << " ";  q++; }
}
```

5.13.5   The Stack Class

The *stack* class provides a restricted subset of container functionality. By default this underlying container type is `vector`. It is a last-in-first-out (LIFO) data structure. The stack class does not allow iteration through its elements. It is a collection of data items organized in a linear sequence, together with the following five operations:

1. *CreateStack* brings a stack into existence

2. *MakeStackEmpty* deletes all items, if any, from the stack

3. *Push* adds an item at one end, called the top, of the stack

4. *Pop* removes the item at the top of a stack and makes it available for use

5. *StackIsEmpty* tests whether a stack contains any items.

The items in a stack might be integers, real numbers, characters, or abstract data types. The stack template has two arguments

```
stack<T,vector<T> > s;
```

The following list gives the operations provided by the STL for the stack class.

```
Insertion and Removal
s.push(value)      push value on front of stack
s.top()            access value at front of stack
s.pop()            remove value from front of stack

Size
s.size()           number of elements in collection
s.empty()          true if collection is empty

Assignment
=

Comparisons
==
!=
<
<=
>
>=
```

The following program shows a simple application of the stack class.

```cpp
// stack1.cpp

#include <iostream>
#include <vector>
#include <stack>
using namespace std;

int main()
{
   stack<char,vector<char> > S, T, U;
   cout << "&S = " << &S << endl;        // => address

   S.push('X');
   S.push('Y');
   S.push('Z');
   cout << "characters pushed onto S: X Y Z\n";

   cout << "size of S = " << S.size() << endl; // => 3

   T = S;
   cout << "after T = S; we have ";
   cout << (S == T ? "S == T" : "S != T") << endl; // => S == T

   U.push('X');
   U.push('W');
   cout << "pushed onto U: X W\n";
   cout << "output now ";
   cout << (S < U ? "S < U" : "S >= U") << endl; // => S >= U

   bool b = (T == S);
   cout << "b = " << b << endl; // => 1 (true)

   return 0;
}
```

5.13.6   The Queue Class

A *queue* is a first-in-first-out (FIFO) data structure. This means that elements are added to the back of the queue and may be removed from the front. Queue is a container adapter, meaning that it is implemented on top of some underlying container type. Queue does not allow iteration through its elements. A queue is a collection of data items organized in a linear sequence, together with the following five operations:

1. *CreateQueue* brings a queue into existence

2. *MakeQueueEmpty* deletes all items, if any, from the queue

3. *EnQueue* adds an item at one end, called the rear, of the queue

4. *DeQueue* removes the item from the other end, called the front, of the queue, and makes it available for use

5. *QueueIsEmpty* tests whether a queue is empty.

The queue template has two arguments

```
queue<T,list<T> > q;
```

The following list gives the operations provided by the STL for the queue class.

```
Insertion and Removal
q.push(value)     push value on back of queue
q.front()         access value at front of queue
q.back()          access value at back of queue
q.pop()           remove value from front of queue

Size
q.size()          number of elements in collection
q.empty()         true if collection is empty
```

The following program shows a simple application of the queue class.

```cpp
// queue1.cpp

#include <iostream>
#include <list>
#include <queue>

using namespace std;

int main()
{
    queue<double,list<double> > Q;
    cout << "&Q = " << &Q << endl;    // => address

    Q.push(3.14);
    Q.push(4.5);
    Q.push(6.9);
    cout << "after pushing 3.14, 4.5, 6.9:\n";
    cout << "Q.front() = " << Q.front() << endl; // => 3.14
    cout << "Q.back() = " << Q.back() << endl;   // => 6.9

    Q.pop();
    cout << "after Q.pop():\n";
    cout << "Q.front() = " << Q.front() << endl; // => 4.5

    cout << "Q.size() = " << Q.size() << endl; // => 2

    queue<double,list<double> > P;
    P = Q;
    cout << "P.front() " << P.front() << endl; // => 4.5
    cout << "P.back() " << P.back() << endl;   // => 6.9

    return 0;
}
```

5.13.7   The Deque Class

A deque (double-ended queue) is an abstract data type in which insertions and deletions can be made at either the front or the rear. Thus it combines features of both
stacks and queues. In practice, the deque has two variations. The first is the input-
restricted deque, where insertions are restricted to one end only, and the second is
the output-restricted deque, where deletions are restricted to a single end.

The following list give the operations provided by the STL for the stack class.

```
Constructors and Assignment
deque<T> d;                   default constructor
deque<T> d(int);              construct with initial size
deque<T> d(int,value);        construct with initial size and initial value
deque<T> d(aDeque);           copy constructor
d = aDeque;                   assignment of deque from another deque
d.swap(aDeque);               swap contents with another deque


Element Access and Insertion
d[i]                          subscript access, can be assignment target
d.front()                     first value in collection
d.back()                      final value in collection
d.insert(iterator,value)      insert value before iterator
d.push_front(value)           insert value at front of container
d.push_back(value)            insert value at back of container


Removal
d.pop_front()                 remove element from front of vector
d.pop_back()                  remove element from back of vector
d.erase(iterator)             remove single element
d.erase(iterator,iterator)    remove range of elements


Size
d.size()                      number of elements currently held
d.empty()                     true if vector is empty


Iterators
deque<T>::iterator itr        declare a new iterator
d.begin()                     starting iterator
d.end()                       stopping iterator
d.rbegin()                    starting iterator for reverse access
d.rend()                      stopping iterator for reverse access
```

The following small program shows an application of the deque class.

```cpp
// deque1.cpp

#include <iostream>
#include <deque>
using namespace std;

int main()
{
   deque<char> D(10,'A');

   deque<char>::iterator i;
   for(i=D.begin(); i != D.end(); ++i)
   cout << *i << " ";   // => A A A A A A A A A A
   cout << endl;

   deque<char> E(D);   // copy constructor

   E.push_front('Z');
   E.push_back('Y');

   for(i=E.begin(); i != E.end(); ++i)
   cout << *i << " ";   // => Z A A A A A A A A A A Y
   cout << endl;

   cout << E.empty() << endl; // => 0 (false)

   cout << E.size() << endl;   // => 12

   char c = E[0];
   cout << "c = " << c << endl;   // => Z

   c = E[5];
   cout << "c = " << c << endl;   // => A

   return 0;
}
```

5.13.8 The Bit Set Class

The bit set class in the STL implements the bitwise operations.

```
Constructors
bitset<N> s                construct bitset for N bits
bitset<N> s(aBitSet)       copy constructor

Bit level operations
s.flip()                   flip all bits
s.flip(i)                  flip position i
s.reset(0)                 set all bits to false
s.reset(i)                 set bit position i to false
s.set()                    set all bits to true
s.set(i)                   set bit position i to true
s.test(i)                  test if bit position i is true

Operations on entire collection
s.any()                    return true if any bit is true
s.none()                   return true if all bits are false
s.count()                  return number of true bits

Assignment
=

Combination with other bitsets
s1 & s2                    bitwise AND
s1 | s2                    bitwise inclusive OR
s1 ^ s2                    bitwise exclusive OR
s == s2                    return true if two sets are the same

Other operations
s << n                     shift set left by one
s >> n                     shift set right by one
s.to_string()              return string representation of set
```

The following small program shows an application of the bitset class.

```cpp
// bitset1.cpp

#include <iostream>
#include <bitset>
#include <string>
using namespace std;

int main()
{
    const unsigned long n = 32;
    bitset<n> s;
    cout << s.set() << endl;        // set all bits to 1

    cout << s.flip(12) << endl;     // flip at position 12

    bitset<n> t;
    cout << t.reset() << endl;      // set all bits to false

    t.set(23);
    t.set(27);

    bitset<n> u;
    u = s & t;
    cout << "u = " << u << endl;

    bitset<n> v;
    v = s | t;
    cout << "v = " << v << endl;

    bitset<n> w;
    w = s ^ t;
    cout << "w = " << w << endl;

    bitset<n> z;
    z = w ^ w;
    cout << "z = " << z << endl;

    cout << "z.to_string() = " << z.to_string();

    return 0;
}
```

5.13.9   The Set Class

The abstract data type set is defined in terms of objects and operations. The objects
are just sets in the mathematical sense with the representation unspecified. The set
and multiset data types in the STL are both template data structures, where the
template argument represents the type of the elements the collection contains. In the
set class each element of a set is identical to its key, and keys are unique. Because
of this, two distinct elements of a set cannot be equal. A multiset differs from a set
only in that it can contain equal elements. The operations in the set class are given
below.

```
Constructors
set<T> s;                        default constructor
multiset<T> m;                   default constructor
set<T> s(aSet);                  copy constructor
multiset<T> m(aMultiset)         copy constructor
s = aSet                         assignment
s.swap(aSet)                     swap elements with argument set


Insertion and Removal
s.insert(value_type)             insert new element
s.erase(value_type)              remove all matching elements
s.erase(iterator)                remove element specified by iterator
s.erase(iterator,iterator)       remove range of values


Testing for Inclusion
s.empty()                        true if collection is empty
s.size()                         number of elements in collection
s.count(value_type)              count number of occurrences
s.find(value_type)               locate value
s.lower_bound(value_type)        first occurrence of value
s.upper_bound(value_type)        next element after value
s.equal_range(value_type)        lower and upper bound pair


Iterators
set<T>::iterator itr             declare a new iterator
s.begin()                        starting iterator
s.end()                          stopping iterator
s.rbegin()                       starting iterator for reverse access
s.rend()                         stopping iterator for reverse access
```

The following program shows an application of the set class.

```cpp
// set2.cpp

#include <iostream>
#include <set>
using namespace std;

int main()
{
    set<int> s;

    s.insert(23);
    s.insert(45);
    s.insert(-1);
    s.insert(-2);
    s.insert(23);
    s.insert(51);

    cout << s.empty() << endl;    // => 0 (false)

    cout << s.size() << endl;      // => 5

    cout << s.count(23) << endl; // => 1

    set<int>::iterator i = s.begin();

    for(i=s.begin(); i != s.end(); ++i)
    cout << *i << " ";

    cout << endl;

    s.erase(45);

    for(i=s.begin(); i != s.end(); ++i)
    cout << *i << " ";

    cout << endl;

    cout << s.size() << endl; // => 4

    return 0;
}
```

5.13.10   The Map Class

The map class (also called dictionary or table) is an indexed collection. The index
values need not be integers, but can be any ordered data values. Therefore a map is
a collection of associations of key value pairs. For maps no two keys can be equal. A
multimap differs from a map in that duplicated keys are allowed. The following table
gives the operations.

```
Constructors
map<T1,T2> m;                        default constructor
multimap<T1,T2> m;                   default constructor
map<T1,T2> m(aMap)                   copy constructor
multimap<T1,T2> m(aMultiMap)         copy constructor
m = aMap                             assignment

Insertion and Removal
m[key]                               return reference to value with key
m.insert(value_type)                 insert given key value pair
m.erase(key)                         erase value with given key
m.erase(iterator)                    erase value at given iterator

Testing for Inclusion
m.empty()                            true if collection is empty
m.size()                             return size of collection
m.count(key)                         count number of elements with given key
m.find(key)                          locate element with given key
m.lower_bound(key)                   first occurrence of key
m.upper_bound(key)                   next element after key
m.equal_range(key)                   lower and upper bound pair

Iterators
map<T>::iterator itr;                declare new iterator
m.begin()                            starting iterator
m.end()                              ending iterator
m.rbegin()                           backwards moving iterator start
m.rend()                             backwards moving iterator end
```

The following two programs show applications of the `map` class.

```cpp
// mmap.cpp

#include <iostream>
#include <map>
#include <string>
using namespace std;

int main()
{
    map<char,int> m1;
    int i;

    for(i=0; i<26; i++)
    {
    m1.insert(pair<char,int>('A'+i,65+i)); // ASCII table
    }

    char ch;
    cout << "enter key: ";
    cin >> ch;

    map<char, int>::iterator p1;
    // find value given key
    p1 = m1.find(ch);
    if(p1 != m1.end())
    cout << "ASCII table number: " << p1 -> second << endl;
    else
    cout << "key not in map.\n";

    map<string,double> m2;
    m2.insert(pair<string, double>("Willi",0.0));
    m2.insert(pair<string, double>("Fritz",1.0));
    m2.insert(pair<string, double>("Charles",2.0));

    map<string,double>::iterator p2;
    p2 = m2.find("Fritz");
    cout << p2 -> second << endl;  // => 1
    cout << p2 -> first << endl;   // => Fritz

    return 0;
}
```

```cpp
// mystl.cpp
#include <iostream>
#include <map>
#include <string>
using namespace std;

typedef map<string, string> vocabulary;

void main(void)
{
    string wrd, def;
    vocabulary voc;
    vocabulary::iterator itr;
    cout << "Populating the dictionary" << endl;
    while(true)
    {
        cout << "Enter word: ";
        cout.flush();
        cin >> wrd;
        if((wrd == "Q") || (wrd == "q")) break;
        cout << "Definition: ";
        cout.flush();
        cin >> def;
        voc[wrd] = def;
    }
    if(voc.empty())
    {
        cout << "Empty vocabulary!" << endl;
        exit(0);
    }
    cout << "Vocabulary populated with ";
    cout << voc.size() << " elements." << endl;
    cout << "Looking up words in the dictionary" << endl;
    while(true)
    {
        cout << "Enter word: ";    cin >> wrd;
        if((wrd == "Q") || (wrd == "q")) break;
        itr = voc.find(wrd);
        if(itr != voc.end()) cout << itr->second << endl;
        else cout << "not found!" << endl;
    }
}
```

5.13.11   The Algorithm Class

Algorithms act on the contents of containers. They include functions for initializing, sorting, searching, and transforming the contents of containers. The functions can also be applied to arrays of basic data types. The functions are:

```
sort, find, copy, merge, reverse, replace
```

The following code shows an application of some of the functions:

```cpp
// algorith1.cpp

#include <iostream>
#include <algorithm>
#include <vector>
using namespace std;

int main()
{
   const int n = 5;
   double x[n];
   x[0] = 3.13; x[1] = 4.5; x[2] = 1.2; x[3] = 0.8; x[4] = 0.1;

   sort(x,x+n);
   cout << "array after sorting: \n";
   double* p;
   for(p=x; p != x+n; p++)
   cout << *p << " ";
   cout << endl;

   double y;
   cout << "double value to be searched for: ";
   cin >> y;

   double* q = find(x,x+n,y);
   if(q == x+n)
   cout << "not found\n";
   else
   {
   cout << "found";
   if(q == x)
   cout << " as the first element";
   else
```

```cpp
    cout << " after " << *--q;
    }
    cout << endl;

    reverse(x,x+n);
    double* r;
    cout << "reversed array: " << endl;
    for(r=x; r != x+n; r++)
    cout << *r << " ";
    cout << endl;

    int a[6] = { 4, 5, 10, 20, 30, 84 };
    int b[5] = { 7, 9, 14, 35, 101 };
    int c[11];
    merge(a,a+6,b,b+5,c);
    int i;
    for(i=0; i<11; i++)
    cout << "c[" << i << "] = " << c[i] << endl;

    char str[] = "otto";
    int length = strlen(str);
    replace(str,str+length,'t','l');
    cout << str << endl;  // => ollo

    vector<int> v(4);
    v[0] = 1; v[1] = 6; v[2] = 3; v[3] = 2;
    vector<int> w(5);
    w[0] = 8; w[1] = 10; w[2] = 3; w[3] = 12; w[4] = 5;

    sort(v.begin(),v.end());
    sort(w.begin(),w.end());

    vector<int> z(9);
    merge(v.begin(),v.end(),w.begin(),w.end(),inserter(z,z.begin()));

    for(i=0; i<9; i++)
    cout << "z[" << i << "] = " << z[i] << endl;

    return 0;
}
```

5.14 Recursion

Recursion plays a central role in computer science. For example a string and a linear linked list are recursive structures. A recursive function is one whose definition includes a call to itself. A recursion needs a stopping condition. Of course, it is not allowed to use the main function in a recursive call.

We consider five examples. In the first example we show how multiplication can be implemented using recursion. The second example shows an implementation of division using subtraction and recursion. The length of a string can be found using recursion. At every step we have to test whether we have reached the end of the string indicated by the null character '\0'. In the fourth example we consider a linked list. All methods use recursion. Finally we give an example for mutual recursion.

```cpp
// recursion1.cpp
// multiplication of two numbers using recursion

#include <iostream.h>

unsigned long mult(unsigned long a,unsigned long b)
{
   unsigned long result = 0;
   if(a == 0) return 0;
   if(b == 0) return 0;
   if(a == 1) return b;
   if(b == 1) return a;
   else result = mult(a,b-1) + a;
   return result;
}

int main()
{
   unsigned long n, m;
   cout << "enter a non-negative integer n = ";
   cin >> n;
   cout << "enter a non-negative integer m = ";
   cin >> m;
   unsigned long result = mult(n,m);
   cout << "result = " << result << endl;

   return 0;
}
```

```cpp
// recursion2.cpp
//
// integer division of two numbers using recursion
// b is the divisor
// a is the dividend

#include <iostream.h>
#include <stdlib.h>

unsigned long divide(unsigned long a,unsigned long b)
{
   unsigned long result = 0;
   if(b > a) return 0;
   if(b == 0) { cout << "division by 0 not allowed"; exit(0); }
   if(b == 1) return a;
   else
   result = divide(a-b,b) + 1;

   return result;
}

int main()
{
   unsigned long n, m;
   cout << "enter a non-negative integer n = ";
   cin >> n;
   cout << "enter a non-negative integer m = ";
   cin >> m;

   unsigned long result = divide(n,m);
   cout << "result = " << result << endl;

   return 0;
}
```

The program determines the length of a string recursively. A string consists of contiguous characters in memory, ending with the NULL character '\0'. Conceptually, we can think of a string as either the NULL string, consisting of just the NULL character, or as a character followed by a string. This definition of a string describes it as a recursive data structure. Thus we can use this to code basic string-handling functions recursively.

```cpp
// recursion3.cpp

#include <iostream.h>
#include <string.h>     // for strcpy

int length(char *s)
{
    if(*s == '\0')
        return 0;
    else
        return (1 + length(s+1));
}

int main()
{
    char* st = "willi hans";
    int res1 = length(st);
    cout << "The length of the string st is: " << res1; // => 10
    cout << endl;

    char* empty = "\0";
    int res2 = length(empty);
    cout << "The length of the string empty is: " << res2; // => 0
    cout << endl;

    char* z = NULL;
    z = new char[4];        // allocating memory
    strcpy(z,"oli");
    int res3 = length(z);
    cout << "The length of the string z is: " << res3; // => 3
    delete [] z;

    return 0;
}
```

The following program shows how to implement a linked list recursively.

```cpp
// rlist.h

#ifndef RLIST_HEADER
#define RLIST_HEADER

#include <assert.h>

template <class T>
class RList
{
 public:
   RList();
   RList(const RList&);
   ~RList();

   RList &operator=(const RList&);
   int operator==(const RList&);

   void Insert(const T&);
   int Search(const T&);
   int Delete(const T&);

   T Head(void);
   RList* Tail(void);
   int Empty(void);
   RList* Reverse(RList*);
 private:
   T head;
   RList* tail;
   int empty;
};

template <class T>
RList<T>::RList()
{
 empty = 1;
}

template <class T>
RList<T>::RList(const RList<T>& RL)
{
  empty = RL.empty;
```

```
  head = RL.head;
  tail = new RList<T>(*RL.tail);
}

template <class T>
RList<T>::~RList()
{
 if(!empty)
  delete tail;
}

template <class T>
RList<T> &RList<T>::operator=(const RList<T> &RL)
{
 if(this == &RL) return;
 if(!empty) delete tail;
 empty = RL.empty;
 head = RL.head;
 tail = new RList<T>(*RL.tail);
}

template <class T>
int RList<T>::operator==(const RList<T> &RL)
{
 if(empty&&RL.empty) return 1;
 if(this == &Rl) return 1;
 if(head != RL.head) return 0;
 return (*tail == *RL.tail);
}

template <class T>
void RList<T>::Insert(const T &toInsert)
{
 if(empty)
 {
  head = toInsert;
  tail = new RList<T>;
  empty=0;
 }
 else
  tail->Insert(toInsert);
}

template <class T>
```

```
int RList<T>::Search(const T &toSearch)
{
 if(empty) return 0;
 else if(head == toSearch) return 1;
 else return tail -> Search(toSearch);
}

template <class T>
int RList<T>::Delete(const T &toDelete)
{
 if(empty) return 0;
 else if(head==toDelete)
 {
  head = tail -> head;
  empty = tail -> empty;
  tail -> Delete(tail -> head);
  if(empty) delete tail;
  return 1;
 }
 else return tail -> Delete(toDelete);
}

template <class T>
T RList<T>::Head(void)
{
 assert(!empty);
 return head;
}

template <class T>
RList<T> *RList<T>::Tail(void)
{
 assert(!empty);
 return tail;
}

template <class T>
int RList<T>::Empty(void)
{
 return empty;
}

template <class T>
RList<T> *RList<T>::Reverse(RList<T> *RL)
```

```
{
 if(RL->Empty())
 {
  RList<T> *temp;
  temp = new RList<T>;
  return temp;
 }
 else
 {
  RList<T> *R;
  R = Reverse(RL->Tail());
  (*R).Insert(RL->Head());
  return R;
 }
}
#endif
```

An appliction is as follows:

```
//rlisteg.cpp

#include <iostream.h>
#include "rlist.h"

int main(void)
{
 RList<int> L;
 int i;
 for(i=1; i<=8; i++) L.Insert(i);

 RList<int>* LX = &L;
 cout << "The initial RList is: " << endl;
 while(!LX -> Empty())
 {
  cout << LX -> Head() << ' ';
  LX = LX -> Tail();
 }
 cout << endl << endl;

 RList<int>* R = L.Reverse(&L);
 RList<int>* LP = R;

 while(!LP -> Empty())
 {
```

```
  cout << LP -> Head() << ' ';
  LP = LP -> Tail();
 }
 cout << endl << endl;

 cout << "what happened to the initial list: "<< endl;
 LP = &L;
 while(!LP -> Empty())
 {
  cout << LP -> Head() << ' ';
  LP = LP -> Tail();
 }
 cout << endl;

 cout << "remove some items: "<< endl;
 L.Delete(1);
 L.Delete(4);
 L.Delete(8);
 LP = &L;
 while(!LP -> Empty())
 {
  cout << LP -> Head() << ' ';
  LP = LP -> Tail();
 }
 cout << endl;

 cout << "is 3 in the list: " << L.Search(3) << endl;
 cout << "is 4 in the list: " << L.Search(4) << endl;

 return 0;
}
```

The output is

```
The initial RList is:
1 2 3 4 5 6 7 8

8 7 6 5 4 3 2 1

what happened to the initial list:
1 2 3 4 5 6 7 8
remove some items:
2 3 5 6 7
is 3 in the list: 1
is 4 in the list: 0
```

In *mutual recursion* we have two functions which call each other. As an example we consider an implementation of sine and cosine. The identities

$$\sin(x) \equiv 2\sin(x/2)\cos(x/2)$$

$$\cos(x) \equiv \cos^2(x/2) - \sin^2(x/2) \equiv 2\cos^2(x/2) - 1.0$$

are used. Both sine and cosine call themselves and are therefore recursive. The sine function calls the cosine function. For the cosine function we have two options. For speed we select the second option where cosine calls only itself.

```cpp
// sincos.cpp

#include <iostream.h>
#include <math.h>

double sine(double,double);   // forward declaration
double cosine(double,double); // forward declaration

void main()
{
   double x = 3.14159;
   double eps = 0.001;
   double res1 = sine(x,eps);
   cout << "res1 = " << res1 << endl;
   double res2 = cosine(x,eps);
   cout << "res2 = " << res2 << endl;
}

double sine(double x,double eps)
{
   double s;
   if(fabs(x) < eps) { s = x*(1.0 - x*x/6.0); }
   else s = 2.0*sine(x/2.0,eps)*cosine(x/2.0,eps);
   return s;
}

double cosine(double x,double eps)
{
   double c;
   if(fabs(x) < eps) { c = 1.0 - x*x/2.0; }
   else c = 2.0*cosine(x/2,eps)*cosine(x/2,eps) - 1.0;
   return c;
}
```

5.15 Summary

This chapter presented the basic programming constructs and tools available in C++. We began by describing pointers and references in C++: why, when and how they can be used. Different constructs in C++ were described, they include

- class
- constructors and destructor
- copy constructor and assignment operator
- type conversion
- operators overloading
- lass template
- function template
- friendship
- inheritance
- virtual functions
- Standard Template Library
- recursion.

In each section, examples were used for illustration. They showed the ability to support the notion of object-orientation, which includes

- encapsulation
- message-passing
- inheritance
- polymorphism, etc.

Based on the features provided in the language, programmers are able to write powerful object-oriented programs.

In the following chapter, a collection of useful classes will be constructed using C++, where most of the features described here will be used.

Chapter 6

Classes for Computer Algebra

In this chapter we introduce the basic building classes of our symbolic system. The chapter deals with many structures in mathematics as well as some common data structures in computer science. The description of the classes are arranged in such a way that primitive structures like very long integer, rational class are placed earlier in the chapter than the more sophisticated structures like vector, matrix class, etc. There are thirteen classes presented in this chapter:

(1) **Verylong** provides the integer numbers abstract data type without upper and lower bound.
(2) **Rational** provides the rational numbers abstract data type.
(3) **Complex** provides the complex numbers abstract data type.
(4) **Quaternion** provides the quaternion abstract data type.
(5) **Derive** provides the exact differentiation class.
(6) **Vector** provides the vector data structure.
(7) **Matrix** provides the matrix data structure.
(8) **Array** provides the array data structure.
(9) **String** provides the string abstract data type.
(10) **BitVector** provides the bit vector field abstract data type.
(11) **List** provides the linked list abstract data type.
(12) **Polynomial** provides the polynomial abstract data type.
(13) **Set** provides the set abstract data type for finite sets.

In each class, the basic ideas and the theory of the class are explained, followed by the abstraction. Different parts of the class, like data fields, constructors, operators and member functions, etc. are also described. Only short examples are given in each section. More advanced applications will be presented in Chapter 8.

6.1　The Verylong Integer Class

6.1.1　Abstraction

The integer data type is implemented internally in most programming languages. However, the effective range is limited due to the nature of the registers of the CPU. For a 4-byte integer, the effective range is between -2^{31} to $2^{31}-1$ or $-2,147,483,648$ to $2,147,483,647$. This range is usually not sufficient for an elaborate computation which requires very large positive integer numbers or negative integer numbers. The purpose of this section is to break the limitation of the built-in integer data type by introducing the concept of a *verylong integer* class.

For a data type to be able to store an arbitrary long integer number, we have to figure out a storage method in memory that imposes no limitation on representing an integer number. Using a string of characters to represent a very long integer is one possibility. Unlike the built-in integer type which could only be stored in 4 bytes of memory, we go beyond this limit. By using a string to represent an integer number, we could in principle make the string as long as possible (subject to availability of memory). This has the implication that an integer number could be represented to any number of digits.

The string that stores the very long number contains only character digits '0', '1', '2', ... , '9'. Each digit in the string represents a decimal digit of the number. For example, the string "123" represents the value 123 in decimal. With this representation, the arithmetic operations could be implemented using the usual manipulation algorithms. Although this is not the only possible representation and it may not be the best way, some other representations such as binary representation may need some less straightforward algorithm for implementing the arithmetic operations. Since simplicity is one of our primary goals, we choose the decimal representation.

An *abstract data type* (ADT) defines not only the representation of the data (for example, the string of characters of integers in the case of the Verylong class) but also the operations which may be performed on the class. However, both the data representation and the implementation details of the operations should be hidden. The user only needs to know the behaviours of the ADT and the public interfaces. It is generally a good idea to strive for complete but minimal class interfaces. This applies to the Verylong class as well. In the following we summarize the behaviours of the Verylong class ADT:

- We have to create new instances of Verylong number abstraction easily.

- We have to use the arithmetic operators such as +, -, *, /, %, ++, -- to manipulate the instances of the Verylong number.

- We have to assign a Verylong number value to a Verylong variable using the operator =.

- The modification forms of assignment, such as +=, -=, *=, /=, %=, have to be supported.

- The relational operators >, >=, <, <=, ==, != should be available.

- We have to convert instances of Verylong number to other standard data types like int, double and char * when necessary.

- Some common functions like absolute value functions abs(), integer power function pow(), integer square root function sqrt() and double division operator div() have to be included as well.

- We have to perform input and output operations using the Verylong numbers.

6.1.2 Data Fields

There are three data fields in the Verylong class, namely vlstr, vlen, vlsign. Following the philosophy of *information hiding* the data fields are declared as private which makes them inaccessible from outside the class. To access or manipulate the private data of the class, one has to use the member functions or operators available.

Below is a description of each data field in the class:

- The variable vlstr stores a string of characters consisting of integers which represents the very long integer number. The string could in principle be stored as the usual ordering or stored in the reversed order. We have chosen to store it in the reversed order to facilitate some manipulations.

- The variable vlen contains the length of the character integer without counting the sign bit.

- The variable vlsign stores 0 or 1 which indicates a positive or negative number, respectively.

6.1.3 Constructors

This section shows how a Verylong number is created. As with all data types, the simplest way to create a Verylong number is through the declaration statement. For example,

```
Verylong x;
```

It creates a new variable named x. This simple statement actually invokes the default constructor provided in the class. During the construction, the Verylong number is initialized to zero internally. However, one may think about initializing the Verylong number to a specific value other than zero. C++ allows the constructors to be overloaded with multiple definitions. Different constructors could be differentiated by different argument lists in the declaration statement. The following statement, for example, creates a new variable named y, and the variable y is initialized to the value 3:

```
Verylong y(3);
```

Again, there exists some problem with this specification. What happens if the user wants to initialize a value that exceeds the built-in integer type? One solution to this problem is to provide a constructor that reads in a string of character integer as its argument. With this implementation, one could declare a `Verylong` number yet initialize it to any possible value. For example,

`Verylong u("123")`	initializes the variable u to the value of 123.
`Verylong v("1234567890123")`	initializes the variable v to a value that exceeds the bound of the built-in integer type.
`Verylong w("-567890")`	initializes a negative integer number.

Below is a brief description of the constructors available in the class:

- `Verylong(const char*)` takes in a string of character integer as argument. It checks and assigns the sign of the integers, allocates memory and stores them internally. If there is no argument during the construction, it is initialized to zero.

- `Verylong(int)` takes an integer as argument, converts it to a string of character integer and stores it internally.

- The copy constructor is crucial for the class. Although the C++ compiler automatically generates one if it is omitted, the generated copy constructor may not be correct whenever it involves dynamic allocation of data fields as in this case. The assignment operator would simply copy the pointers. This is not what we want. Therefore the programmer is responsible to define a proper copy constructor to ensure that the duplication of instances are correct.

- The destructor simply releases memory that is no longer in use.

6.1.4 Operators

When we multiply two matrices A*B, the method used to perform the operations is quite different from multiplying two floating point numbers. In C++, arithmetic operators can be overloaded. The only requirement is that a new definition must not be ambiguous. This means that the definition must not require arguments that match any existing definition.

In the case of the `Verylong` class, various operators have been overloaded. They are

```
++, --, -(unary), +, -, *, /, %, =,
+=, -=, *=, /=, %=, ==, !=, <, <=, >, >=.
```

In the following, we describe the functions and the algorithms used for each operator:

- The assignment operator = is used to assign a Verylong number to another. Its implementation is similar to the copy constructor but their functions are different.

- Just like any other built-in data type, the increment and decrement operators are overloaded in the class. However, they can be used in two different ways — prefix and postfix. In order to overload these operators, we need to know how to distinguish between them. As shown in the following, the operator with no parameter is for prefix usage and the operator with an int parameter is for postfix usage:

```
class Verylong
{
    ...
    Verylong operator ++ ();    // prefix:  ++Verylong
    Verylong operator ++ (int); // postfix: Verylong++
    ...
}
```

When implementing the two functions, we should remember that:
(1) For prefix use, change the value and then use it.
(2) For postfix usage, use the value and then change it.

This explains the structure of these functions:

```
// Prefix increment operator
Verylong Verylong::operator ++ ()
{
    return *this = *this + one;
}

// Postfix increment operator
Verylong Verylong::operator ++ (int)
{
    Verylong result(*this);
    *this = *this + one;
    return result;
}
```

- The addition operator + adds two Verylong integers.

  The exclusive or (XOR) operator ˆ is used to determine the signs of each argument which in turn determine how the operations are to be carried out.

Suppose we are evaluating u+v, where u, v are instances of the **Verylong** class, we perform the following:

Step 1. If u and v are of different sign then

      - if u is positive then return u-|v| else return v-|u|

      - in both cases, the result is evaluated using the subtraction operator.

    else get digit by digit from each operand and add them together using the usual addition arithmetic.

Step 2. Finally determine the correct sign and return the result.

- The subtraction operator - subtracts one **Verylong** integer from another. Similar to addition, the exclusive or operator ˆ is used to determine the signs of each argument. Suppose we are evaluating u-v, where u, v are instances of the **Verylong** class, we perform the following:

  Step 1. if u and v are of different sign then

        - if u is positive then return u+|v| else return -(v+|u|)

        - in both cases, the result is evaluated using the addition operator.

      else get digit by digit from each operand and subtract them using the usual subtraction arithmetic.

  Step 2. Finally determine the correct sign and return the result.

Note that the addition and subtraction operators invoke each other during the manipulation.

- Suppose we are evaluating u*v, where u, v are instances of the **Verylong** class. The multiplication is carried out using the usual method:

  For each digit in v, we multiply it by u using a private member function **multdigit()**. The summation of these results gives the product of u and v.

- For the *division operator* /, the algorithm used is the usual long division. Consider the expression u/v. First we make sure that the divisor v is non-zero. A zero value is return if u<v. The rest of the operations involves finding the quotient digit by digit. Finally, we assign the correct sign to the value and return it.

- The *modulo operator* % is calculated using u - v*(u/v).

- There are 6 relational operators defined in the class. For ==, we compare the signs and the contents of the two numbers. If they are both equal to each other, then the two numbers must be equal. The != could be calculated using !(u==v).

To check if u<v is true, we do the following:

- Compare the signs of the two numbers u and v; a positive number is always greater than a negative number.

- If both numbers have the same sign, compare the lengths of each number taking their signs into consideration.

- If both numbers have the same length, reverse the string and compare their values using the built-in function strcmp().

- Return the boolean value — 1 or 0 (True/False).

The rest of the relational operators could be constructed based on the less than operator <:

- <= could be constructed by (u<v or u==v)

- > could be constructed by (!(u<v) and u!=v)

- >= could be constructed by (u>v or u==v)

6.1.5 Type Conversion Operators

In C++, a floating point number could be assigned to an integer number and vice versa. The type conversion is done implicitly. In the case of the Verylong class, we would like to include these properties as well. There are two types of conversion:

1. Convert a Verylong number to a built-in data type.
   This type of conversion is accomplished by the conversion operator provided in C++. Three conversion operators are used in the class:

   • operator int() converts the Verylong number into a built-in integer type if it is within the valid range, otherwise an error message is reported.

   • operator double() converts the Verylong number to a double precision floating point number where applicable.

   • operator char*() converts the Verylong number to a pointer to character type. This conversion is useful when we apply the routines in the library <string.h> (see Chapter 11).

2. Convert a built-in data type to a Verylong number.
   This type of conversion is carried out by the constructors of the class. The constructors read the data type of the arguments and perform the appropriate transformation which converts a built-in data type to a Verylong number.

6.1.6   Private Member Functions

Some operations on the data fields should not be visible to the user. Member functions like this are declared as `private`. There are 3 private member functions in the class. Their behaviours are described as follows:

- `multdigit(int num)` multiplies this `Verylong` number by num where num is an integer ranged between 0 and 9. This function is invoked during the multiplication of two `Verylong` numbers.

- `mult10(int num)` multiplies this `Verylong` number by 10^{num}.

  e.g. `v.mult10(5)` is equivalent to v * 100000.

- `strrev(char *s)` reverses the order of the string s and returns it.

6.1.7   Other Functions

For a class to be useful, one must include a complete set of public interfaces. A `Verylong` class could not be considered as complete if the following functions were omitted:

- `abs(const Verylong&)` returns the absolute value of a `Verylong` number.

- Integer square root function `sqrt(const Verylong&)`

  Given a positive integer b, the integer square root of b is given by a provided

$$a^2 \leq b < (a+1)^2.$$

  For example, the integer square root of 105 is 10 since $10^2 \leq 105 < 11^2$.

  There are many ways to find an integer square root of a positive integer. The method we use here is based on the identity $(a + b)^2 \equiv a^2 + 2ab + b^2$. The algorithm is as follows:

  Step 1. Start from the rightmost digit towards the left and split the number into 2 digits each. The number of segments is equal to the number of digits of the integer square root.

  Step 2. Get the first digit a of the result by taking the integer square root of the first segment. Record the result immediately.

  Step 3. Subtract a^2 from the first segment and get the remainder.

  Step 4. Divide the remainder by $2a$ to obtain the second digit of the root b. Record the result.

  Step 5. Subtract $2ab$ and b^2 from the remainder in the appropriate position.

  Step 6. If all the digits required have been obtained, return the final result.

  Step 7. Go to Step 4 to find the next few digits of the integer square root.

For the purpose of illustration, consider the following example: $\sqrt{394384} = ?$

$$
\begin{array}{r}
3\;9\;4\;3\;8\;4 \\
-\;3\;6 \\
\hline
3\;4\;3 \\
-\;\;2\;4 \\
\hline
4 \\
\hline
9\;9\;8\;4 \\
-\;9\;9\;2 \\
\hline
6\;4 \\
\hline
\underline{0}
\end{array}
$$

$\cdots\cdots$ $a \to 6$ $\quad\cdots\cdots$ First digit

$\cdots\cdots$ $34 \div 2 * 6 \to 2$ $\quad\cdots\cdots$ Second digit

$\cdots\cdots$ $2a * b = 2 * 6 * 2 = 24$

$\cdots\cdots$ $b^2 = 4$

$\cdots\cdots$ $998 \div 2 * 62 \to 8$ $\quad\cdots\cdots$ Third digit

$\cdots\cdots$ $2\,ab * c = 2 * 62 * 8 = 992$

$\cdots\cdots$ $c^2 = 64$

Therefore, sqrt(394384) = 628.

- pow(const Verylong&, const Verylong&)

Suppose we want to compute x^{29}, we could simply start with x and multiply by x twenty-eight times. However it is possible to obtain the same answer with only seven multiplications: start with x, square, multiply by x, square, multiply by x, square, square, multiply by x, forming the sequence

$$
x \to x^2 \to x^3 \to x^6 \to x^7 \to x^{14} \to x^{28} \to x^{29}.
$$

This sequence could be obtained by its binary representation 11101: replace each "1" by the pair of letters SX, replace each "0" by S, we get SX SX SX S SX and remove the leading SX to obtain the rule SXSXSSX, where "S" is interpreted as squaring and "X" is interpreted as multiplying by x. This method can readily be programmed. However it is usually more convenient to do so from right to left. Here, we present an algorithm based on a right-to-left scan of the number:

Consider for positive N:

Step 1. Set $N \leftarrow n, Y \leftarrow 1, Z \leftarrow x$

Step 2. If N is odd, $Y \leftarrow Y \times Z, N \leftarrow \lfloor N/2 \rfloor$
    If $N = 0$ return the answer Y
    else $N \leftarrow \lfloor N/2 \rfloor$

Step 3. $Z \leftarrow Z \times Z$ and goto Step 2

where $\lfloor x \rfloor$ denotes the *floor function*. It is defined as the largest integer value smaller than x. Let us consider the steps in the evaluation of x^{29}:

	N	Y	Z
After Step 1.	29	1	x
After Step 3.	14	x	x^2
After Step 3.	7	x	x^4
After Step 3.	3	x^5	x^8
After Step 3.	1	x^{13}	x^{16}
After Step 3.	0	x^{29}	–

- `div(const Verylong&,const Verylong&)`

  The usual division operator / performs integer division. In many cases, however, we need the floating point value of the quotient of `Verylong` numbers. The function `div(u,v)` is used to perform floating point division and the return type of the function is `double`. In this function, we perform the following:

  - First, we ensure that the denominator is not equal to 0.
  - Next, we find the appropriate scale factor to start with.
  - Then, we find the quotient digit by digit.
  - The division is completed if there is no remainder or 16 significant digits have been obtained.

6.1.8   Streams

The `>>` and `<<` operators are used for input and output, respectively. These operators work with all the built-in data types. For the `Verylong` class we overload the stream operators. For example, one could read a value from an input stream into a `Verylong` variable named x and then display the value.

```
#include <iostream.h>
#include "Verylong.h"

void main()
{
    Verylong x;
    cin >> x;
    cout << "The value entered is " << x << endl;
}
```

The implementation for these functions are quite straightforward:

- `istream& operator >> (istream&,Verylong&)`
  It reads in a sequence of characters of integers from the stream (e.g. keyboard), assuming a maximum string length of 1000 characters.

- `ostream& operator << (ostream&,const Verylong&)`
  It displays the value of the `Verylong` number.

6.2 The Rational Number Class

6.2.1 Abstraction

In mathematics a number system is built up level by level. In this section, we construct a Rational number class, which is the natural extension of integer **Z**. The mathematics of *rational numbers* has been described in Section 2.3.

Here, we encounter again the problem faced in the last section, namely, the integral data type in C++ has limited range. In order to solve this problem, instead of using int, we use the Verylong to represent the numerator and denominator of a Rational number. However, we have decided to extend it further; we make use of the class template feature provided in C++ to implement the class. This allows the users to select the data type that suits their purposes.

We have to specify the behaviours of the Rational number ADT:

- It is a **template** class for which the data type of the numerator and denominator are to be specified by the users.

- Creation of a new instance of Rational number is simple.

- The Rational number is reduced and stored in its simplest form, using the greatest common divisor algorithm.

- Arithmetic operators such as +, -, *, / are available.

- Assignment and modification forms of assignment =, +=, -=, *=, /= are available.

- The relational operators >, >=, <, <=, ==, != are available.

- Conversion to the type double is supported.

- Functions that return the numerator, denominator, and fractional part of Rational numbers are available.

- Input and output operations with Rational numbers are supported.

6.2.2 Template Class

Rational numbers with numerator and denominator of type int could not represent the whole class of rational numbers due to the limitation on the data type int. This is possible only if the numerator and denominator can represent the whole range of the integer. The Verylong class developed in the previous section in principle allows an arbitrarily long integer number. Therefore, we need to incorporate the Verylong number into the Rational class somehow.

The template in C++ provides *parametrized types*. With this feature, the same code could be used with respect to different types where the type is a parameter of the code body. Template classes give us the ability to reuse code in a simple type safe manner that allows the compiler to automate the process of type instantiation.

We have developed the `Rational` number as a template class. The reason is that the data items, numerator and denominator, could be of type `int` or `Verylong` as specified by the user. With this desirable feature, one can perform extensive computation without worrying that a number could possibly run out of range.

6.2.3 Data Fields

The `Rational` class declares two fields of data type T. This means that the user has to specify the actual data type represented in order to use the class. The first field p maintains the numerator, while the second field q maintains the denominator of the `Rational` number. Note that in order to enforce the concept of *data hiding*, both fields are declared as `private`. Therefore they are accessible only within the class.

6.2.4 Constructors

To declare a `Rational` number, we proceed as follows:

```
// To declare a Rational number u of int-type which is initialized to 0
Rational<int> u;

// To declare v and initialize it to 5
Rational<int> v(5);

// To declare w and initialize it to 2/3
Rational<int> w(2,3);

// To declare x of Verylong-type and initialize it to 127762/2384623
Rational<Verylong> x(127762,2384623);

// To declare y and initialize it to a value that exceeds the int-bound
Rational<Verylong> y("32872134727762","23489729384879822384623");
```

There are three constructors in the class which cater for different ways of construction of a `Rational` number:

1. The default constructor declares a `Rational` variable and it is initialized to zero.

2. `Rational(const T N)` declares a `Rational` variable and initializes it to N.

3. Rational(const T N, const T D) declares a Rational variable and initializes it to N/D.

One may argue that it is possible to combine the three constructors into one by putting Rational(const T = 0, const T = 1). However, the compilers we use complain about this construct because it is a class template. To be on the safe side, we decide to maintain three constructors.

During the construction, a private member function gcd() is invoked to reduce the Rational number into its standard form. The copy constructor and destructor provided by the compiler work well in this case, but for completeness we have included them in the class.

6.2.5 Operators

Since the Rational class is a mathematical object, the common operators used in the class are the arithmetic operators. We have included many operators in the class, namely

-(unary), +, -, *, /, =, +=, -=, *=, /=, ==, !=, >, >=, <, <=.

The definitions of some of these operators are as follows:

$$-\frac{a}{b} = \frac{-a}{b},$$

$$\frac{a}{b} + \frac{c}{d} = \frac{a*d+b*c}{b*d}, \qquad \frac{a}{b} - \frac{c}{d} = \frac{a*d-b*c}{b*d},$$

$$\frac{a}{b} * \frac{c}{d} = \frac{a*c}{b*d}, \qquad \frac{a}{b} \div \frac{c}{d} = \frac{a*d}{b*c},$$

$$\frac{a}{b} == \frac{c}{d} \Rightarrow (a == c) \text{ and } (b == d), \qquad \frac{a}{b} < \frac{c}{d} \Rightarrow (a*d) < (b*c).$$

The implementations of these operators are straightforward and can be understood easily.

6.2.6 Type Conversion Operators

A *type conversion operator* allows a data type to be cast into another when needed. A floating point representation of a Rational number is always useful. Here, we have included a conversion operator to the data type double. The conversion operator in the class exists in two forms. One is the general form which allows any

data type to be cast into `double`. The other is only specific to the conversion from
`Rational<Verylong>` to `double`. One may ask why we need an extra conversion
operator for the `Rational<Verylong>` when the other one could do the job as well.
As we see in the following example, there exists a better and more accurate method
for the `Rational<Verylong>` number.

The general method which works for any data type simply returns the `double`-cast
value of the `Rational` number:

```
return double(p)/double(q);
```

This method works quite well for the `Rational<Verylong>` number in general, but it
fails sometimes due to the limited accuracy in `double`. Recall that there is a function
named `div()` which could do the double division for `Verylong` numbers. With this
function we could do a better job:

```
return div(p,q);
```

For the purpose of illustration, consider the following example:

```
// division.cxx

#include <iostream.h>
#include "Verylong.h"

void main()
{
    Verylong P("999"), Q("111"), D("105");

    P = pow(P,D);   // 999^105 = 9.00277e+314 (exceeded double limit)
    Q = pow(Q,D);   // 111^105 = 5.74001e+214
    cout << div(P,Q) << endl;              // 1.56842e+100 - OK
    cout << div(Q,P) << endl;              // 6.37583e-101 - OK
    cout << double(P)/double(Q) << endl;   // NaN
    cout << double(Q)/double(P) << endl;   // NaN
}
```

The statement `double(P)/double(Q)` does not work when P or/and Q exceed the
limit of the data type `double` which is about $\pm 1.7977 \times 10^{308}$ as shown in above.
The word NaN stands for Not A Number. From this example, we conclude that the
definition for the `Rational<Verylong>` is necessary.

6.2.7 Private Member Functions

Private member functions are internal member functions that are only known to the class itself. They are inaccessible from outside the class. There is only one private member function in our class. The private member function

```
gcd(T a,T b)
```

returns the *greatest common divisor* of a and b using the following algorithm:

Step 1. While $b > 0$, do the following:

* $m \leftarrow a \bmod b$
* $a \leftarrow b$ and $b \leftarrow m$

Step 2. Return the answer a.

For example, gcd(4,8) returns 4.

6.2.8 Other Functions

There are only three member functions, other than the arithmetic operators, defined in the class: num(), den() and frac(). They return the numerator, denominator and the fractional part of the Rational number, respectively.

```
template <class T> T Rational<T>::num() const
{ ... }
template <class T> T Rational<T>::den() const
{ .... }
template <class T> Rational<T> Rational<T>::frac() const
{ ... }
```

Notice that they are declared as const functions. This indicates that the functions do not alter the value of the instance. Any value can be declared as constant in C++. A constant variable is bound to a value and its value can never be changed. Therefore, only constant operations can be performed on the value. An application of these methods is given in the following program

```
// testfrac.cpp

#include <iostream.h>
#include "Rational.h"

int main()
{
    Rational<int> r(7,3);
    int n = r.num();
```

```
    int d = r.den();
    Rational<int> f;
    f = r.frac();
    cout << "n = " << n << endl; // => 7
    cout << "d = " << d << endl; // => 3
    cout << "f = " << f << endl; // => 1/3

    return 0;
}
```

6.2.9   Streams

In this section, we describe how the >> and << operators are overloaded to perform the input and output for a Rational number:

- The input stream operator >> is implemented in the way that one could input a fraction like a/b from the keyboard and the class could recognize that a is the numerator whereas b is the denominator. On the other hand, if a non-fraction is entered, the class should be able to recognize that it is an integer (i.e. a rational number with denominator equal to one).

  In order to fulfill the requirement, the function make use of some functions from <iostream.h> library. The manipulator ws clears any leading white space from the input. The function peek() is used to have a sneak preview of the next character and the function get() reads a character from the input stream. No precaution against erroneous input is taken. The users take the full responsibility for handling the function in a proper manner.

- The implementation for the output stream operator << is much simpler. If the denominator of the Rational number is equal to one, then output only the numerator. Otherwise, output the value of the fraction.

6.3 The Complex Number Class

6.3.1 Abstraction

The header file `<complex.h>` is available in the C++ library. However, the data type of the real and imaginary part is `double`. This is not sufficient for applications which require exact manipulation of the complex numbers. This limitation has prompted us to construct another complex number class which uses the `template`. This class allows us to use the `Rational` number class, developed in the previous section, as the real and imaginary part of the complex number. Hence, exact manipulation becomes possible. We describe the abstraction and give an implementation for the `Complex` number class. An example is also given at the end of this section to demonstrate how the class can be used.

Since the `Complex` ADT is a numeric type, it should have all the properties that a number should have. The behaviours of the class are as follows:

- It is a `template` class for which the data type of the real and imaginary parts are to be specified by the user.

- There are three simple ways to create a `Complex` number.

- Arithmetic operators such as +, -, *, / are available.

- Assignment and modification forms of assignment =, +=, -=, *=, /= are available.

- Six relational operators >, >=, <, <=, ==, != are available.

- Six member functions are supported, they are:

  1. `realPart()` returns the real part of a `Complex` number.
  2. `imagPart()` returns the imaginary part of a `Complex` number.
  3. `magnitude()` returns the magnitude of a `Complex` number.
  4. `argument()` returns the argument of a `Complex` number.
  5. `conjugate()` returns the conjugate of a `Complex` number.
  6. `negate()` returns the negative of a `Complex` number.

- Input and output operations with `Complex` numbers are supported.

6.3.2 Template Class

As in the `Rational` number class, the users may want to have a better control over the coefficients of a `Complex` number. This can be achieved by implementing the class as `template`. Below are some possible ways to use the `Complex` number class:

- `Complex<int>`

  $z = x + iy$ x, y are of data type `int`

- `Complex<double>`

  $z = x + iy$ x, y are of data type `double`

- `Complex<Rational<int> >`

  $z = a/b + i\, c/d$ a, b, c, d are of data type `int`

- `Complex<Rational<Verylong> >`

  $z = a/b + i\, c/d$ a, b, c, d are of data type `Verylong`

Note that a space between the two `>`'s that ended the template names is always required. If the space between the `>`'s is omitted, the compiler will complain about it.

6.3.3   Data Fields

Since there are two different ways of representing a complex number, we have to choose one to represent the data field internally. We use the Cartesian form of the complex number because this representation is simpler. However, the choice is left to the readers. No matter which representation has been chosen, the users should be unaware of which is used internally. Complex numbers in Cartesian form have two components, i.e. the real and imaginary part. This is represented by the variables `real` and `imag` in the class.

6.3.4   Constructors

The most important functions of a class are the constructors. We provide three different means to construct a `Complex` number:

1. The default constructor declares a `Complex` variable without specifying its initial value. It is set to zero automatically.

2. The constructor which contains one argument declares a `Complex` number with its imaginary part equal to zero and the real part is initialized to the value specified in the argument.

3. The constructor which contains two arguments declares a `Complex` number with the real and imaginary part initialized to the values specified.

The following examples illustrate how a `Complex` number could be constructed in different ways:

```
// Declare a Complex number u of int-type
// which is initialized to 0
Complex<int> u;

// Declare a Complex number v of double-type
// which is initialized to 0
Complex<double> v;

// Declare w and initialize it to 2
Complex<int> w(2);

// Declare x and initialize it to 2 + 3i
Complex<int> x(2,3);

// Declare and initialize y
// with coefficient of Rational<int>-type
Complex<Rational<int> > y(Rational<int>(1,2), Rational<int>(3,4));

// Declare and initialize z
// with coefficient of Rational<Verylong>-type
Complex<Rational<Verylong> > z(Rational<Verylong>("12345","345123"),
    Rational<Verylong>("987659876543","82567834536572693487"));
```

As in the Rational class, one may tend to believe that the constructor with default argument

```
Complex(const T = 0,const T = 0)
```

could replace the three constructors mentioned above. However, for the reason of generality we keep them separated.

The copy constructor and destructor are trivial in this class. However, we have included them for completeness.

6.3.5 Operators

Here we discuss the behaviours of the arithmetic operators available in the class. Operators that have been overloaded include

$$-(\text{unary}), +, -, *, /, =, +=, -=, *=, /=, <, <=, >, >=, ==, !=.$$

The definitions and operations for these operators are as follows:

$$(a + ib) + (c + id) = (a + c) + i(b + d)$$
$$(a + ib) - (c + id) = (a - c) + i(b - d)$$

$$(a + ib) \times (c + id) \quad = \quad (ac - bd) + i(ad + bc)$$

$$\frac{(a + ib)}{(c + id)} \quad = \quad \frac{ac + db}{c^2 + d^2} + i\left(\frac{bc - ad}{c^2 + d^2}\right)$$

$$(a + ib) < (c + id) \quad \Rightarrow \quad (a^2 + b^2) < (c^2 + d^2)$$
$$(a + ib) == (c + id) \quad \Rightarrow \quad (a == c) \text{ and } (b == d)$$

The implementations of these operators are quite straightforward and the readers should have no difficulty of understanding them.

6.3.6   Type Conversion Operators

We have

$$\mathbf{Z} \subset \mathbf{Q} \subset \mathbf{R} \subset \mathbf{C}$$

i.e. the integers are a subset of the rationals, which are a subset of the reals, which are a subset of the complex numbers. Thus the set of the complex numbers is the superset of all the other sets including the built-in data types. Therefore, we do not define any type conversion function to any built-in data type, because by doing so the complex number would suffer from loss of information.

6.3.7   Other Functions

Other than the operators described above, below are some further properties on the Complex number $a + ib$ that we would like to obtain:

- The real part and imaginary part of a complex number.

- The magnitude of a complex number which is defined as $\sqrt{a^2 + b^2}$.

- The argument which is defined as $\tan^{-1}(b/a)$ (principle branch).

- The complex conjugate of $a + ib$ which is equal to $a - ib$.

- The negative of a complex number $a + ib$ which is equal to $-a - ib$.

They are readily obtained by using the six member functions defined in the class:

`realPart()`, `imagPart()`, `magnitude()`, `argument()`, `conjugate()`, `negate()`.

The following program shows the usage of the six member functions:

```
// scomplex.cxx

#include <iostream.h>
#include "MComplex.h"

int main()
{
    Complex<int> c(3,4);   // declares a complex number 3 + 4i

    cout << c.realPart()  << endl;      // 3
    cout << c.imagPart()  << endl;      // 4
    cout << c.magnitude() << endl;      // 5
    cout << c.argument()  << endl;      // 0.927295
    cout << c.conjugate() << endl;      // (3,-4i)
    cout << c.negate()    << endl;      // (-3,-4i)

    return 0;
}
```

For the Complex class to be complete, we still have to include all the special mathematical functions that can be applied to the Complex numbers, for example

sin(), cos(), exp(), cosh(), sinh(), sqrt(), log().

The implementations for these functions are left as exercises for the readers.

6.4   The Quaternion Class

6.4.1   Abstraction

The *quaternion* described in Section 2.7 is a higher level mathematical structure compared with the basic numeric structures like rational and complex numbers. It is therefore based on the basic structures. The behaviours of the Quaternion ADT can be summarized as follows:

- It is a template class for which the underlying data type for the coefficients of each component is to be specified.

- The construction of instances of Quaternion is simple.

- Arithmetic operators for Quaternion are overloaded. For example, (unary)-, +, -, *, /.

- The assignment operator = is available.

- Operations like finding the magnitude, conjugate and inverse of a Quaternion are available.

- Input and output stream operations are supported.

6.4.2   Template Class

The coefficients of a Quaternion could be of type int, double or Verylong, etc. A class like this is best implemented in template form. By doing so, we could avoid code duplication for different data type of coefficients. In the following, we list some possible ways to use the Quaternion class:

- Quaternion<int>, Quaternion<double> declare Quaternion with coefficients of built-in type int and double, respectively.

- Quaternion<Verylong> declares a Quaternion with coefficients of user-defined type Verylong.

- Quaternion<Rational<int> >, Quaternion<Rational<Verylong> > declare Quaternion with coefficients of type Rational.

- Quaternion<Complex<int> >, Quaternion<Complex<Verylong> >, Quaternion<Complex<Rational<int> > >, Quaternion<Complex<Rational<Verylong> > > are different ways to declare Quaternion with coefficients of type Complex.

Note that it is interesting to couple four user-defined types successfully to form a new data type.

6.4.3 Data Fields

To define a quaternion uniquely, we have to specify the coefficients of the four components. This is exactly what we need to maintain in the data fields of the `Quaternion` class. The entries r, i, j, k represent the coefficients of $1, I, J, K$ respectively.

6.4.4 Constructors

The construction of a `Quaternion` is simple. What the users have to do is to provide the four coefficients and a `Quaternion` is created. The users could also opt for not providing any coefficient whereby the default constructor would be invoked and the coefficients would be initialized to zero. Below are some examples of how instances of `Quaternion` could be constructed:

```
// To declare u of int-type that is initialized to 0
Quaternion<int> u;

// To declare v of double-type and initialize to 1 + 2I - 3J + 4K
Quaternion<double> v(1,2,-3,4);

// To declare w and initialize to 1/2 - 2/3 I + 3/4 J - 4/5 K
Quaternion<Rational<int> > w(Rational<int>(1,2),Rational<int>(-2,3),
                             Rational<int>(3,4),Rational<int>(-4,5));
```

The copy constructor and destructor are trivial.

6.4.5 Operators

Suppose q and p are two arbitrary quaternions,

$$q + p, \qquad q - p, \qquad q * p, \qquad q/p$$

are overloaded as the addition, subtraction, multiplication, division of q and p respectively; whereas $-q$ is the negative of q. The mathematics has been described in Section 2.7.

6.4.6 Other Functions

In this section, we would like to demonstrate the usage of the functions `conjugate()`, `inverse()`, `magnitude()` and the normalization operator (`~`):

```
// squater.cxx

#include <iostream.h>
#include "Quatern.h"
```

```
void main()
{
   Quaternion<double> Q1(3,4,5,6),
                      Q2 = Q1.conjugate(),
                      Q3 = Q1.inverse();
   double Mag  = Q1.magnitude(), Magz = (~Q1).magnitude();

   cout << "Q1 = "                       << Q1 << endl;
   cout << "Q2 = Conjugate of Q1 = "  << Q2 << endl;
   cout << "Q3 = Inverse of Q1 = "    << Q3 << endl;
   cout << "Mag  = Magnitude of Q1 = " << Mag << endl;
   cout << "Magz = Magnitude of normalized Q1 = " << Magz << endl;
   cout << endl;

   cout << "Q1 * Q2 = " << Q1 * Q2 << endl;
   cout << "Q2 * Q1 = " << Q2 * Q1 << endl;
   cout << "Mag^2 = Square of magnitude = " << Mag * Mag << endl;
   cout << endl;

   cout << "Q1 * Q3 = " << Q1 * Q3 << endl;
   cout << "Q3 * Q1 = " << Q3 * Q1 << endl;
}
Results
=======
Q1 = (3,4,5,6)
Q2 = Conjugate of Q1 = (3,-4,-5,-6)
Q3 = Inverse of Q1 = (0.0348837,-0.0465116,-0.0581395,-0.0697674)
Mag  = Magnitude of Q1 = 9.27362
Magz = Magnitude of normalized Q1 = 1

Q1 * Q2 = (86,0,0,0)
Q2 * Q1 = (86,0,0,0)
Mag^2 = Square of magnitude = 86

Q1 * Q3 = (1,0,0,0)
Q3 * Q1 = (1,0,0,0)
```

6.4.7   Streams

The input and output stream functions in the class are straightforward. The input stream function simply reads in the four coefficients of each component. For the output stream function, the quaternion $q = a_1 * 1 + a_I * I + a_J * J + a_K * K$ is formatted and exported as (a_1, a_I, a_J, a_K).

6.5 The Derive Class

6.5.1 Abstraction

So far we have been dealing with numeric types; here we are going to specify a somehow quite different abstraction. The `Derive` class provides an operator which applies to a numeric type, and the result of the operation is again the numeric type. The behaviours of the ADT are as follows:

- The data type of the result is the template `T` which is to be specified by the users.

- The construction of an expression is simple, using arithmetic operators such as (unary)-, +, -, *, /.

- The member function `set()` is used to specify the point (a number) where the derivative is taken.

- The value of the derivative can be obtained using the function `df()`.

- Output operation with the derivative is supported.

6.5.2 Data Fields

There are only two data fields in the class: one being the variable u, which stores the value of the point where the derivative is evaluated, whereas the other variable du stores the derivative value of u.

6.5.3 Constructors

The constructors of the class are as follows:

- The default constructor declares an independent `Derive` variable. The dependent variable is declared using the assignment operator or copy constructor.

- `Derive(const T num)` declares a constant number num.

- The private constructor `Derive(const T,const T)` is used to define the values of the point and its derivative.

- The copy constructor and assignment operator are trivial in this case, and perform no more than member-wise copying.

- The destructor is trivial.

As an example, the declaration of $y = 2x + 1$ requires the following statements:

```
Derive<int> x;          // This declares the independent variable x
Derive<int> y = 2*x + 1; // ...  and the dependent variable.
```

6.5.4   Operators

The arithmetic operators

(unary)-, +, -, *, /

are overloaded to perform operations that obey the *derivative rules* as described in Section 2.9.

6.5.5   Member Functions

The followings member functions are available in the class:

- The function set(const T) operates on the independent variable. It is used to specify the value of u which is the point where the derivative of f is evaluated.

- The function df(const Derive &x) returns the value of the derivative evaluated at x.

- The output stream operator << returns the value of u.

Let us consider the function $f(x) = 2x^3 + 5x + 1$. Suppose we intend to evaluate the value of $df(x = 2)/dx$, we do the following:

```
// sderive1.cxx

#include <iostream.h>
#include "Derive.h"

void main()
{
   Derive<int> x;
   x.set(2);
   Derive<int> y = 2*x*x*x + 5*x + 1;

   cout << "The derivative of y at x = " << x << " is "
        << df(y)<< endl;
}
Result
======
The derivative of y at x = 2 is 29
```

Consider another function $g(x) = x^2 + 3/x$. Suppose we want to evaluate the value of $dg(x = 37/29)/dx$, then

```
// sderive2.cxx

#include <iostream.h>
#include "Derive.h"
#include "Rational.h"

void main()
{
    Derive<Rational<int> > x;
    x.set(Rational<int>(37,29));

    Derive<Rational<int> > c(3);
    Derive<Rational<int> > y = x*x + c/x;

    cout << "The derivative of y at x = " << x << " is "
         << df(y)<< endl;
}
```
```
Result
======
The derivative of y at x = 37/29 is 28139/39701
```

Notice that after the declaration of the independent variable x, we always set the value where the derivative is taken. This is important and has to be done before the declaration of the dependent variable y. Otherwise, unpredictable results will be obtained.

6.5.6 Possible Improvements

The major drawback of this class is the inflexibility in specifying the function f. For example, to specify the expression $y = x^5 + 2x^3 - 3$, it requires a long statement like y = x*x*x*x*x + 2*x*x*x - 3. What happens if the function required is of the order of x^{100}?

Other drawbacks include the fact that the derived function f' is not known. Only the value of $f'(a)$ can be evaluated. This imposes a strict restriction on the usefulness and applications of the class.

These shortcomings can be overcome by implementing a more elaborate class which is shown in the next chapter – the Symbolic Class. The class not only solves all the problems mentioned above, it also includes many more features.

6.6   The Vector Class

6.6.1   Abstraction

A *vector* is a common mathematical structure in linear algebra and vector analysis (see Section 2.6). This structure could be constructed using arrays. However, C and C++ arrays have some weaknesses. They are effectively treated as pointers. It is therefore useful to introduce a `Vector` class as an abstract data type. With bound checking and mathematical operators overloaded (for example vector addition), we built a comprehensive and type-safe `Vector` class. On the other hand, it could replace the array supported by C and C++ as a collection of data objects.

The `Vector` class is a structure that possesses many interesting properties for manipulation. The behaviour of the `Vector` ADT is summarized as follows.

- It is best implemented as a template class, because it is a *container class* whereby the data items could be of any type.

- The construction of a `Vector` is simple.

- Arithmetic operators such as +, -, *, / with `Vector` and numeric constants are available.

- The assignment and modification forms of assignment =, +=, -=, *=, /= are overloaded. We could also copy one `Vector` to another by using the assignment operator.

- The subscript operator [] is overloaded. This is useful for accessing individual data items in the `Vector`.

- The equality (==) and inequality (!=) operators check if the two vectors contain the same elements in the same order.

- Operations such as dot product and cross product for `Vector` are available.

- The member function `length()` returns the size of a `Vector` while `resize()` reallocates the `Vector` to the size specified.

- The `Matrix` class is declared as a `friend` of the class. This indicates that the `Matrix` class is allowed to access the `private` region of the class.

- Input and output stream operations with `Vector` is supported.

- The auxiliary file `VecNorm.h` contains different types of norm operators: $\|v\|_1$, $\|v\|_2$, $\|v\|_\infty$ and the normalization function for the `Vector`.

6.6.2 Templates

A *container class* implements some data structures that "contain" other objects. Examples of containers include arrays, lists, sets and vectors, etc. Templates work especially well for containers since the logic to manage a container is often largely independent of its contents. In this section, we see how templates can be used to build one of the fundamental data structures in mathematics — the Vector class.

The container we implement here is *homogeneous*, i.e. it contains objects of just one type as opposed to a container that contains objects of a variety of types. It also has *value semantics*. Therefore it contains the *object* itself rather than the *reference* to the object.

6.6.3 Data Fields

There are only two data fields in the class:

- The variable size stores the length of the Vector.

- The variable data is a pointer to template type T that is used to store the data items of the Vector. The memory is allocated dynamically so that it fits the need of the application. For data types that require a huge amount of memory, it is advisable to release the memory as soon as it is no longer in use. This is not possible with static memory allocation. Therefore, most array-based data types such as vectors and matrices should use dynamic memory for their data storage.

Note that there is no item in the data fields that records the lower or upper index bound of the vector; this means that the index will run from $0, 1, \ldots, \text{size-1}$. To make a vector that starts from an index other than zero, we may introduce a derived class that inherits all the properties and behaviours of the Vector class and adds an extra data field that indicates the lower index bound of the Vector. It is therefore important to declare the data fields as protected rather than private. This allows the derived classes to access the data fields. The implementation of such a bound vector is left as an exercise for the readers.

6.6.4 Constructors

Whenever an array of n vectors is declared, the compiler automatically invokes the default constructor. Therefore, in order to ensure the proper execution of the class, we need to initialize the data items properly in the default constructor. This includes assigning NULL to data. This step is crucial or else we may run into some run-time problem.

A common error is to assign a Vector to an uninitialized one using the assignment operator =, which first frees the old contents (data) of the left-hand side. But there is no "old" value, some random value in data is freed, probably with disastrous effect. The remedy is to nullify the variable data because deleting a NULL pointer is perfectly valid and has no side effect.

There are, in fact, two more overloaded constructors:

```
Vector(int n)     Vector(int n,T value)
```

The first constructor allocates n memory locations for the Vector, whereas the other initializes the data items to value on top of that.

The copy constructor Vector(const Vector& source) constructs a new Vector identical to source. It will be invoked automatically to make temporary copies when needed, for example for passing function parameters and return values. It could also be used explicitly during the construction of a Vector.

In the following, we list some common ways to construct a Vector:

```
// declare a vector of 10 numbers of type int
Vector<int> u(10);

// declare a vector of 10 numbers and initialize them to 0
Vector<int> v(10,0);

// use a copy constructor to create and duplicate a vector
Vector<int> w(v);
```

Whenever a local Vector object passes out of scope, the destructor comes into play. It releases the array storage in free memory. Otherwise, it will constitute unusable storage, because it is allocated but no pointer points to it.

6.6.5 Operators

Most of the operators applicable to Vector are implemented in the class, namely

```
(unary)+, (unary)-, +, -, *, /,
=, +=, -=, *=, /=, ==, !=, |, %, [] .
```

Suppose u, v, w are vectors and c is a numeric constant, then the available operations are defined as follows:

- u+v, u-v, u*v, u/v adds, subtracts, multiplies or divides corresponding elements of u and v.

- u+=v, u-=v, u*=v, u/=v adds, subtracts, multiplies or divides corresponding elements of v into u.

- u+=c, u-=c, u*=c, u/=c adds, subtracts, multiplies or divides each element of u with the scalar.

- The assignment operator = should be overloaded in the class. Should one omit to define an assignment operator, the C++ compiler will write one. However, one should bear in mind that the code produced by the compiler simply makes a byte-for-byte copy of the data members. In the case where the class allocates memory dynamically, we usually have to write our own assignment operator. This is because the byte-for-byte operation copies only the memory address of the pointer not the memory content. It is dangerous to have multiple pointers pointing at the same memory location without proper management. The same argument applies to the copy constructor.

  Two forms of the assignment operator have been overloaded:

  - u=v makes a duplication of v into u.
  - u=c assigns the constant c to every entry of u.

  Note that the assignment operator is defined such that it returns a reference to the object iteself, thereby allowing constructs like u = v = w.

- u==v returns *true* if u and v contain the same elements in the same order and returns *false* otherwise.

- u!=v is just the converse of u==v.

- We use the symbol | as the *dot product* operator (also called the *scalar product* or *inner product*). It is defined as $u|v = u \cdot v = \sum_{j=1}^{n} u_j v_j$.

- The *vector product* (also called the *cross product*) is operated by % in the class.

- The [] operator allows u[i] to access the ith entry of the Vector u. It must return a reference to, not the value of, the entry because it must be usable on the left-hand side of an assignment. In C++ terminology, it must be an lvalue.

The following shows some examples of the usage of the dot product and cross product of Vector. Suppose A, B, C, D are four vectors in $\mathbf{R}^3$, then

$$A \times (B \times C) + B \times (C \times A) + C \times (A \times B) = 0$$

$$(A \times B) \times (C \times D) = B(A \cdot C \times D) - A(B \cdot C \times D)$$
$$= C(A \cdot B \times D) - D(A \cdot B \times C)$$
$$A \cdot (B \times C) = (A \times B) \cdot C$$

The following excerpt program demonstrates that the identities are obeyed for some randomly selected vectors:

```
// vprod.cxx

#include <iostream.h>
#include "Vector.h"

void main()
{
    Vector<double> A(3), B(3), C(3), D(3);

    A[0] = 1.2; A[1] = 1.3; A[2] = 3.4;
    B[0] = 4.3; B[1] = 4.3; B[2] = 5.5;
    C[0] = 6.5; C[1] = 2.6; C[2] = 9.3;
    D[0] = 1.1; D[1] = 7.6; D[2] = 1.8;

    cout << A%(B%C) + B%(C%A) + C%(A%B) << endl;

    cout << (A%B)%(C%D) << endl;
    cout << B*(A|C%D)-A*(B|C%D) << endl;
    cout << C*(A|B%D)-D*(A|B%C) << endl;

    // precedence of | is lower than <<
    cout << (A|B%C) - (A%B|C) << endl;
}
```

Result
======
[1.42109e-14]
[0]
[0]

[372.619]
[376.034]
[540.301]

[372.619]
[376.034]
[540.301]

[372.619]
[376.034]
[540.301]

0

The small, non-zero value $1.42109e-14$ is due to rounding errors. Thus to obtain the correct result, namely the zero vector, we use the data type `Vector<Rational<int> >`.

6.6.6 Member Functions and Norms

Other than the arithmetic operators, there exist some useful operations for the `Vector` class. Their definitions and properties are listed as follows:

- The function `length()` returns the size of the `Vector`.

- `resize(int n)` sets the `Vector`'s length to n. All elements are unchanged, except that if the new size is smaller than the original, than trailing elements are deleted, and if greater, trailing elements are uninitialized.

- `resize(int n,T value)` behaves similar to the previous function except when the new size is greater than the original, trailing elements are initialized to value.

In the auxiliary file `VecNorm.h`, we implement three different vector norms and the normalization function:

- `norm1(x)` is defined as $\|\mathbf{x}\|_1 := |x_1| + |x_2| + \ldots + |x_n|$.

- `norm2(x)` is defined as $\|\mathbf{x}\|_2 := \sqrt{x_1^2 + x_2^2 + \ldots + x_n^2}$, the return type of `norm2()` is double.

- `normI(x)` is defined as $\|\mathbf{x}\|_\infty := \max\{|x_1|, |x_2|, \ldots, |x_n|\}$.

- The function `normalize(x)` is used to *normalize* a vector x. The normalized form of the vector x is defined as x/|x| where |x| is the 2-norm of x.

In order to have a better understanding about these functions, let us consider some examples:

```
// vnorm.cxx

#include <iostream.h>
#include "Vector.h"
#include "VecNorm.h"

void main()
{
  Vector<int> v;
  v.resize(5,2);
  cout << "The size of vector v is " << v.length() << endl;
```

```
      cout << endl;

      Vector<double> a(4,-3.1), b;
      b.resize(4);
      b[0] = 2.3; b[1] = -3.6; b[2] = -1.2; b[3] = -5.5;

      // Different vector norms
      cout << "norm1() of a = " << norm1(a) << endl;
      cout << "norm2() of a = " << norm2(a) << endl;
      cout << "normI() of a = " << normI(a) << endl;
      cout << endl;
      cout << "norm1() of b = " << norm1(b) << endl;
      cout << "norm2() of b = " << norm2(b) << endl;
      cout << "normI() of b = " << normI(b) << endl;
      cout << endl;

      // The norm2() of normalized vectors a and b is 1
      cout << "norm2() of normalized a = " << norm2(normalize(a)) << endl;
      cout << "norm2() of normalized b = " << norm2(normalize(b)) << endl;
   }
```
Result
======
```
The size of vector v is 5

norm1() of a = 12.4
norm2() of a = 6.2
normI() of a = 3.1
norm1() of b = 12.6
norm2() of b = 7.06682
normI() of b = 5.5
norm2() of normalized a = 1
norm2() of normalized b = 1
```

6.6.7   Streams

The overloaded output stream operator << simply exports all the entries in the vector v and puts them in between a pair of square brackets $[v_0, v_1, v_2, \ldots, v_{n-1}]$.

The input stream operator >> first reads in the size of the Vector followed by the data entries.

6.7 The Matrix Class

6.7.1 Abstraction

Matrices are two-dimensional arrays with a certain number of rows and columns. They are important structures in *linear algebra*. To build a matrix class, we do not have to start from scratch. We make use of the advantages (reusability and extensibility) of object-oriented programming to build a new class based on the Vector class. A vector is a special case of a matrix with the number of columns being equal to one. Thus we are able to define a matrix as a vector of vectors.

```
template <class T> class Matrix
{
    private:
        // Data Fields
        int rowNum, colNum;
        Vector<T>* mat;
        ...
}
```

We have declared the Matrix as a template class. Defining the matrix as a vector of vectors has the advantage that the matrix class methods can use many of the vector operations defined for the vector class. For example, we could perform vector additions on each row of the matrix to obtain the result of a matrix addition. This reduces the amount of code duplication.

To use the matrix class effectively, the users need to familiarize themselves with the behaviours and interfaces of the class. Below, we summarize the properties of the Matrix ADT:

- It is implemented as a template class. This indicates that the data type of the data items could be of any type including built-in types and user-defined types.

- There are several simple ways to construct a Matrix.

- Arithmetic operators such as +, -, * with Matrix and +, -, *, / with numeric constants are available.

- The assignment and modification forms of assignment =, +=, -=, *= and /= are overloaded.

- The vectorize operator is available.

- The *Kronecker product* of two matrices is supported.

- The subscript operator [] is overloaded to access the row vector of the matrix while the parenthesis operator () is overloaded to access the column vector of the matrix.

- The equality (==) and inequality (!=) operators check if two matrices are identical.

- The transpose, trace and determinant of a matrix are implemented.

- The inverse of a square matrix is implemented.

- The member function `resize()` reallocates the memory for row and column vectors according to the new specification provided in the arguments of the function.

- The member functions `rows()` and `cols()` return the number of rows or columns of the matrix, respectively.

- Input (>>) and output (<<) stream operators are supported.

- The auxiliary file `MatNorm.h` contains the three different matrix norms: $||A||_1$, $||A||_\infty$ and $||A||_H$.

6.7.2   Data Fields

The data fields `rowNum` and `colNum` specify the number of rows and columns of the matrix respectively.

`Vector<T>* mat` stores the data items of the matrix. It is declared as a vector of vectors. To allocate the memory space for an $m \times n$ matrix, we have to first allocate a vector of m pointers to `Vector`. Then for each pointer, we allocate again a vector of size n. This would result in a total of $m \times n$ memory space for the matrix. After the initialization has been done properly, the matrix is then operational.

6.7.3   Constructors

There are a couple of ways to construct a `Matrix` in the class. One prime criterion for a matrix to exist is the specification of the number of rows and columns:

- `Matrix()` declares a matrix with no size specified. Such a matrix is not usable. To activate the matrix, we make use of a member function called `resize()`, which reallocates the matrix with the number of rows and columns specified in the arguments of the function.

- `Matrix(int nr,int nc)` declares an nr*nc matrix with the entry values undefined.

- `Matrix(int nr,int nc,T value)` declares an nr*nc matrix with all the entries initialized to `value`.

- `Matrix(const Vector<T>& v)` constructs a matrix from a vector v. It is understood that the resultant matrix will contain only one column.

- The copy constructor duplicates a matrix. It is invoked automatically by the compiler when needed and it can be invoked by the users explicitly as well.

- The destructor releases the unused memory back to the free memory pool.

Below are some examples on how to declare a `Matrix`:

```
// declare a 2-by-3 matrix of type int
Matrix<int> m(2,3)

// declare a 3-by-4 matrix and initialize the entries to 5
Matrix<int> n(3,4,5)

// duplicate a matrix using the copy constructor
Matrix<int> p(n);

// construct a matrix q from a vector v
Vector<double> v(3,0);
Matrix<double> q(v);
```

6.7.4 Operators

There are many matrix operators implemented in the class, namely

```
(unary)+, (unary)-, +, -, *, /,
=, +=, -=, *=, /=, [], (), ==, !=.
```

Some of the operators are overloaded with more than one meaning! The users are advised to read the documentation carefully before using the class.

In the following, we discuss the behaviour and usage of the operators. Suppose A, B are matrices, v is a vector and c is a numeric constant,

- The operations A+B, A-B and A*B add, subtract and multiply two matrices according to their normal definitions.

- The operations A+c, A-c, A*c and A/c are defined as A+cI, A-cI, A*cI and A/cI respectively where I is the identity matrix.

- The operations c+A, c-A, c*A and c/A have similar definitions as above.

- A=B makes a duplication of B into A whereas A=c assigns the value c to all the entries of the matrix A.

- A+=B, A-=B and A*=B are just the modification forms of assignments which perform two operations in one shot. For example, A+=B is equivalent to A = A+B.

- A+=c, A-=c, A*=c and A/=c are just the modification forms of assignments. For example, A+=c is equivalent to A = A+cI.

- The function vec(A) is used to create a vector that contains elements of the matrix A, one column after the other. Suppose

$$A = \begin{pmatrix} 2 & x & a \\ 0 & 3 & -3 \end{pmatrix}$$

then

$$\text{vec(A)} = \begin{pmatrix} 2 \\ 0 \\ x \\ 3 \\ a \\ -3 \end{pmatrix}.$$

- The *Kronecker product* of two matrices is described in Section 2.12. The function kron(A,B) is used for calculating the Kronecker product of the matrices A and B. Note that $A \otimes B \neq B \otimes A$ in general (if $A \otimes B$ and $B \otimes A$ are of the same size) and $(A \otimes B)(C \otimes D) = (AC) \otimes (BD)$ (if A is compatible with C and B is compatible with D).

- The subscript operator [] is overloaded to access a specific row vector of a matrix. For example, A[i] returns the ith row vector of matrix A.

- The parenthesis operator () is overloaded to access a specific column vector of a matrix. For example, B(j) returns the jth column vector of matrix B.

- The equality (==) and inequality (!=) operators compare whether the individual entries of two matrices match each other in the right order.

The precedence of == and != are lower than the output stream operator <<. This means that a pair of brackets is required when the users write statements that resemble the following:

```
cout << (u != v) << endl;
cout << (u == v) << endl;
```

otherwise, the compiler may complain about it.

6.7.5 Member Functions and Norms

Many useful operations have been included in the class. Their properties are described as follows:

- The `transpose()` of an $m \times n$ matrix A is the $n \times m$ matrix A^T such that the ij entry of A^T is the ji entry of A.

- The `trace()` of an $n \times n$ matrix A is the sum of all the diagonal entries of A, $\mathrm{tr}A := a_{11} + a_{22} + \cdots + a_{nn}$.

- `determinant()`: The method for evaluating the determinant of a matrix has been described in Section 2.13. The method employed depends on whether the matrix is symbolic or numeric. Since our system is meant to solve symbolic expressions, we use the Leverrier's method for solving the determinant.

- `inverse()` is used to obtain the inverse of an invertible matrix. The methods used for finding the inverse of a matrix are different for numeric and symbolic matrices. For a numeric matrix, we can use the LU decomposition [40] and backward substitution routines, whereas for a symbolic matrix we use Leverrier's method. This method can also be used for a numeric matrix.

- `resize()` reallocate the number of rows and columns according to the new specification provided in the arguments of the function.

- `rows()` and `cols()` return the number of rows and columns of the matrix, respectively.

In the auxiliary file `MatNorm.h`, we implement three different matrix norms:

- `norm1(A)` is defined as the maximum value of the sum of the entries in column vectors,

$$\|A\|_1 := \max_{1 \leq j \leq n} \left\{ \sum_{i=1}^{n} |a_{ij}| \right\}$$

- `normI(A)` is defined as the maximum value of the sum of the entries in row vectors,

$$\|A\|_\infty := \max_{1 \leq i \leq n} \left\{ \sum_{j=1}^{n} |a_{ij}| \right\}$$

- normH(A) is the *Hilbert–Schmidt norm*, defined as

$$||A||_H := [\operatorname{tr} A^* A]^{1/2} = [\operatorname{tr} A A^*]^{1/2} = \sqrt{\sum_{i=1}^{n} \sum_{j=1}^{m} |a_{ij}|^2}$$

Example 1

In this example, we demonstrate the usage of the Kronecker product of matrices. We declare and define four matrices and then calculate the Kronecker product.

```
// kron.cxx

#include <iostream.h>
#include "Matrix.h"

void main()
{
    Matrix<int> A(2,3), B(3,2), C(3,1), D(2,2);

    A[0][0] = 2; A[0][1] = -4; A[0][2] = -3;
    A[1][0] = 4; A[1][1] = -1; A[1][2] = -2;

    B[0][0] = 2; B[0][1] = -4;
    B[1][0] = 2; B[1][1] = -3;
    B[2][0] = 3; B[2][1] = -1;

    C[0][0] =  2;
    C[1][0] =  1;
    C[2][0] = -2;

    D[0][0] = 2; D[0][1] =  1;
    D[1][0] = 3; D[1][1] = -1;

    cout << kron(A,B) << endl;
    cout << kron(B,A) << endl;
    cout << kron(A,B)*kron(C,D) - kron(A*C,B*D) << endl;
}
```

Result
======

```
[4 -8 -8 16 -6 12]
[4 -6 -8 12 -6 9]
[6 -2 -12 4 -9 3]
[8 -16 -2 4 -4 8]
[8 -12 -2 3 -4 6]
[12 -4 -3 1 -6 2]

[4 -8 -6 -8 16 12]
[8 -2 -4 -16 4 8]
[4 -8 -6 -6 12 9]
[8 -2 -4 -12 3 6]
[6 -12 -9 -2 4 3]
[12 -3 -6 -4 1 2]

[0 0]
[0 0]
[0 0]
[0 0]
[0 0]
[0 0]
```

Example 2

In this example, we demonstrate that

$$\text{tr}(AB) = \text{tr}(BA) \qquad \text{and} \qquad \text{tr}(AB) \neq \text{tr}(A)\text{tr}(B)$$

in general.

```
// trace.cxx

#include <iostream.h>
#include "Matrix.h"

void main()
{
   Matrix<int> A(3,3), B(3,3,-1);
   A[0][0] = 2; A[0][1] = -1; A[0][2] =  1;
   A[1][0] = 1; A[1][1] = -2; A[1][2] = -1;
   A[2][0] = 3; A[2][1] =  2; A[2][2] = 2;

   cout << "A =\n" << A << endl;
   cout << "B =\n" << B << endl;
   cout << "tr(A) = " << A.trace() << endl;
   cout << "tr(B) = " << B.trace() << endl;
   cout << "tr(AB) = " << (A*B).trace() << endl;
   cout << "tr(BA) = " << (B*A).trace() << endl;
   cout << "tr(A)tr(B) = " << A.trace() * B.trace() << endl;
}
```
```
Result
======
A =
[2 -1 1]
[1 -2 -1]
[3 2 2]
B =
[-1 -1 -1]
[-1 -1 -1]
[-1 -1 -1]

tr(A) = 2
tr(B) = -3
tr(AB) = -7
tr(BA) = -7
tr(A)tr(B) = -6
```

Example 3

In this example, we demonstrate the usage of the determinant function.

```
// deter.cxx

#include <iostream.h>
#include "Matrix.h"

void main()
{
Matrix<double> A(2,2);

A[0][0] = 1.0; A[0][1] = 2.0;
A[1][0] = 3.0; A[1][1] = 4.0;

cout << A;
cout << "Determinant of the matrix = " << A.determinant() << endl;
cout << endl;

for(int i=3; i<5; i++)
{
A.resize(i,i,i);
cout << A;
cout << "Determinant of the matrix = " << A.determinant() << endl;
cout << endl;
}
}
```
Result
======
[1 2]
[3 4]
Determinant of the matrix = -2

[1 2 3]
[3 4 3]
[3 3 3]
Determinant of the matrix = -6

[1 2 3 4]
[3 4 3 4]
[3 3 3 4]
[4 4 4 4]
Determinant of the matrix = 8

6.8   Array Classes

6.8.1   Abstraction

Array is a common data structure in computer programming. Although C and C++ provide built-in array data structures, there are some weaknesses. For example, the bounds of the array are not checked to prevent possible run-time errors. It is therefore useful to introduce an **Array** class as an abstract data type. With bound checking and some simple mathematical operators overloaded (for example array addition), we built a comprehensive and type-safe **Array** class. On the other hand, it could replace the array supported by C and C++ as a collection of data objects.

In this section, we implement one-, two-, three- and four-dimensional arrays. Every higher-dimensional array makes use of the member functions and operators of the lower-dimensional one. Their close relationship shows up transparently in the data fields:

```
template <class T> class Array1          template <class T> class Array2
{                                        {
   private:                                 private:
   // Data Fields                           // Data Fields
   int n_data;                              int rows, cols;
   T *data;                                 Array1<T> *data2D;
   ...                                      ...
};                                       };


template <class T> class Array3          template <class T> class Array4
{                                        {
   private:                                 private:
   // Data Fields                           // Data Fields
   int rows, cols, levs;                    int rows, cols, levs, blks;
   Array2<T> *data3D;                       Array3<T> *data4D;
   ...                                      ...
};                                       };
```

In doing so, we have greatly simplified the job of construction and the code is more concise and informative. Once again, we have demonstrated the power of object-oriented programming in code reusability and extensibility.

The applications for one- and two- dimensional **Array** are common. The reasons we extended the **Array** class to three and four dimensional is because some applications for tensor fields need it. For example, the *Christoffel symbols* require a three-dimensional **Array** and for the *curvature tensor*, we need a four-dimensional **Array**.

There are four classes of `Array` described in this section, we have named them as `Array1`, `Array2`, `Array3` and `Array4` which stand for one-, two-, three- and four-dimensional `Array`, respectively. Although they are different in terms of definitions and functionalities, they share similar interfaces. Therefore, the behaviours described below apply to all four classes, except for some minor differences which will be pointed out as the description goes on.

- It is implemented as a template class, because templates work very well for container classes.

- The construction of an `Array` is simple.

- Simple arithmetic operators such as +, -, * are available.

- The assignment and modification forms of assignment =, +=, -=, *= are overloaded.

- We could copy one `Array` to another by using the assignment operator =.

- The subscript operator [] is overloaded. It provides individual element access and returns a reference to the indexed element.

- The equality (==) and inequality (!=) operators check if the two `Arrays` contain the same elements in the same order.

- For `Array` with dimensionality D, the member function `size(int index)` returns the number of elements in the `index` th dimension of the `Array`, where $index = 0, \ldots, D - 1$. For example, `size(0)` returns the number of elements for the zeroth dimension for one-, two-, three- and four-dimensional arrays.

- `resize()` reallocates the `Array` to the size specified.

- Input and output stream operations with `Array` is supported.

6.8.2 Data Fields

There are only two data fields in each class. The dimensions of the `Array` and a pointer to the data items.

- The one-dimensional array class `Array1` maintains only an integer variable `n_data`, which specifies the size of the array. It also maintains a pointer to a contiguous array of Ts.

  The memory will be allocated dynamically so that it fits the need of the application. For data types that require a huge amount of memory, it is advisable to release the memory as soon as it is no longer in use. This is not possible with static memory allocation. Therefore, array-based data types should use dynamic memory for their data storage.

- The two-dimensional array class Array2 requires two integers — rows and cols to specify its size, and a pointer to Array1 as well. To allocate memory space for an $m \times n$ array, we first allocate an array of m pointers to Array1 followed by an allocation of an array of size n to each pointer.

- The three- and four-dimensional array classes Array3 and Array4 require three and four integer variables, respectively, to specify their sizes. Array3 contains a pointer to Array2 whereas Array4 maintains a pointer to Array3. These pointers will be used to allocate the memory space required to store the data items when necessary. The procedures for the memory allocation for these pointers follow the same logic as the two-dimensional array described above.

6.8.3   Constructors

The specification for the default constructor is important especially when the data fields of a class involve some pointers or references. It is important to initialize the data items in the class including NULLifying the pointers. This would prevent some unexpected problems which may occur later on. Every class has two other overloaded constructors:

```
Array1(int)               Array1(int,T value)
Array2(int,int)           Array2(int,int,T value)
Array3(int,int,int)       Array3(int,int,int,T value)
Array4(int,int,int,int)   Array4(int,int,int,int,T value)
```

The first set of constructors allocates an appropriate amount of memory for the Array while the second set initializes all the data items to value in addition. The copy constructor and destructor must be properly implemented for a class which contains pointers in the data fields.

In the following we give some examples of the usage of the constructors:

```
// declare u as a one-dimensional array of size 20,
// the entries are uninitialized.
Array1<int> u(20);

// declare v as a two-dimensional array of size 2x3,
// the entries are initialized to zero.
Array2<double> v(2,3,0.0);

// declare w as a three-dimensional array of double,
// with size 2x3x2, all its entries are initialized to 2.0
Array3<double> w(2,3,2,2.0);

// declare x as a four-dimensional array of size 1x2x3x4,
```

```
// all the entries in the array are uninitialized.
Array4<double> x(1,2,3,4);

// copy constructor is invoked to duplicate the array.
Array3 y(w);
```

6.8.4 Operators

Some common operators are overloaded in the class, namely (unary)+, (unary)-, +, -, *, =, +=, -=, *=, ==, !=, [], << and >>. Suppose u, v, w are arrays and c is a constant, then the operations are defined as follows:

- +u has no effect on u while -u negates each element of u.

- u+v, u-v adds or subtracts corresponding elements of u and v.

- u*c multiplies each element of u by the scalar c.

- u+=v, u-=v adds or subtracts corresponding elements of v into u.

- u*=c multiplies the scalar c into each element of u.

- u=v makes a duplicate copy of v and stores into u. Note that the assignment operator returns a reference to the object itself. It means that multiple duplication like u = v = w is allowed.

- u=c overwrites all the entries in u with the constant c.

- u==v returns *true* if u and v are identical, otherwise returns *false*.

- u!=v is just the converse of u==v.

- The [] operator is overloaded to perform individual element accessing. The construct u[i] will return a reference to the ith entry of the array. Since a reference value is returned, it can be used on either side of an assignment expression.

- The output stream operator << simply exports all the entries and encloses them with a pair of square brackets.

  An instance u of **Array1** is formatted as

$$\begin{bmatrix} u_0 & u_1 & \cdots & u_{n\_data-1} \end{bmatrix}$$

  An instance v of **Array2** is formatted as

$$\begin{bmatrix} v_{00} & v_{01} & \cdots & v_{0\ \text{cols}} \\ v_{10} & v_{11} & \cdots & v_{1\ \text{cols}} \\ \vdots & & \cdots & \\ v_{\text{rows}\ 0} & v_{\text{rows}\ 1} & \cdots & v_{\text{rows}\ \text{cols}} \end{bmatrix}$$

The output for **Array3** is less obvious; it can be considered as a cube of data entries to be output. The result is formatted in such a way that every layer of the cube is output accordingly, separated by an empty line. An instance of **Array3** w(2,3,2) will be printed as

$$\begin{bmatrix} w_{000} & w_{001} \end{bmatrix}$$
$$\begin{bmatrix} w_{010} & w_{011} \end{bmatrix}$$
$$\begin{bmatrix} w_{020} & w_{021} \end{bmatrix}$$

$$\begin{bmatrix} w_{100} & w_{101} \end{bmatrix}$$
$$\begin{bmatrix} w_{110} & w_{111} \end{bmatrix}$$
$$\begin{bmatrix} w_{120} & w_{121} \end{bmatrix}$$

The idea for exporting an instance of **Array3** is extended to **Array4**. Here, a four-dimensional array can be visualized as a collection of three-dimensional blocks of **Array3**. Therefore, in printing an instance of **Array4**, we print a block of **Array3** followed by another until all the fourth dimensional entries are printed. As an example, we print an instance of **Array4** x(2,2,2,3)

$$\begin{bmatrix} x_{0000} & x_{0001} & x_{0002} \end{bmatrix}$$
$$\begin{bmatrix} x_{0010} & x_{0011} & x_{0012} \end{bmatrix}$$

$$\begin{bmatrix} x_{0100} & x_{0101} & x_{0102} \end{bmatrix}$$
$$\begin{bmatrix} x_{0110} & x_{0111} & x_{0112} \end{bmatrix}$$

$$\begin{bmatrix} x_{1000} & x_{1001} & x_{1002} \end{bmatrix}$$
$$\begin{bmatrix} x_{1010} & x_{1011} & x_{1012} \end{bmatrix}$$

$$\begin{bmatrix} x_{1100} & x_{1101} & x_{1102} \end{bmatrix}$$
$$\begin{bmatrix} x_{1110} & x_{1111} & x_{1112} \end{bmatrix}$$

Each layer in **Array3** is separated by an empty line whereas each block in **Array4** is separated by two empty lines.

- The input stream operator >> is just the reverse of the output stream operator. It reads in the sizes of each dimension of the arrays, followed by reading in the individual data entries until all the entries have been input.

6.8.5 Member Functions

The member functions available for each array class include:

- `int size(int = 0) const;`
  Returns the number of data entries in each dimension for each class. For example, `size(0)` returns the size for `Array1`, or the rows number for `Array2`, `Array3` and `Array4`. `size(2)` returns the levs number for `Array3` and `Array4` as it is not defined for both `Array1` and `Array2`.

  Note that default arguments have been used for the function; this indicates that if an argument is omitted, the size of the zeroth dimension for each array will be returned.

- `void resize(int);`
  It is an important function for the array classes. The following statement

  ```
  Array1<int> u;
  ```

  declares a variable u but no memory space is allocated. Such a variable is not usable. When the statement

  ```
  u.resize(100);
  ```

  is executed, the compiler allocates 100 memory spaces for u. After this the variable becomes operational. If we wish to fill the array with some initial values upon memory allocation, another form of the function could be used:

  ```
  void resize(int n,T value);
  ```

  In addition to the memory allocation, all the entries are initialized to `value`. The logic for `Array2`, `Array3` and `Array4` are similar, except that `resize()` requires two, three or four arguments to specify the sizes of the arrays, respectively.

  The function `resize()` can actually be used to change the sizes of arrays in use. An expansion of array size would result in the old storage being copied into the new array with the expanded region left undefined or initialized to `value` depending on which form of `resize()` is used. When `resize()` shrinks an array only the part of the old storage that will fit is copied into the new array and the extra entries are "truncated".

The following program demonstrates the usage of the function `resize()`:

```
// aresize.cxx

#include <iostream.h>
#include "Array.h"

void main()
{
    Array1<double> M;
    M.resize(3);
    M[0] = 5.0;  M[1] = 8.0; M[2] = 4.0;
    cout << "M = \n" << M << endl; cout << endl;

    Array2<double> A;
    A.resize(2,3);
    A[0][0] = 1.0;  A[0][1] = 3.0;  A[0][2] = 2.0;
    A[1][0] = 4.0;  A[1][1] = 5.0;  A[1][2] = 6.0;

    A.resize(3,4,2);
    cout << "A = \n" << A << endl;
    A.resize(2,2);
    cout << "A = \n" << A << endl;
    A.resize(4,4,9);
    cout << "A = \n" << A << endl;
}
```

Result
======
M =
[5 8 4]

A =
[1 3 2 2]
[4 5 6 2]
[2 2 2 2]

A =
[1 3]
[4 5]

A =
[1 3 9 9]
[4 5 9 9]
[9 9 9 9]
[9 9 9 9]

6.9 The String Class

In traditional C, a string is represented as a pointer to char. This means a string can be considered as an array of type char. In this representation, the end-of-string is denoted by '\0'. This convention has a major drawback. The '\0' requires one byte to store and is part of the string. This means the string "abc" requires 4 bytes of memory. The inconsistency between the length of the string and the memory required for the string has caused much confusion to many users. It is also a pitfall. Therefore, to avoid possible error and improve efficiency on the string operations, it is best to implement it as an abstract data type. A string class has been included into the C++ standard.

6.9.1 Abstraction

The behaviour of the class is summarized as follows:

- The construction of a String is easy.

- An instance of String can be duplicated using the assignment operator.

- The subscript operator [] is used to access individual characters in the String.

- The addition operator + is used to concatenate two strings.

- The relational operators >, >=, <, <=, ==, != are included.

- Instances of String can be converted to const char * when necessary.

- reverse() reverses the order of the String.

- length() returns the length of the String.

- Input and output operations with String are supported.

6.9.2 Data Fields

There are two data fields in the class:

- datalength stores the size of the largest possible String.

- data contains a pointer to char. When it is allocated to some appropriate memory space, it will be used to store the characters of the String.

6.9.3 Constructors

Strings are constructed and initialized as follows:

- `String x; String y=0; String z="";`

  set x, y and z to an empty `String`. Note that 0 and "" may be used as NULL string.

- `String x('A'), y('#');`

  set x to A and y to # respectively.

- `String x("abc"); String y="abc";`

  set x and y to the `String` "abc".

- `String x(int n);`

  allocates n memory spaces to x and initializes it to a NULL `String`.

- `String u(x); String v=x;`

  invoke the copy constructor to duplicate x into u and v.

- The destructor is invoked to free unused memory back to the pool.

6.9.4 Operators

Some common operators have been included in the class, namely

     `=, [], +, ==, !=, >, >=, <, <=, >>, <<.`

The properties are described as follows:

- The assignment operator = makes a duplicate of `String`.

- The subscript operator `[]` gives permission to access individual characters in the `String`.

- The concatenation operator + combines two strings into one.

- The six relational operators `==, !=, >, >=, <, <=` compare the ordering of `String` lexicographically using the `strcmp()` function.

- The input and output stream operators provide means to import and export instances of `String`.

- The type conversion operator will automatically convert the `String` into the built-in type `const char*` when needed.

6.9.5 Member Functions

There are only two member functions implemented in the class: length() and reverse():

- length() returns the size of the String.

- reverse() reverses the order of the String and returns it.

Most of the operations and member functions are self-explanatory and can be understood easily. The conversion operator to built-in type const char* plays an important role in the class. It allows all the functions in the <string.h> library to be used with the class. With this feature, duplication of codes may be avoided although some of the functions have been re-implemented in a better form. We demonstrate the usage of the class in the following examples:

```
// sstring.cxx

#include <iostream.h>
#include "MString.h"

void main()
{
    String S1("Children");
    cout << "The string S1 is " << S1 << endl;
    cout << "The reverse of S1 is " << S1.reverse() << endl;
    cout << "The length of S1 is " << S1.length() << endl;
    cout << endl;

    String S2 = "Mother";
    String S3(S2);
    S3[0] = 'F';
    S3[1] = 'a';

    cout << "The string S2 is " << S2 << endl;
    cout << "The string S3 is " << S3 << endl;
    cout << endl;

    // Concatenation of strings
    cout << S2 + " and " + S3 << endl;
    cout << S3 + " and " + S2 << endl;
    cout << endl;

    // Comparing strings
    cout << (S2 == S3) << endl;
    cout << (S2 != S3) << endl;
```

```
        cout << (S2 < S3) << endl;
        cout << (S2 > S3) << endl;
    }
Result
======
The string S1 is Children
The reverse of S1 is nerdlihC
The length of S1 is 8

The string S2 is Mother
The string S3 is Father

Mother and Father
Father and Mother

0
1
0
1
```

6.9.6   Type Conversion Operator

The class has one type conversion operator, namely, the conversion to the built-in string in the C language (`char*`). This operator does the conversion automatically when needed. It could also be invoked explicitly using the *cast* (see Section 5.5). This conversion becomes useful when we need the library routines from `<string.h>`, such as `strlen()`, `strcpy()`, `strtok()`, etc.

6.9.7   Possible Improvements

It is obvious that the class may be improved and extended. For example, substring extraction and modification is not implemented. Other features worth implementing include searching and matching of `Strings`. For example, the substring, which provides the ability to access a subportion of a string, can be overloaded using the parentheses operator `()`. The operator takes two arguments, where the first specifies the left index of the substring and the second specifies the right index. The following example illustrates this concept:

```
    String s("Diamond Ring");

    cout << s(0,4)  << endl;  // output => Diamo
    cout << s(8,11) << endl;  // output => Ring
```

The implementation of this extension is left as an exercise for the readers.

6.10 Bit Vectors

A bit vector is simply a vector of 0 or 1 values. Since bits are the most fundamental values stored in computers, we can implement a bit vector using any data type in principle. Here, we selected unsigned char to store the bits. Since an unsigned char is composed of 8 bits, we can store 8 bit values in each character. In other words, to implement a bit vector we use a Vector of unsigned char to store the bit values.

Bits 0-7	Bits 8-15	Bits 16-23	Bits 24-31
00110101	11101011	10101100	01001011
Vector 0	Vector 1	Vector 2	Vector 3

By reusing the Vector data structure in the internal representation, we avoid the necessity of redeveloping an existing structure. We also need not be concerned with issues such as dynamic memory allocation or deletion. Hence, the programmers can concentrate on features available only to the new structure and the potential for introducing new errors is greatly reduced.

6.10.1 Abstraction

For a bit vector to be useful, it has to be easy to use and to access individual elements. Below is a summary of the abstraction of the BitVector class:

- A BitVector could be constructed easily by specifying the required size of the vector.

- size() returns the size of the bit vector.

- set(), clear(), test() and flip() do the operations on a particular bit as specified in the arguments.

- The output stream operator is supported.

6.10.2 Data Fields

The class maintains two data fields — bsize and data. bsize stores the size of the bit vector whereas data is a Vector of unsigned char used to store the bit values. Here, we do not need to be concerned about the issue of dynamic memory allocation and deletion as these operations will be taken care of by the Vector class, which has been developed and tested carefully.

6.10.3   Constructors

There are two possible means to construct a BitVector:

- BitVector(unsigned int num) specifies the size num of the bit vector and allocates a proper amount of vectors to accommodate. An unsigned char is made up of 8 bits, hence it computes the smallest multiple of 8 larger than or equal to num. This constructor will initialize the bit vector to all zero values.

- BitVector(unsigned int num,unsigned int value) does the same as the previous constructor except the initial values of the bit vector are set according to value as specified in the argument list.

- The copy constructor and destructor simply invoke the code provided by the Vector class.

6.10.4   Member Functions

Suppose v, u are BitVector of size N and M is a positive integer, then

- v.size() returns the size of the bit vector as specified by the user.

- v.reset(M) re-specifies the size of the BitVector to M and initializes all the bit values to zero.

- v.reset(M,flag) does the same as the previous except the initialization of the bit values follows the flag values $(0/1)$.

- v.set(i) sets the ith bit value.

- v.clear(i) clears the ith bit value.

- v.test(i) checks whether the ith bit value is set or clear.

- v.flip(i) flips the ith bit value $(0 \rightarrow 1$ or $1 \rightarrow 0)$.

- v.unionSet(u) forms the union set of v and u.

- v.intersectSet(u) forms the intersection set of v and u.

- v.differenceSet(u) forms the difference set of v and u.

- v.subset(u) checks if u is a subset of v.

- v==u checks if v is equal to u.

- cout << v puts v on the output stream.

In the operations

v.subset(u), v.unionSet(u), v.intersectSet(u), v.differenceSet(u)

and v==u, the size of the bit vectors u, v must be equal to each other.

In the following, we demonstrate the usage of these member functions:

```
// sbitvec.cxx

#include <iostream.h>
#include "Bitvec.h"

void main()
{
    // Set up a bit vector of size=30 with all bits set to 1
    BitVector B(30,1);

    // Duplicate a bit vector A
    BitVector A(B);

    // Bits operations
    A.clear(5); A.clear(20); A.clear(29);
    B.clear(5); B.clear(10); B.clear(12); B.clear(29);
    B.flip(0); B.flip(1);

    cout << "The size of the bit vector A is " << A.size() << endl;
    cout << endl;

    cout << "      A      = " << A << endl;
    cout << "      B      = " << B << endl;
    cout << "-----------------------------------------------" << endl;

    // Set operations
    cout << "A union B      = " << A.unionSet(B)      << endl;
    cout << "A intersect B  = " << A.intersectSet(B)  << endl;
    cout << "A difference B = " << A.differenceSet(B) << endl;
    cout << endl;

    cout << "Is A a subset of B ? " << A.subset(B) << endl;
    cout << "Is A equals to B   ? " << (A==B)      << endl;
}
```

```
Result
======
The size of the bit vector A is 30

        A     = 111110111111111111110111111110
        B     = 001110111101011111111111111110
--------------------------------------------------
A union B     = 111110111111111111111111111110
A intersect B = 001110111101011111110111111110
A difference B = 110000000010100000000000000000


Is A a subset of B ? 0
Is A equals to B    ? 0
```

6.10.5 Private Member Functions

Two auxiliary functions do their jobs quietly without being known by the users — byteNumber() and mask(). They are declared as private member functions.

- byteNumber(i) returns the respective byte number of the ith bit index. It is obtained by dividing the index value i by 8 because each byte is made up of 8 bits. For example, suppose the index value is 19. Its corresponding bit pattern is 00010011. Dividing the value by 8 is equivalent to shifting the bit pattern to the right by 3 places, obtaining 00000010 which corresponds to 2 numerically. Thus, the 19th bit is located at byte number 2.

- mask(i) returns the masked value of the selected byte where the ith index locates. The masked value is represented by a '1' bit for the specified bit and '0' elsewhere. It is obtained by taking the three least significant bits of the selected byte, evaluating their value v, and shifting left the numeric one by v places. For example, consider the selected byte as 19 = 00010011. The three least significant bits are 011 = 3. The masked value is thus obtained by shifting the value one by 3 spaces, yielding 00001000.

6.11 The Linked List Class

6.11.1 Abstraction

Suppose we have a collection of elements, where the number of elements is not known in advance, or varies over a wide range. It is natural to keep the data in a *linked list*. Basically, a linked list maintains all its elements in a chain. Each component of the chain holds just a single value and a pointer. The pointer either points at the next item of the list, or points at NULL if it happens to be the last element in the chain.

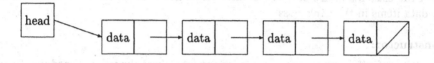

Figure 6.1: *Schematic diagram of a linked list*

The operations of the linked list abstraction are as follows:

- *Insertion:* A new element can be added to the beginning of a linked list.

- *Removal:* A data item may be removed from the beginning of a linked list.

- *Access:* The value of the first element may be enquired.

- *Duplication:* The linked list may be duplicated when required.

- *Size:* It is possible to check if a linked list is empty.

- *Inclusion:* It is possible to check if an element exists in a linked list.

It seems that the operations available in the linked list data structure are limited. Operations such as insertion or removal of an element from an arbitrary location are not possible. This problem will be overcome later with the introduction of the *iterator* mechanism.

The linked list data structure consists of two major classes:

- The List class. This class contains the public interface of the linked list data structure. The interface includes the constructor and all the operations listed above.

- The Link class. The operations on the list values are built on top of this class. It contains the data areas as well as the chain to the next link.

6.11.2   The List Class

The linked list data structure is declared in this class. All the interfaces and operations are defined in the class. However, the class maintains only one pointer to the first link. The data values of the list are stored in the Link class. The class is implemented as a template form so that the linked list may hold elements of different data types, such as character, string, or other abstract data types.

Data Fields:

The List class maintains only one pointer called head. It points to the first link of the data items in the Link class.

Constructor:

- The default constructor creates an empty list, where the pointer head points to NULL.

- The copy constructor duplicates a list when necessary.

- The destructor deletes all the data items in a list. It invokes a member function called deleteAllNodes().

Member functions:

The member functions listed here provide the necessary interface of the class:

- first_Node() returns the first element of a linked list.

- duplicate() makes a duplicate copy of a linked list.

- is_Empty() checks if a list is empty.

- add_data(T value) adds the data item value to the beginning of a list.

- deleteAllNodes() frees the memory space occupied by a list. Hence the linked list ceases to exist after this operation.

- is_Include(T value) checks if a list contains the data item value.

- delete_First() removes the first data item from a linked list.

- The assignment operator (=) allows a linked list to be assigned to another.

Friend class:

The class ListIterator is declared as a friend of this class. The ListIterator class offers a convenient way to access and manipulate the data items in the linked list. It will be discussed in a later section.

6.11.3 The Link Class

The Link class is a *facilitator class*. It is used only to facilitate the operations of the List class. Its existence is not known to the users of the linked list data structure. Therefore, most of the functions in the class, including the constructor, are declared as private.

Data Fields:

There are two data members in the class — data and next. The variable data stores the data item itself whereas the variable next contains a pointer to the next link.

Constructor:

The constructor

```
Link(const T value,Link<T>* next)
```

takes two arguments. The first argument takes the value of the data item, whereas the second argument is a pointer to the next link.

The copy constructor Link(const Link<T>&) copies the data fields as usual.

Member functions:

- Link<T>* insert(const T val);
  This is the only public member function in the class. It is used for inserting a data item after the current location.

- Link<T>* duplicate() const;
  It makes a duplication of the whole chain of links.

Friend classes:

Since the classes List and ListIterator are declared as friend, they are able to access the private region of the class.

6.11.4   The List Iterator

This class provides an easy means to access and manipulate the list. This includes looping over the elements of a list, removing arbitrary elements from a list, and adding elements at an arbitrary location within a list.

Basically, the idea to maintain an iterator arises because we want a method to examine each value in turn in a data structure. Usually, this process is performed using a loop. However, difficulty arises when we want a loop that provides users with access to individual values easily, yet preventing the inner details of the implementations of the data structures from being exposed.

On the other hand, creating an iterator that is separated from the actual data structure has several advantages:

- The responsibilities of an iterator are different from the data structure itself; separating them is in line with the design of object-oriented programming.

- This implementation allows more than one different style of iterator to operate on a structure. For example, in the traversal of a binary tree structure, we may have different iterators such as preorder, inorder and postorder.

- Finally, this implementation allows more than one iterator to operate simultaneously on the same data structure.

The ListIterator class is a public derived class of an abstract base class Iterator. The abstract base class declares the interface of an iterator abstraction. However, the implementations of these operations are left for the derived classes of this base class.

```
#ifndef MITERATOR_H
#define MITERATOR_H

// Iterator class
template <class T> class Iterator
{
public:
    virtual int  init() = 0;          // Initialization
    virtual int  operator !() = 0;  // Check if a current element exists
    virtual T    operator ()() = 0; // Return current element
    virtual int  operator ++() = 0; // Increment operator
    virtual void operator = (const T) = 0; // Assignment operator
};
#endif
```

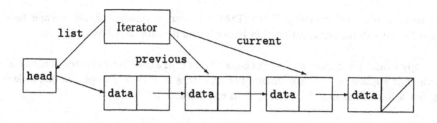

Figure 6.2: *Schematic diagram of a list iterator*

In general, an iterator has the following five operations:

- *Initialization:* This operation initializes the iterator to point at the beginning of the data structure.

- *Termination:* This operation checks if the iterator has reached the end of the iterated data structure.

- *Current Value:* This operation returns the current value of the data item which the iterator points at.

- *Increment:* This operation increases the pointer of the iterator to the next item of the data structure.

- *Assignment:* This operation assigns a new value to the current data item of the data structure.

In the following, we describe the design of the ListIterator class.

Data Fields:

The class has three data fields:

- Link<T>* current maintains a pointer that points at the "current" node of the linked list.

- Link<T>* previous maintains a pointer that points at the "previous" node, i.e., the node that is located immediately prior to the "current" node of the linked list.

- Link<T>& list maintains a reference to a linked list. A reference is used instead of a pointer because the linked list to be iterated, once assigned, should not be changed. A reference provides such a restriction. Furthermore, a reference cannot be assigned to a NULL value.

Constructor:

The constructor `ListIterator(List<T>&)` takes an argument of a reference to a linked list. It establishes a connection between the iterator and the linked list.

The copy constructor `ListIterator(const ListIterator<T>&)` creates another iterator that applies on the same linked list. An iterator can only apply on one linked list, but a linked list may have more than one iterator.

Member function:

- `int init()` sets the iterator to the first element in the list.

- `T operator() ()` returns the value of the current element.

- `int operator !()` checks if the iterator of the linked list has reached NULL or the end of the list.

- `int operator ++()` increases the current pointer to the next node.

- `void operator = (const T value)` assigns the current element to `value`.

- `delete_Current()` removes the current element from the list.

- `add_Before(const T value)` adds a new element to the list before the current node.

- `add_After(const T value)` adds a new element to the list after the current node.

Next, we consider some examples on the linked list class. The first example applies the basic operations available in the `List` class. It is not trivial to print the content of a linked list or to access an arbitrary element in the list. However, the task becomes relatively simple with the use of the list iterator. This point is demonstrated in the second example, which constructs a linked list while maintaining the data items in increasing order.

Example 1

In this example, we apply some basic linked list operations on a list of numbers.

```cpp
// num1.cxx

#include <iostream.h>
#include "MList.h"

void main()
{
    int i;
    int N = 10;
    List<int> A;

    cout << "Is the linked list A empty ? " << A.is_Empty() << endl;
    cout << endl;

    for(i=0; i<N; i++)
    {
        A.add(i);
        cout << "The first element of linked list A is "
             << A.first_Node() << endl;
    }
    cout << endl;

    List<int> B(A);    // copy constructor

    do
    {
        cout << "The first element of linked list B is "
             << B.first_Node() << endl;
        B.delete_First();
    } while(!B.is_Empty());
    cout << endl;

    // Inclusion test
    int p = 3, q = 15;

    cout << "Does " << p << " exist in linked list A ? "
         << A.is_Include(p) << endl;
    cout << "Does " << q << " exist in linked list B ? "
         << B.is_Include(q) << endl;
```

```
        cout << endl;

        A.deleteAllNodes();
        B.deleteAllNodes();

        cout << "Is the linked list A empty ? " << A.is_Empty() << endl;
        cout << "Is the linked list B empty ? " << B.is_Empty() << endl;
    }
Result
======
Is the linked list A empty ? 1

The first element of linked list A is 0
The first element of linked list A is 1
The first element of linked list A is 2
The first element of linked list A is 3
The first element of linked list A is 4
The first element of linked list A is 5
The first element of linked list A is 6
The first element of linked list A is 7
The first element of linked list A is 8
The first element of linked list A is 9

The first element of linked list B is 9
The first element of linked list B is 8
The first element of linked list B is 7
The first element of linked list B is 6
The first element of linked list B is 5
The first element of linked list B is 4
The first element of linked list B is 3
The first element of linked list B is 2
The first element of linked list B is 1
The first element of linked list B is 0

Does 3 exist in linked list A ? 1
Does 15 exist in linked list B ? 0

Is the linked list A empty ? 1
Is the linked list B empty ? 1
```

Example 2

In this example, we use the list iterator to construct a sorted linked list. With the use of the list iterator, accessing and manipulating a linked list becomes an easy task.

```
// num2.cxx

#include <iostream.h>
#include "MList.h"

template <class T>
void Insert(List<T>& LL,T value)
{
    ListIterator<T> LI(LL);
    LI.init();
    while(!LI && LI() < value) ++LI;
    LI.add_Before(value);
}

void main()
{
    List<int> A;
    ListIterator<int> P(A);

    Insert(A,10);
    Insert(A,1);
    Insert(A,-15);
    Insert(A,7);
    Insert(A,12);
    Insert(A,5);
    Insert(A,-7);
    Insert(A,0);
    Insert(A,15);

    cout << "The linked list A is" << endl;

    for(P.init(); !P; ++P) cout << P() << " ";
}
```

Result
======
The linked list A is
-15 -7 0 1 5 7 10 12 15

6.12 The Polynomial Class

6.12.1 Abstraction

The behaviour of the class is specified as follows:

- It is a **template** class for which the data type of the polynomial coefficients are to be specified by the user.

- Creation of a new instance of **Polynomial** is simple.

- Arithmetic operators such as +, -, *, / are available.

- Assignment and modification forms of assignment =, +=, -=, *=, /=, %= are available.

- The relational operators == and != are available.

- **Polynomial** variables can be raised to an integer power with the ^ operator.

- A **Polynomial** can be used as a function and can be evaluated for a given value for the polynomial's variable.

- Symbolic differentiation and integration of a **Polynomial** is possible.

- A monomial class is provided to allow the symbolic specification of a polynomial and represent the polynomial variable.

- The monomial class also supports the operators defined for the **Polynomial** class to provide a full symbolic specification of polynomials.

- The monomial class implements the coefficient and power of the polynomial variable and the sum of monomials is a polynomial.

- Output operations with **Polynomial** are supported.

The monomial class is called the **Polyterm** class. Since polynomials consist of a sum of polynomial terms it makes sense to implement a polynomial using a linked list.

6.12.2 Template Class

The **Polynomial** class is a **template** class so that the user can select a data type for the polynomial coefficients and evaluation. For example

```
Polynomial<Rational<int> >
```

can be used to create a polynomial with rational coefficients and

```
Polynomial<Polynomial<double> >
```

can be used to create a two-variable polynomial with coefficients of type `double`. Thus the `template` class provides multi-variable polynomials.

6.12.3 Data Fields

The polynomial's variable is required to effectively use the class. The variable is implemented as a simple monomial with the necessary functionality to create polynomials. The class uses a linked list of monomials to form the polynomial.

6.12.4 Constructors

There are three ways to construct a `Polynomial`:

1. Construct a `Polynomial` by specifying the symbol string of the polynomial's variable (the only argument).

2. Construct a `Polynomial` by specifying a monomial which is used as the symbol of the polynomial's variable. This is the easiest way to construct a `Polynomial` since the polynomial can be built from the specified monomial.

3. Construct a constant polynomial from an integer.

There is also a copy constructor. The monomial class `Polyterm` is also a `template` class taking the same template parameter as the `Polynomial` class for which it will be used. Although more functionality is available the most important function of the `Polyterm` class is the constructor

```
Polyterm<T>(char *name)
```

(which creates a monomial with symbol string name) and the mathematical operators. Some examples are:

```
// Declare a Polynomial p1 with integer coefficients and
// symbolic variable "x"
Polynomial<int> p1("x");

// Declare a Polynomial p2 with integer coefficients and
// symbolic variable "x"
Polyterm<int> x("x");
Polynomial<int> p2(x);

// Declare a monomial x as a polynomial variable
// Declare a Polynomial q with double coefficients and
// symbolic variable "x" and initialize to x^2+2x+1
```

```
Polyterm<double> x("x");
Polynomial<double> q=(x^2)+2.0*x+1;

// Declare a monomial y as a polynomial variable
// Declare a Polynomial r with polynomial coefficients
// and symbolic variable "y" and initialize to q(x)*(y^2+y);
Polyterm<Polynomial<double> > y("y");
Polynomial<Polynomial<double> > r=q*((y^2)+y);
```

6.12.5 Operators

The `Polynomial` class overrides the following operators:

$$-(\text{unary}), \ +, \ -, \ *, \ /, \ \%, \ \hat{\ }, \ =, \ +=, \ -=, \ *=, \ /=, \ \%=, \ ==, \ !=, \ ().$$

Care must be taken with the `^` operator since the precedence is lower than `+,-` and `*`. Most operators have their usual meaning except for `^` which raises a polynomial to a positive integer power and `()` which is used to evaluate the polynomial at a point. Most operators can be used with the monomial class `Polyterm`. Addition and subtraction of `Polyterms` are defined to create `Polynomials`.

6.12.6 Type Conversion Operators

A constant polynomial can be created from an integer using the `Polynomial(int)` constructor (typecast).

6.12.7 Private Member Functions

The `Polynomial` class has a private member function `tidy(void)`. Its only purpose is to remove terms with zero coefficients from a polynomial. This is to save memory resources but also makes output of `Polynomials` easier to read.

6.12.8 Other Functions

A template function `_poly_power(T,int)` is used to raise an instance of `T` to a positive integer power. The function `Diff(Polynomial)` differentiates a polynomial with respect to its variable. The function `Int(Polynomial)` integrates a polynomial with respect to its variable. The constant of integration is assumed to be zero.

6.12.9 Streams

The output stream operator `<<` first checks if the linked list for the polynomial is empty and if so outputs zero. Otherwise the operator outputs each term in the linked list using the overloaded operator `<<` for `Polyterm` which outputs the coefficient, variable and exponent of the `Polyterm`.

6.12.10 Example

As an example we create a single and a multivariable polynomial and differentiate, integrate and square them.

```cpp
// pexample.cpp

#include <iostream.h>
#include "poly.h"
#include "rational.h"

int main(void)
{
 Polyterm<double> x("x");
 Polyterm<Polynomial<double> > y("y"); //multivariable term

 Polynomial<double> p=(x^3)+2.0*(x^2)+7.0;

 cout << "p(x)=" << p <<endl;
 cout << "Diff(p)=" << Diff(p) << endl;
 cout << "Int(p)=" << Int(p) << endl;
 cout << "p(x)^2=" << (p^2) << endl << endl;

 // multivariable polynomial
 // differentiation and integration are with respect to y
 Polynomial<Polynomial<double> > q=p+(4.0*p)*(y^2);
 cout << "q(y)=" << q << endl;
 cout << "Diff(q)=" << Diff(q) << endl;
 cout << "Int(q)=" << Int(q) << endl;
 cout << "q(y)^2=" << (q^2) << endl;
 return 0;
}
Results
=======
p(x)=x^3+(2)x^2+(7)
Diff(p)=(3)x^2+(4)x
Int(p)=(0.25)x^4+(0.666667)x^3+(7)x
p(x)^2=x^6+(4)x^5+(4)x^4+(14)x^3+(28)x^2+(49)

q(y)=((4)x^3+(8)x^2+(28))y^2+(x^3+(2)x^2+(7))
Diff(q)=((8)x^3+(16)x^2+(56))y
Int(q)=((1.33333)x^3+(2.66667)x^2+(9.33333))y^3+(x^3+(2)x^2+(7))y
q(y)^2=((16)x^6+(64)x^5+(64)x^4+(224)x^3+(448)x^2+(784))y^4
       +((4)x^6+(16)x^5+(16)x^4+(56)x^3+(112)x^2+(196))y^2
       +(x^6+(4)x^5+(4)x^4+(14)x^3+(28)x^2+(49))
```

6.13   The Set Class

6.13.1   Abstraction

The behaviour of the class is specified as follows:

- It is a **template** class for which the data type of the elements of the finite set are to be specified by the user.

- Creation of a new instance of **Set** is simple.

- The assignment operator = is overloaded.

- The relational operator == is available.

- The operators + and * are overloaded for union and intersection of finite sets.

- Output operations with **Set** are supported.

6.13.2   Template Class

The **Set** class is a **template** class so that the user can select a data type for the polynomial coefficients and evaluation. For example

Set<char>

can be used to create a **Set** with rational coefficients and

Set<string>

can be used to create a finite **Set** with strings as elements.

6.13.3   Data Fields

The data fields contain the elements of the linked list the **Set** class is based on.

6.13.4   Constructors

The constructor

Set(const T)

is used to construct the list. The copy constructor and destructor are implemented.

6.13.5 Operators

The Set class overrides the following operators:

+, *, =, ==

Note that the * has higher precedence then the + operator.

6.13.6 Member Functions

The Set class has a public member function `cardinality()` which finds the number of elements in the finite set.

6.13.7 Streams

The output stream operator << is overloaded and first checks if the linked list for the finite set is empty and if so outputs (). Otherwise it displays the finite set.

6.13.8 Example

The following example shows a simple use of the Set class.

```
// setapp.cpp

#include <iostream>
#include <string>
#include "set.h"
using namespace std;

void main(void)
{
  Set<char> c1, c2, c3;
  Set<char> setA('X');
  Set<char> setB('Y');
  Set<char> setC('Z');
  Set<char> setD('Y');

  cout << setA << "\n";

  c1 = setA + setB;
  cout << "c1 = " << c1 << endl;

  c2 = setA + setB + setC + setD;
```

```
  cout << "c2 = " << c2 << endl;

  c3 = c1 * c2;
  cout << "c3 = " << c3 << endl;

  int size = c2.cardinality();
  cout << "size of c2 = " << size << endl;

  Set<string> s1, s2;
  Set<string> setS1("willi");
  Set<string> setS2("hans");
  s1 = setS1 + setS2;
  cout << "s1 = " << s1 << endl;

  s2 = setS1 * setS2;
  cout << "s2 = " << s2 << endl;
}
```

Results
========
```
(X)
c1 = (X,Y)
c2 = (X,Y,Z)
c3 = (X,Y)
size of c2 = 3
s1 = (hans,willi)
s2 = ()
```

The Standard Template Library described in Chapter 5 also includes a set class.

6.14 Summary

This chapter presented a collection of useful classes written in C++. For each class, the abstraction, data fields, constructors, operators and member functions etc. were described in great detail.

Among the classes introduced in this chapter:

- The `Verylong`, `Rational`, `Complex` and `Quaternion` classes form the basic abstract data types.

- The `Vector`, `Matrix`, `Array`, `Derive` and `Polynomial` classes are higher level mathematical structures.

- Last but not least, the `String`, `BitVector`, `List` and `Set` classes are useful data structures which will become apparent in later chapters.

The applications of these classes will be described in Chapter 8.

In fact, there are a large number of other classes which could be added to the classes described above. An example is a class for multivariate polynomials. Other examples are classes for Lie algebras, finite groups and Galois fields, etc.

In Chapter 9, we introduce a polymorphic linked list which is capable of storing different data types (basic data types and abstract data types) in a single linked list. This special linked list forms the basic data structure for the programming language LISP.

6.14 Summary

Chapter 7

The Symbolic Class

Computer algebraic systems which perform symbolic manipulations have proved useful in many respects and they have become indispensable tools for research and scientific calculation. However, most of the software systems available are independent systems and the transfer of mathematical expressions from them to other programming environments such as C is rather tedious, time consuming and error prone. It is therefore useful to use a high level language that provides all the necessary tools to perform the task elegantly. This is the aim of this chapter.

In the next few sections we construct, step by step using object-oriented techniques, a computer algebra system — **SymbolicC++**. The system can be used in many areas of study. We describe the structures, functions and special features of the system. Examples are also included to demonstrate the usage of the functions.

The computer algebra system is built upon the concept of classes in object-oriented programming. Therefore, it inherits all the advantages and flexibilities of object-oriented programming. Some major plus points of object-oriented programming are modularity, reusability and extensibility. These features are very important especially for computer algebra systems because mathematics is built up level by level. For example, complex numbers are built upon real numbers, rational numbers are built upon integers and so on. From this point of view, the mathematical hierarchy is parallel with the philosophy of object-oriented programming. It seems to indicate that object-oriented programming is a natural choice for developing a computer algebra system. On the other hand, reusability and extensibility play a crucial role in clean coding which produce a more robust system.

The programs in this chapter can be found in MSymbol.h. This header file makes use of Mall.h and MList.h. Mall.h contains classes of the basic data structures, such as Term, Magnitude, Function, etc. MList.h is a linked list data structure, which has been described in Chapter 6. The linked list is used to store the dependent list of symbolic variables.

277

7.1   Object-Oriented Design

We first define what is meant by an *expression*. There are a number of ways to define an expression. Here, we consider it as a recursive data structure — that is, expressions are defined in terms of themselves. Suppose an expression can only use +, -, *, / and parentheses, then it can be defined with the following rules:

expression → term [+ term] [- term]
term   → factor [* factor] [/ factor]
factor   → variable, number or (expression).

The square brackets designate an optional element, and → means produces. These rules are usually called the *production rules* of the expression. Based on these rules, we could construct classes that work closely among themselves to represent mathematical expressions in a simple and clear form.

7.1.1   The Expression Tree

For the symbolic system discussed here, an expression is organized in a tree-like structure. Every node in the tree may represent a variable, a number, a function or a root of a subtree. A subtree may represent a term or an expression. There are two types of node defined in the system — Sum and Product. A Sum node is used to represent a variable, a number, a function or a root of a tree/subtree. A Product node is used to represent a term which is composed of the product or quotient of two factors. This representation conforms to the production rules we mentioned above.

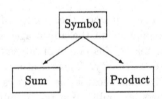

Both the Sum and Product classes are derived from a class called Symbol. They have, therefore, inherited the characteristics from the base class which describes the symbolic quantities. However, all the algebraic manipulations are done in Sum and Product which are interrelated.

The Sum class is the most important class as it defines a symbolic variable as well as a numeric number. It can be added or multiplied by another instance of Sum to form an expression, which is also represented by the Sum class. All the operators, functions and interfaces are defined in the class. The Sum class also defines special functions like sin(x) and cos(x) which will be described in detail in later sections.

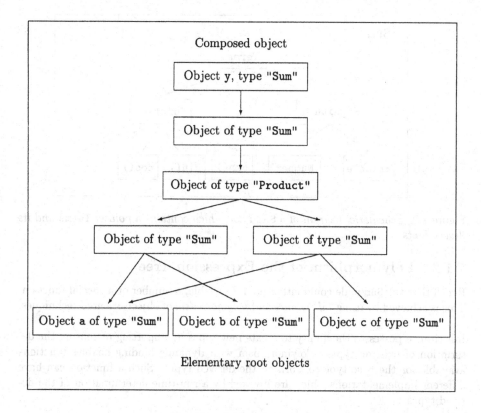

Figure 7.1: *Schematic diagram of the expression* y = (a+b)*(a+c)

On the other hand, the Product class describes terms that are multiplied or divided. It is a facilitator class and its existence is not known by the users.

As an example, consider the expression

$$y = (a+b)*(a+c).$$

The object y has a structure as given in Figure 7.1. Notice that the symbolic variables a, b and c are represented by Sum nodes. They are the leaf nodes of the expression tree. The expression a+b is composed of another Sum node which points at the variables a and b. a+c is constructed similarly. The Product node is used to multiply (a+b) and (a+c) which is in turn pointed to by another Sum node to form a tree as an expression. A Sum node y points at the final expression.

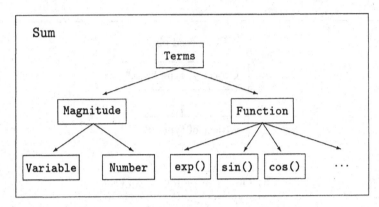

Figure 7.2: *Schematic diagram of a* Sum *node which contains a pointer* Terms *and its descendants*

7.1.2 Polymorphism of the Expression Tree

Recall that the Sum node could either hold a variable, a number or a special function. This is achieved by using *inheritance* and *polymorphism* of object-oriented techniques.

Inheritance possesses the ability to create new types by importing or reusing the description of existing types. Polymorphism with dynamic binding enables functions available for the base type to work on the derived types. Such a function can have different implementations which are invoked by a run-time determination of the derived types.

For our case here, the Sum node maintains a pointer to a class called Terms. It has two derived classes — Magnitude and Function. The Magnitude has two more derived classes Variable and Number, whereas Function has many derived classes. Each derived class of Function corresponds to a special function and more functions can be added if required (Figure 7.2). Making use of *virtual functions*, the properties and behaviours of the derived classes can be determined during run-time.

The class Terms is an *abstract base class*, as are its descendants Magnitude and Function:

```
template <class T>
class Terms
{
public:
    virtual ~Terms();
    virtual char type() const = 0;
    virtual String varName() const = 0;
    virtual void oprint(ostream&) const = 0;
```

```
    };

    template <class T>
    class Magnitude : public Terms<T>
    {
    public:
        virtual ~Magnitude();
        virtual T val() const = 0;
        virtual void set(const T) = 0;
    };

    template <class T>
    class Function : public Terms<T>
    {
    public:
        virtual ~Function();
        virtual double f(const T&) const = 0;
    };
```

The member functions in Terms, Magnitude and Function are *pure virtual functions*. Any class that declares or inherits a pure virtual function is an abstract base class. An attempt to create an object of an abstract base class will cause a compile time error. Therefore, all the pure virtual functions have to be overridden in the derived class that is declarable. An abstract base class is used to declare an *interface* without declaring a full set of implementations for that interface. That interface specifies the abstract operations supported by all objects derived from the class; it is up to the derived classes to supply implementations for those abstract operations.

The interfaces declared by the Terms class include the methods:

- type() returns the type information of the class, whether it is a Variable, a Number or different types of Function.

- varName() returns the name of the Variable or the function name. A Number does not have a variable name.

- oprint(ostream& os) places the name of a Variable, the numerical value of a Number or the function name on the output stream os.

Magnitude declares the following interfaces:

- val() returns the numerical value of a Variable or a Number.

- set(const T num) assigns the numerical value num to a Variable or a Number.

The class Function has only one extra interface, that is, f(const T& a) which returns the numerical value of the function evaluated at a.

Let us consider the derived classes one by one:

1. The class Variable: As the name suggests, it holds the information of a symbolic variable and its corresponding numerical value. It has two data fields — name and value. The variable name stores the symbolic name for the Variable whereas value stores the corresponding numerical value of name. The constructor reads in the variable name and its numerical value. The member functions override the base class definitions:

   - type() returns a character 'V' to indicate that it is a Variable class.
   - varName() returns name, the symbolic name of the instance.
   - oprint() puts name on the output stream.
   - val() returns value, the numerical value of the instance.
   - set(const T num) assigns the numeric constant num to the data field value.

2. The Number class defines a template for all available built-in or user defined numeric data types. For example,

   > Number<double>

   declares a user-defined double type which possesses all the original properties of the built-in double type. It also possesses some extra features that are useful for the symbolic system which will be discussed later in the section.

   The class contains only one data field

   > T data .

   It stores the value of type T. The constructor reads in the numeric value via its argument. The member functions here override the base class definitions:

   - type() returns a character 'N' to indicate that it is a Number class.
   - varName() returns a NULL string because it has no meaningful definition in the class.
   - oprint() puts data on the output stream.
   - val() returns data, the numerical value of the class.
   - set(const T num) assigns the numeric constant num to the data field data.

In addition to the member functions declared as in the interfaces of the abstract base class, the class has two extra member functions:

- The assignment operator = makes sure that instances of Number are assigned correctly.

- The conversion operator to type T is an important function. It is the gateway to all the properties that are possessed by the data type T. It converts instances of Number to the corresponding type T when necessary, hence enabling all the functions that are applicable to type T to be applicable to the class Number.

3. The derived classes of Function declare the special functions

   exp(x), sin(x), cos(x), sinh(x), cosh(x), sqrt(x)

Although many functions can be included here, they share some common properties. Each class contains only one data member Fname, which is used to store the function name. It also overrides all the interfaces declared by its ancestors:

- type() returns a character 'e' to indicate that it is an exponential function exp(). Other characters are returned to indicate different functions. However, only lower case characters are returned as part of the convention used in the system.

- varName() returns Fname which is the function name.

- oprint() puts Fname on the output stream.

- f(const T& x) returns the numerical value of the function. Note that the return type of the function is double, since we have made use of the built-in functions provided in the <math.h> library.

7.2   Data Fields and Types of Symbol

Recall that Sum is a public derived class of the Symbol class. Here, Symbol is also an abstract base class. It contains many pure virtual functions which wait to be overridden by its derived class. Next we describe how Symbol, Sum and Product co-operate among themselves as an expression tree to build a symbolic system.

The Symbol class has four data fields:

- branches stores the number of branches descended from this node.

- Symbol<T> **ep is a pointer to a pointer of Symbol<T>. This variable, after proper memory allocation, will contain an array of pointers to Symbol<T>. The size of the array is equivalent to branches, as these pointers will point at the descendants of this node.

- T *fac_exp is a pointer to T. After it is allocated to an array of size branches, it stores the multiplication factor for a Sum node or the power factor for a Product node.

- int *next_var is a pointer to int. It will be allocated to an array of size branches. The array contains only 0 or 1, which is used to indicate whether the decendent of the tree is of type 'V'. This information is essential for the destructor of the class.

There are six different types of nodes in an expression tree. Five of them are represented by a Sum node and one is by the Product node. When the function type() is invoked, different characters are returned to indicate the type of the tree node:

- 'S' is the root of an expression tree or subtree. For example, a+b or a+b*c. It does not have a variable name.

- 'V' is a Variable node, e.g. a, b. It is usually a leaf of an expression tree. Every Variable has a name associated with it.

- 'N' is a Number node, e.g. 2.0, 4/5. It is also a leaf of an expression tree.

- 'B' is a 'bound' node in the sense that it is a Variable but it is not a leaf of an expression tree. For example y=a+b, where y is a bound node.

- 'P' is a Product node, which is used to represent the product of two terms or expressions. e.g. a*b. It is always preceded by a Sum node.

- The last type of Sum denotes special functions; different characters will be returned to indicate different functions. For example, exp(x) function returns an 'e', sin(x) function returns a 'z' and cos(x) function returns a 'c'. Only lower case characters are returned for such special functions.

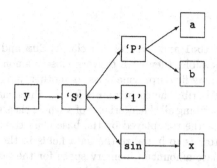

Figure 7.3: *Schematic diagram of the expression* y = a*b + 1 + sin(x)

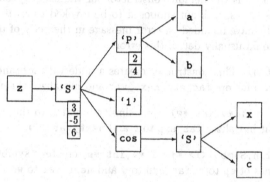

Figure 7.4: *Schematic diagram of* z = 3*a^2*b^4 - 5 + 6*cos(x+c)

Consider the expression

$$y = a*b + 1 + sin(x)$$

and its corresponding expression tree in Figure 7.3. Note that all six types of nodes have been used in this expression: y is 'B'-type; a, b, x are 'V'-type; '1' is 'N'-type, 'sin' is 'Function'-type and the 'S' and 'P'-types are used as intermediate connecting nodes for different purposes.

Another example is given in Figure 7.4. The expression is

$$z = 3*a^2*b^4 - 5 + 6*cos(x+c)$$

We notice that expressions which contain only +, -, *, / and parentheses can be represented uniquely by the expression tree. The smaller boxes attached to the 'S' and 'P' nodes are fac_exp which represent the multiplication factors or the power factors respectively. The '1' node is a special node which we discuss later. The expression that follows the special function is the argument of the function.

7.3   Constructors

We have a base class **Symbol** and two derived classes **Sum** and **Product**. For the symbolic system to be possible, they have to work closely among themselves. Since the logic of the program may flow from one class to another and due to the nature of the data structures, we describe the constructors for the three classes all together at this stage rather than describing all the interfaces of a whole class followed by another. We start off by describing the role played by the base class constructors. The main purpose of the constructors is to initialize the data fields in the class including the allocation of an appropriate amount of memory space for the pointers. This step is important for ensuring the proper operation of the classes.

- **Symbol()**: This is the default constructor for the class **Symbol**. Since it is an abstract base class, it is not supposed to be invoked to create any instance at any time. We have included an error message in the body of the function. The compiler should usually detect this error.

- **Symbol(int n)**: This constructor creates a node with n branches. It allocates memory space for **ep**, **fac_exp**, **next_var** and initializes them.

- **Symbol(int n,T *v,int *w)** has a similar function to the previous constructor, except it initializes **fac_exp** to v and **next_var** to w.

- **Symbol(int n,Symbol<T> **s,T *v,int *w)** creates a symbolic node and initializes entries of **ep** to s, **fac_exp** to v and **next_var** to w.

- The destructor simply releases unused memory back to the pool.

All the constructors are declared as **protected**. These constructors are only visible and accessible to the member functions and derived classes but not to the users or functions outside the class.

Consider the constructors in the Sum class. The class has a data field **Terms<T> *data** and it is a public derived class of **Symbol**. A constructor of **Sum** will always invoke the constructor of **Symbol** first, followed by the initialization of **data** and other operations:

- The default constructor declares a "zero" node. The "zero" node and "one" node are two special nodes in the classes. They have the following node structures:

A number n other than "one" or "zero", is represented by putting n in the fac_exp of the Sum node as a multiplication factor as shown in the structure below:

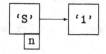

- Sum(const T n) creates a number node with numerical value equal to n as described above.

- Sum(const T n,char) creates a leaf node of type 'N' with its numerical value equal to n. It is a private constructor which is used internally.

- Sum(char ftype,char) creates a 'Function' node according to the ftype provided by the caller. For example, if ftype = 'e', an exp() function is created. It is also a private constructor.

- Sum(int n,int) creates a Sum node which has n branches.

- Sum(String name,int) is used to declare a symbolic variable name.

- Sum(int i,Symbol<T> **newep,T *new_fe,int *new_nx,String nm) creates a Sum node with all its data fields and base class structures initialized according to the items in the argument.

- The copy constructor duplicates the Sum node and the tree expression rooted on this node.

- The destructor releases memory that is no longer in use back to the free memory pool.

The constructors for the Product class are much simpler because the Product has only one data field Comm, which indicates the commutativity of the multiplication operator *.

- Product(int n) creates a Product node with n branches.

- Product(int i, Symbol<T> **newep, T *new_fe, int *new_nx) creates a Product node with the base class structure initialized as specified in the argument.

- The copy constructor and destructor for Product are trivial.

- The data field Comm is declared as `static`. It is therefore shared by all instances of the class and is stored uniquely in one place. Since a static member is independent of any particular instance, it is accessed in the form Product<T>::Comm. We have specified the algebra system to be commutative by default and it is initialized by the following statement:

```
template <class T> int Product<T>::Comm = 1;
```

where 1 specifies a commutative system and 0 specifies otherwise.

7.4   Operators

In this section, we describe the arithmetic and mathematical operators available in the system:

```
(unary)+, (unary)-, +, -, *, /, =, +=, -=, *=, /=, ==, !=,
power(), df(), Int(), expand(), coeff()
```

Suppose x, y are symbolic variables of type Sum and c is a numerical constant of type T, the operators do the following:

- x+y, x+c, c+x add symbolic variables, expressions or numerical constants.

- x-y, x-c, c-x subtract symbolic variables, expressions or numerical constants.

- x*y, x*c, c*x multiply symbolic variables, expressions or numerical constants.

- x/y, x/c, c/x divide symbolic variables, expressions or numerical constants.

- x=y assigns a symbolic variable/expression y to x.

- x=c assigns a numerical constant c to x.

- +=, -=, *= and /= are just the modification forms of the assignment operator.

- ==, != compare the equality of two expressions. For example, x+y==y+x returns *true* and x!=x returns *false*.

  The result actually depends on the commutativity of the * operator as well. For example, a*b==b*a will return *true* if it is commutative, otherwise it is *false*.

- power(const Sum<T> &s,int n) raises expression s to power n. For example, y=power(x,5) has the structures shown on the next page.

  For the other case when the power is equal to zero, e.g. y=power(x,0), the function simply returns a constant value '1'.

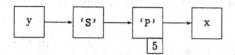

- `power(T x,T n)` evaluates x^n using the algorithm discussed in Section 6.1. Note that the function works only for numeric powers. Terms such as x^x cannot be expressed using the `power()` function. An alternative way to express the term making use of the exponential function is given by

$$x^x \equiv \exp(x * \ln(x)).$$

- `df(const Sum<T>& y,const Sum<T>& x)` evaluates the partial derivative of y with respect to x. The evaluation of `df(y,x)` proceeds as follows:

   1. Suppose y is expressed in terms of x, then each term of y which contains the variable x is differentiated according to the standard rules.

   2. If z is another variable then its derivative with respect to x is zero, unless z has been declared previously as dependent on x, in which case the derivative is `df(z,x)`.

Consider a polynomial P of degree n,

$$P(x) = a_n x^n + a_{n-1} x^{n-1} + \cdots + a_k x^k + \cdots + a_2 x^2 + a_1 x + a_0.$$

Since the derivative of the monomial $a_k x^k$ is $k a_k x^{k-1}$, we obtain by using `diff()` in the function `df()`:

$$\frac{dP}{dx} = n a_n x^{n-1} + (n-1) a_{n-1} x^{n-2} + \cdots + k a_k x^{k-1} + \cdots + 2 a_2 x + a_1.$$

In the process, we have made use of several derivative rules.

   – The function `Sum<T>::diff()` manipulates expressions following the properties listed below:

   1. $\dfrac{dx}{dx} = 1$, $\dfrac{dc}{dx} = 0$ where c is a numerical constant.

   2. $\dfrac{dy}{dx} = \begin{cases} \dfrac{dy}{dx} & \text{if } y \text{ is dependent on } x \\ 0 & \text{otherwise} \end{cases}$

3. The differentiation operator is a *linear operator*, i.e.

$$\frac{d}{dx}[a * u(x) + b * v(x)] = a * \frac{du(x)}{dx} + b * \frac{dv(x)}{dx}$$

where a, b are numerical constants, x is an independent variable and $u(x)$, $v(x)$ are functions of x.

4. *Chain-rule*:

$$\frac{d}{dx}f(u(x)) = \frac{df}{du} * \frac{du}{dx}$$

where f is a function of u which is dependent on x.

5. Integration is the inverse of differentiation

$$\int \frac{d}{dx}y(x)dx = y(x).$$

– The `Product<T>::diff()` implements the *product rule* for differentiation

$$\frac{d}{dx}(u_1(x) * u_2(x)) = \frac{du_1}{dx} * u_2 + u_1 * \frac{du_2}{dx}$$

and the derivative rule for monomials

$$\frac{d}{dx}(a * (u(x))^n) = n * a * u^{n-1} * \left(\frac{du}{dx}\right)$$

where a is a constant.

The differentiation operator needs knowledge of the dependency between various variables. Such dependency may be declared by the function `depend()`. For example, `y.depend(x); y.depend(z);` declares that y is dependent on x and z. After such declarations are made, `df(y,x)` would be evaluated to itself `df(y,x)`, instead of 0 as prior to the declarations.

If, for some reason, the dependency needs to be removed, then `nodepend()` can be used. For example, given the above dependencies, `y.nodepend(x)` indicates that y is no longer dependent on x, although it remains dependent on z.

Sometimes, we need the information about the dependencies on various variables. We could check them using `isdepend()`. For example, `y.isdepend(z)` would return *true* for the above dependencies whereas `y.isdepend(x)` would return *false*.

All the information regarding the dependencies between variables is stored using a linked list `MList.h`. Note that a linked list data structure is chosen because the number of dependent variables varies for different symbolic variables. It is therefore difficult to preassign the memory space required for such a case.

- `df(const Sum<T>& y,const Sum<T>& x,int n)` evaluates the nth derivative of y with respect to x.

- `Int(const Sum<T> &y,const Sum<T> &x)` evaluates simple integrals of y with respect to x, neglecting constants of integration.

Consider a polynomial P of degree n,

$$P(x) = a_n x^n + a_{n-1} x^{n-1} + \cdots + a_k x^k + \cdots + a_2 x^2 + a_1 x + a_0.$$

Since the integral of the monomial $a_k x^k$ is $a_k x^{k+1}/(k+1)$, we obtain by using `integrate()` in the function `Int()`:

$$\int P(x)dx = \frac{a_n}{n+1}x^{n+1} + \frac{a_{n-1}}{n}x^n + \cdots + \frac{a_k}{k+1}x^{k+1} + \cdots + \frac{a_2}{3}x^3 + \frac{a_1}{2}x^2 + a_0 x.$$

In the process, we have made use of several integration rules.

- The function `Sum<T>::integrate()` manipulates expressions following the properties listed below:

1. For c a numerical constant and x an independent variable

$$\int c\, dx = c * x, \qquad \int x\, dx = \frac{x^2}{2}.$$

2.

$$\int y\, dx = \begin{cases} \int y\, dx & \text{if } y \text{ is dependent on } x \\ y * x & \text{otherwise} \end{cases}$$

3. The integration operator is a *linear operator*, i.e.

$$\int [a * u(x) + b * v(x)] \, dx = a * \int u(x) \, dx + b * \int v(x) \, dx$$

where a, b are numerical constants, x is an independent variable and $u(x)$, $v(x)$ are functions of x.

4. Integration is the inverse operation of differentiation:

$$\int \frac{d}{dx} y(x) \, dx = y(x).$$

The function Product<T>::integrate() implements the integration rules for monomials

$$\int a * x^n dx = \begin{cases} \frac{a}{n+1} x^{n+1} & n \neq -1 \\ \ln(x) & n = -1 \end{cases}$$

where a is a constant.

The integration operator sometimes needs knowledge of the dependency between various variables. Such dependency may be declared by the function depend(). For example, y.depend(x); declares that y is dependent on x. After such a declaration is made, Int(y,x) would be evaluated to itself Int(y,x), instead of y*x as prior to the declaration. The description of df() expands on this discussion.

- Int(const Sum<T> &y, const Sum<T> &x, int n) evaluates the nth integral of y with respect to x.

- expand() expands the product of expressions that may be raised to a power n. It is composed of two major functions — dxpand() and mxpand(). dxpand() expands expressions using the distributive law, whereas mxpand() uses the *binomial theorem* to perform the binomial expansion

$$(a + b)^n = {}^nC_0 a^n + {}^nC_1 a^{n-1} b + {}^nC_2 a^{n-2} b^2 + \cdots + {}^nC_{n-1} ab^{n-1} + {}^nC_n b^n$$

For a multinomial expansion, where the expression is a sum of more than two components, we further extend the idea of the binomial theorem by grouping terms into only two sets of terms. However, we have to decide how to split it into the a and b pieces. There are two obvious ways: either cut the expression in half, so that a and b will be of equal size, or split off one component at a time. It can be shown that the latter method is more efficient in most cases. In other words, an expression $t_1 + t_2 + t_3 + \cdots + t_k$ will be treated as the sum $a + b$ where $a = t_1$ and $b = t_2 + t_3 + \cdots + t_k$, i.e.

$$(t_1 + t_2 + t_3 + \cdots + t_k)^n = [t_1 + (t_2 + t_3 + \cdots + t_k)]^n$$

and the binomial expansion is applied.

- `coeff()` returns the symbolic or numeric coefficients of an expression. It is overloaded in three different forms:

    - `Sum<T> coeff(const Sum<T>& s) const;`
      returns the coefficient of the term `s` in the expression.

    - `T coeff(int) const;`
      returns the constant term in the expression.

    - `Sum<T> coeff(const Sum<T>& s,int n) const;`
      returns the coefficient of the term s^n in the expression.

- `put(const Sum<T>& s1,const Sum<T>& s2)` replaces all the terms `s1` in an expression by another term `s2`. The function returns a non-zero integer if a replacement is successful, otherwise a zero is returned.

The function proceeds by checking through each term of the expression. If a match with `s1` is found, it replaces `s1` with `s2`. This is a useful function that aids the simplification of expressions. For example, an expression with a mixture of $\sin^2 x$ and $\cos^2 x$ terms such as

$$10 * \sin^2 x + 5 * \cos^2 x$$

may be simplified to

$$5 * \sin^2 x + 5$$

by replacing all the $\cos^2 x$ term by $1 - \sin^2 x$ and combining like terms.

Example 1

In this example, we consider the function power().

```
// expand.cxx

#include <iostream.h>
#include "MSymbol.h"

void main()
{
   Sum<int> a("a",0), b("b",0), c("c",0), y, z;

   y = power(7,3);            cout << " y = " << y << endl;
   y = power(a,0);            cout << " y = " << y << endl;
   y = power(a,3);            cout << " y = " << y << endl;
   cout << endl;

   y = power(a+b-c,3);        cout << " y = " << y << endl;
   cout << endl;

   y = (a+b)*(a-c);           cout << " y = " << y << endl;
   cout << endl;

   y = a+b;
   z = power(y,4);            cout << " z = " << z << endl;
}
/*
Results
=======
y = 343
y = 1
y = a^(3)

y = a^(3)+3*a^(2)*b-3*a^(2)*c+3*a*b^(2)-6*a*b*c+3*a*c^(2)+b^(3)
    -3*b^(2)*c+3*b*c^(2)-c^(3)

y = a^(2)-a*c+b*a-b*c

z = a^(4)+4*a^(3)*b+6*a^(2)*b^(2)+4*a*b^(3)+b^(4)
*/
```

Example 2

In this example, we consider the derivatives of

$$y(x) = \frac{1}{1-x} + 2x^3 - z$$

with respect to x. Notice that

$$\frac{dz}{dx} = 0$$

because the system assumes no dependency for any two variables unless specified otherwise. In the second part of this example, we consider the derivatives of the expression

$$u(v) = \frac{1}{3}v^{3/5} - \frac{2}{7}v^{1/5} + \frac{1}{6}$$

with respect to v. Note that the coefficients and the degrees of the variable v are rational numbers.

```
// derivatv.cxx

#include <iostream.h>
#include "Rational.h"
#include "MSymbol.h"

void main()
{
    int i;
    Sum<int> x("x",0), z("z",0), y;

    y = 1/(1-x) + 2*x*x*x - z;
    cout << "y = " << y << endl;

    for(i=0; i<8; i++)
    {
        y = df(y,x);
        cout << "y = " << y << endl;
```

```
       }
       cout << endl;

       Sum<Rational<int> > v("v",0), u;
       u = Rational<int>(1,3) * power(v,Rational<int>(3,5))
           - Rational<int>(2,7) * power(v,Rational<int>(1,5))
           + Rational<int>(1,6);
       cout << "u = " << u << endl;

       for(i=0; i<8; i++)
       {
          u = df(u,v);
          cout << "u = " << u << endl;
       }


   }
/*
Results
=======
y = (1-x)^(-1)+2*x^(3)-z
y = (1-x)^(-2)+6*x^(2)
y = 2*(1-x)^(-3)+12*x
y = 6*(1-x)^(-4)+12
y = 24*(1-x)^(-5)
y = 120*(1-x)^(-6)
y = 720*(1-x)^(-7)
y = 5040*(1-x)^(-8)
y = 40320*(1-x)^(-9)

u = 1/3*v^(3/5)-2/7*v^(1/5)+1/6
u = 1/5*v^(-2/5)-2/35*v^(-4/5)
u = -2/25*v^(-7/5)+8/175*v^(-9/5)
u = 14/125*v^(-12/5)-72/875*v^(-14/5)
u = -168/625*v^(-17/5)+144/625*v^(-19/5)
u = 2856/3125*v^(-22/5)-2736/3125*v^(-24/5)
u = -62832/15625*v^(-27/5)+65664/15625*v^(-29/5)
u = 1696464/78125*v^(-32/5)-1904256/78125*v^(-34/5)
u = -54286848/390625*v^(-37/5)+64744704/390625*v^(-39/5)
*/
```

Example 3

In this example, we investigate the properties of the df() operator. By default, any two variables are assumed to be independent of each other. Dependency can be created by depend() and removed by nodepend().

```
// depend.cxx

#include <iostream.h>
#include "MSymbol.h"

void main()
{
    Sum<int> a("a",0), b("b",0), u("u",0), v("v",0),
             x("x",0), z("z",0), y("y",0);

    cout << "System assumes no dependency by default" << endl;
    cout << "df(y,x) => " << df(y,x) << endl;

    cout << "y is dependent on x" << endl;
    y.depend(x);
    cout << "df(y,x) => " << df(y,x) << endl;
    y = sin(x*x+5) + x; cout << "y = " << y << endl;
    cout << "df(y,x) => " << df(y,x) << endl;
    cout << endl;

    cout << "u depends on x" << endl;
    u.depend(x);
    cout << "df(cos(u),x) => " << df(cos(u),x) << endl;

    cout << "u depends on x, and x depends on v" << endl;
    x.depend(v);
    cout << "df(cos(u),v) => " << df(cos(u),v) << endl;

    // example
    y = cos(x*x+5)-2*sin(a*z+b*x); cout << "y = " << y << endl;
    cout << "df(y,v) => " << df(y,v) << endl;
    cout << endl;

    cout << "check dependency" << endl;
    cout << "y.isdepend(x) -> " << y.isdepend(x) << endl;
    cout << "y.isdepend(v) -> " << y.isdepend(v) << endl;
    cout << "y.isdepend(z) -> " << y.isdepend(z) << endl;
    cout << endl;
```

```
        cout << "remove dependency" << endl;
    x.nodepend(v);
    cout << "df(y,v) => " << df(y,v) << endl;
    cout << endl;

    cout << "derivative of constants gives zero" << endl;
    cout << "df(5,x) => " << df(5,x) << endl; '
    cout << endl;

    // renew the variable y
    y.clear();
    y.depend(x);
    y.depend(z);
    cout << "df(y,x)+df(y,x) => " << df(y,x)+df(y,x) << endl;
    cout << "df(y,x)*df(y,x) => " << df(y,x)*df(y,x) << endl;
    cout << endl;

    cout << "Another example," << endl;
    x.clear();
    y.clear();
    u.clear();
    y.depend(u);
    u.depend(v);
    v.depend(x);
    y = df(u,x)*sin(x);
    cout << "y = " << y << endl;
    u = 2*v*x;
    cout << "let u = " << u << " then," << endl;
    cout << "y => " << y << endl;
    cout << "df(y,x) => " << df(y,x) << endl;
    }
/*
Results
=======
System assumes no dependency by default
df(y,x) => 0

y is dependent on x
df(y,x) => df(y,x)
y = sin(x^(2)+5)+x
df(y,x) => 2*x*cos(x^(2)+5)+1

u depends on x
```

```
df(cos(u),x) => -df(u,x)*sin(u)
u depends on x, and x depends on v
df(cos(u),v) => -df(u,v)*sin(u)
y = cos(x^(2)+5)-2*sin(a*z+b*x)
df(y,v) => -2*x*df(x,v)*sin(x^(2)+5)-2*b*df(x,v)*cos(a*z+b*x)

check dependency
y.isdepend(x) -> 1
y.isdepend(v) -> 1
y.isdepend(z) -> 0

remove dependency
df(y,v) => 0

derivative of constants gives zero
df(5,x) => 0

df(y,x)+df(y,x) => 2*df(y,x)
df(y,x)*df(y,x) => df(y,x)^(2)

Another example,
y = df(u,x)*sin(x)
let u = 2*v*x then,
y => (2*df(v,x)*x+2*v)*sin(x)
df(y,x) => (2*df(df(v,x),x)*x+4*df(v,x))*sin(x)+(2*df(v,x)*x+2*v)*cos(x)
*/
```

Examples 4, 5 and 6 demonstrate how different forms of coeff() could be used to extract coefficients of expressions. Example 7 illustrates the use of put() and how identities can be used for simplification.

Example 4

```
// coeff1.cxx

#include <iostream.h>
#include "MSymbol.h"

void main()
{
    Sum<int> a("a",0), b("b",0), c("c",0);

    cout << (2*a-3*b*a-2+c).coeff(a) << endl;
    cout << (2*a-3*b*a-2+c).coeff(b*a) << endl;
    cout << (2*a-3*b*a-2+c).coeff(b) << endl;
    cout << endl;

    cout << (b*b+c-3).coeff(a,0) << endl;
    cout << (-b).coeff(a,0) << endl;
}
/*
Results
=======
2-3*b
-3
-3*a

b^(2)+c-3
-b
*/
```

Example 5

```
// coeff2.cxx

#include <iostream.h>
#include "MSymbol.h"

void main()
{
    Sum<int> a("a",0), b("b",0), c("c",0), d("d",0),
            y, z;

    y = -a*5*a*b*b*c + 4*a - 2*a*b*c*c + 6 - 2*a*b
            + 3*a*b*c - 8*c*c*b*a + 4*c*c*c*a*b - 3*b*c;
    cout << "y = " << y << endl; cout << endl;

    z = a;          cout << y.coeff(z) << endl;
    z = b;          cout << y.coeff(z) << endl;
    z = c;          cout << y.coeff(z) << endl;
    z = a*a;        cout << y.coeff(z) << endl;
    z = a*b;        cout << y.coeff(z) << endl;
    z = d;          cout << y.coeff(z) << endl;
    cout << endl;

    // find coefficients of the constant term
    z = a;          cout << y.coeff(z,0) << endl;
    cout << y.coeff(0) << endl;
}
/*
Results
=======
y = -5*a^(2)*b^(2)*c+4*a-10*a*b*c^(2)+6-2*a*b+3*a*b*c+4*c^(3)*a*b-3*b*c

4-10*b*c^(2)-2*b+3*b*c+4*c^(3)*b
-10*a*c^(2)-2*a+3*a*c+4*c^(3)*a-3*c
-5*a^(2)*b^(2)+3*a*b-3*b
-5*b^(2)*c
-10*c^(2)-2+3*c+4*c^(3)
0

6-3*b*c
6
*/
```

Example 6

```
// coeff3.cxx

#include <iostream.h>
#include "Rational.h"
#include "MSymbol.h"

void main()
{
    int i;
    Sum<Rational<int> > x("x",0),w("w",0),p("p",0);
    w = Rational<int>(0);

    for(i=-5; i<=5; i++)
       w += Rational<int>(6+i,6-i)*power(p,5-i)*power(x,i);

    cout << "w = " << w << endl; cout << endl;

    cout << "And the coefficients are" << endl;
    for (i=-5; i<=5; i++)
        cout << w.coeff(x,i) << endl;
}
/*
Results
=======
w = 1/11*p^(10)*x^(-5)+1/5*p^(9)*x^(-4)+1/3*p^(8)*x^(-3)
   +1/2*p^(7)*x^(-2)+5/7*p^(6)*x^(-1)+p^(5)+7/5*p^(4)*x+2*p^(3)*x^(2)
   +3*p^(2)*x^(3)+5*p*x^(4)+11*x^(5)

And the coefficients are
1/11*p^(10)
1/5*p^(9)
1/3*p^(8)
1/2*p^(7)
5/7*p^(6)
p^(5)
7/5*p^(4)
2*p^(3)
3*p^(2)
5*p
11
*/
```

Example 7

```
// put.cxx

#include <iostream.h>
#include "MSymbol.h"

void main()
{
    Sum<int> a("a",0), b("b",0), c("c",0), w, y;

    // Test (1)
    y = (a+b)*(a+sin(power(cos(a),2)))*b;
    cout << "y = " << y << endl;
    y.put(b,c+c); cout << "y = " << y << endl;
    y.put(cos(a)*cos(a),1-power(sin(a),2));
    cout << "y = " << y << endl; cout << endl;

    // Test (2)
    y = sin(power(cos(a),2)) + b; cout << "y = " << y << endl;
    y.put(cos(a)*cos(a),1-power(sin(a),2));
    cout << "y = " << y << endl; cout << endl;

    // Test (3)
    a.depend(c); b.depend(c);
    y = 2*a*df(b*a*a,c) + a*b*c; cout << "y = " << y << endl;
    y.put(a,cos(c)); cout << "y = " << y << endl;
    w = df(b,c); y.put(w,exp(a)); cout << "y = " << y << endl;
}
/*
Results
=======
y = a^(2)*b+a*sin(cos(a)^(2))*b+b^(2)*a+b^(2)*sin(cos(a)^(2))
y = 2*a^(2)*c+2*a*sin(cos(a)^(2))*c+4*c^(2)*a+4*c^(2)*sin(cos(a)^(2))
y = 2*a^(2)*c+2*a*sin(1-sin(a)^(2))*c+4*c^(2)*a+4*c^(2)*sin(1-sin(a)^(2))

y = sin(cos(a)^(2))+b
y = sin(1-sin(a)^(2))+b

y = 2*a^(3)*df(b,c)+4*a^(2)*b*df(a,c)+a*b*c
y = 2*cos(c)^(3)*df(b,c)-4*cos(c)^(2)*b*sin(c)+cos(c)*b*c
y = 2*cos(c)^(3)*exp(a)-4*cos(c)^(2)*b*sin(c)+cos(c)*b*c
*/
```

Example 8

In this example, we integrate the expression

$$y(x) = (1 - x)^2 + \cos(x) + xe^x + c + z$$

with respect to x. We declare z to be dependent on x. Since the system assumes no dependency for any two variables we can use c as a symbolic constant. We also make use of

$$\int xe^x = xe^x - \int e^x dx.$$

```cpp
// integ.cpp

#include <iostream.h>
#include "MSymbol.h"
#include "rational.h"

void main()
{
 int i;
 Sum<Rational<int> > x("x",0), c("c",0), z("z",0), y;
 Sum<Rational<int> > one(Rational<int>(1));
 z.depend(x);
 y=(one-x)*(one-x)+cos(x)+x*exp(x)+c+z;
 cout << "y = " << y << endl;

 for(i=0; i<3; i++)
 {
 y = Int(y,x);
 y.put(Int(x*exp(x),x), x*exp(x) - exp(x));
 cout << "y = " << y << endl;
 }
 cout << endl;
}
```

Results
=======
```
y = 1-2*x+x^(2)+cos(x)+x*exp(x)+c+z
y = x-x^(2)+1/3*x^(3)+sin(x)+x*exp(x)-exp(x)+c*x+Int(z,x)
y = 1/2*x^(2)-1/3*x^(3)+1/12*x^(4)-cos(x)+x*exp(x)-2*exp(x)
    +1/2*c*x^(2)+Int(Int(z,x),x)
y = 1/6*x^(3)-1/12*x^(4)+1/60*x^(5)-sin(x)+x*exp(x)-3*exp(x)
    +1/6*c*x^(3)+Int(Int(Int(z,x),x),x)
```

7.5 Functions

We have described the availability of special functions such as sine and cosine in our symbolic system. All the operations available can be applied to these functions, e.g. the differentiation operator df().

A function is stored in the expression tree as a leaf node, i.e. it has no descendant. However, a function has to operate on an argument. We need to find an appropriate place to store the argument. Since an argument is attached to every function, it should also belong to the expression tree. We have decided to make it like an expression tree rooted at the function itself. Consider the function sin(x+y) and its tree structure in Figure 7.5.

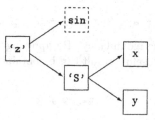

Figure 7.5: *Schematic diagram of the function* sin(x+y)

In the diagram, the 'z' node denotes a sin() function with its data points at the sin class. Its argument is pointed to by ep which is the root of the argument expression tree. With this representation, we have successfully implemented the function as a component of the expression tree. Below, we listed the functions available in the symbolic system and the operations associated with them. For the time being, there are only seven special functions implemented in the class:

exp(), sinh(), cosh(), sin(), cos(), sqrt(), ln()

Member functions that are associated with them include:

- The differentiation rules:

$$\frac{d}{dx}\exp(u) = \frac{du}{dx}\exp(u), \quad \frac{d}{dx}\sinh u = \frac{du}{dx}\cosh u, \quad \frac{d}{dx}\cosh u = \frac{du}{dx}\sinh u,$$

$$\frac{d}{dx}\sin u = \frac{du}{dx}\cos u, \quad \frac{d}{dx}\cos u = -\frac{du}{dx}\sin u, \quad \frac{d}{dx}\sqrt{u} = \frac{du}{dx}\frac{1}{2\sqrt{u}}, \quad \frac{d}{dx}\ln u = \frac{du}{dx}\frac{1}{u}.$$

- The value for special arguments:

$$\exp(0) = 1, \quad \sinh(0) = 0, \quad \cosh(0) = 1, \quad \sin(0) = 0,$$
$$\cos(0) = 1, \quad \sqrt{1} = 1, \quad \sqrt{0} = 0, \quad \ln(1) = 0.$$

7.6   Simplification

Simplification plays a vital role in a symbolic system. It produces an equivalent but simpler form of a given expression. The following are some examples of simplification of algebraic expressions:

```
2+5              ->  7
3*x-2*x          ->  x
u+0              ->  u
sin(2*x-x-x)     ->  0
```

where the symbol -> means "simplifies to" or "transforms to". Consider another expression,

```
8*a*a + b*c*(a + b - a - 2*b + b) + 2 - power(a,2)*3 + cos(x - x)
```

There exists some obvious redundancy. By applying the algebraic simplification, a much more "clean" and "useful" result can be obtained, such as the simplified form of the expression above,

```
5*a^2 + 3
```

Expression simplification is based on the existence of simplification rules for each type of mathematical expression. For example, simplification of the expression above involved the following algebraic manipulations:

```
    8*a*a + b*c*(a + b - a - 2*b + b) + 2 - power(a,2)*3 + cos(x - x)
→   8*a^2 + b*c*0 + 2 - 3*a^2 + cos(0)
→   5*a^2 + 0 + 2 + 1
→   5*a^2 + 3
```

In fact, there is no finite set of simplification rules that could simplify all kinds of algebraic expressions. This means that it is hard to find an algorithm that can completely simplify an arbitrary algebraic expression. However, with a suitable finite set of simplification rules, very good results can be achieved in practice.

7.6.1   Canonical Forms

Mathematical expressions may exist in several different but equivalent symbolic representations. A *canonical form* is a designated or "standard" representation for such expressions. For example, polynomials are usually represented by a series of terms with decreasing exponents, with each term preceded by a numerical coefficient. Following the "standard" convention, the conversion below would be appropriate:

$$-5*x^3 + x^6*8 + 4 - 2*x^2 \quad \rightarrow \quad 8*x^6 - 5*x^3 - 2*x^2 + 4.$$

A conversion to canonical form may increase or decrease the complexity of an expression, or simply result in the rearrangement of terms. Quite often, such a conversion is to improve the readability. Nevertheless, the most important application is the determination of two distinct symbolic representations for their "equivalence". Therefore, we need some rules that permit equivalent transformation from one representation to another. However, it is difficult to obtain the complete set of correct rules.

When a canonical form of an expression is carefully defined, transformation to the canonical form of two equivalent expressions will produce identical, or nearly identical representations. This will greatly reduce the complexity for comparison.

A mathematical expression is represented by a tree structure for our symbolic system. The simplification process consists of two major parts: One being the transformation to the canonical form, and the other being the reduction to simplified form based on some known rules. For example, a*(-1) could be rewritten as -a. Its symbolic simplification is as follows:

```
              a*(-1)
[Step 1]  →  -(a*(1))
[Step 2]  →  -(a*1)
[Step 3]  →  -(a)
[Step 4]  →  -a
```

The process involves three steps of canonical transformations [Step 1, 2, 4] and one step of rule-based simplification [Step 3] (Figure 7.6).

This example illustrates that a canonical form is a way of restricting the forms of expression which are allowed. Hence the simplification rules need only deal with a restricted class of expressions.

7.6.2 Simplification Rules and Member Functions

In this section, we describe in greater detail the simplification rules and the transformation to canonical forms. Suppose e, e_1, e_2, e_3, e_4 are arbitrary algebraic expressions and c, c_1, c_2 denote positive numerical constants. Some sample rules of simplification for each operator are listed in Figure 7.7.

The subtraction operator (-) has similar properties to the addition operator (+) while the division operator (/) has similar properties to the multiplication operator (*). Meanwhile, the transformation rules to canonical forms are given in Figure 7.8.

Finally, individual terms are grouped, summed or multiplied if they belong to the same class. For example,

```
2*a + 3*a          ->   5*a
exp(z) + exp(z)    ->   2*exp(z)
p^(-2) * p^(5)     ->   p^(3)
cos(x) * cos(x)    ->   (cos(x))^2
```

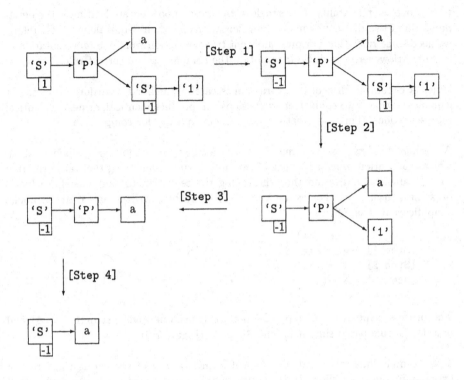

Figure 7.6: *Schematic diagram of the simplification process for* a*(-1)

and the final simplified expression is obtained. One may wonder with this limited set of simplification rules, could an expression be simplified to its simplest possible form? The answer to this question is not obvious. However, transformation to canonical form does play an important role in the process as it reduces the number of rule-based simplifications and also the complexity of the process.

In our canonical form for product expressions with numeric operands, only the first operand may be numeric. This implies that

$$(e * c) \rightarrow (c * e)$$

which is just the application of commutative law for multiplication. Hence the following rules are equivalent and they are not shown in Figure 7.7:

$$e * 1 \longleftrightarrow 1 * e$$

Operations	Functions
$c_1 + c_2 \rightarrow c$	Sum::shrink_4()
$e + 0 \rightarrow e$	Sum::shrink_3()
$c_1 * c_2 \rightarrow c$	Sum::shrink_1()
$0 * e \rightarrow 0$	Product::shrink_3()
$1 * e \rightarrow e$	Product::shrink_1()
$0^c \rightarrow 0$	Product::shrink_3()
$1^c \rightarrow 1$	Product::shrink_1()
$e^0 \rightarrow 1$	Product::gather()

Figure 7.7: *Simplification rules*

Operations	Functions
$(c_1 * e_1) * (c_2 * e_2) \rightarrow c * (e_1 * e_2)$ where $c = c_1 * c_2$	Sum::shrink_1()
$((e_1 + e_2) + e_3) + e_4 \rightarrow e_1 + e_2 + e_3 + e_4$	Sum::shrink_2()
$((e_1 * e_2) * e_3) * e_4 \rightarrow e_1 * e_2 * e_3 * e_4$	Product::shrink_2()
$e_1 + (e_2) \rightarrow e_1 + e_2$	Sum::shrink_5()

Figure 7.8: *Canonical conversion rules*

$$e * 0 \longleftrightarrow 0 * e$$
$$e * c \longleftrightarrow c * e$$

Similarly, the commutative law for addition may be employed:

$$(e + c) \rightarrow (c + e).$$

This eliminates many unnecessary addition rules. Another important canonical conversion rule is the outward propagation multiplication rule. For example,

$$(c_1 * e_1) * (c_2 * e_2) \longleftrightarrow (c_1 * c_2) * (e_1 * e_2)$$
$$(c_1 * e_1)/(c_2 * e_2) \longleftrightarrow (c_1/c_2) * (e_1/e_2)$$

These rules do not result in simplifications, but convert expressions to canonical forms which may result in simplifications later.

7.7   Commutativity

So far we have assumed the commutative law holds for multiplication of symbols. However, in some applications this is not necessarily true. For example,

$$A * B \neq B * A$$

in general if A, B are matrices. This leads us to think about the need for a non-commutative operator. In fact, a large branch of mathematics called Lie Algebra involves operators which are non-commutative. Therefore, non-commutative operators must be built into a symbolic system.

For our computer algebra system, we have made the system commutative by default. To specify a non-commutative algebra, use the function `Commutative(T,int t)`. The first argument specifies the type with which the underlying field is concerned. This is the type as specified during the declaration of the variables. The second argument t specifies the choice of commutativity. If `t = 0`, non-commutativity prevails. Otherwise, commutativity prevails.

Consider the following program:

```
// commute.cxx

#include <iostream.h>
#include "MSymbol.h"

void main()
{
   Sum<int> a("a",0), b("b",0), y;

   cout << "Commutative Algebra" << endl;
   cout << "===================" << endl;

   // The system is commutative by default
   y = a*b*a;          cout << " y = " << y << endl;
   y = a*b-b*a;        cout << " y = " << y << endl;

   y = power(a-b,2);   cout << " y = " << y << endl;
   cout << endl;

   cout << "Non-commutative Algebra" << endl;
   cout << "=======================" << endl;

   Commutative(int(),0);
```

```
        y = a*b*a;              cout << " y = " << y << endl;
        y = a*b-b*a;            cout << " y = " << y << endl;

        y = power(a-b,2);       cout << " y = " << y << endl;
        cout << endl;

        cout << "Commutative Algebra" << endl;
        cout << "===================" << endl;

        Commutative(int(),1);
        cout << " y = " << y << endl;
    }
/*
Result
======
Commutative Algebra
===================
  y = a^(2)*b
  y = 0
  y = a^(2)-2*a*b+b^(2)
Non-commutative Algebra
=======================
  y = a*b*a
  y = a*b-b*a
  y = a^(2)-a*b-b*a+b^(2)
Commutative Algebra
===================
  y = a^(2)-2*a*b+b^(2)
*/
```

From the result obtained, we notice that the function could be invoked at any time
within a program to specify or change the status of commutativity. The last two
display statements show that by switching the mode from non-commutative to commutative, expressions will be evaluated automatically. This rule does not apply the
other way round.

7.8   Symbolic and Numeric Interface

When a symbolic algebraic expression is evaluated, its numerical algebraic value is also needed. Our system handles the situation in the following way:

1. Starting with the algebraic values of the parts, all variables and operators with an argument list have the algebraic values as they were last assigned, or if never assigned, are taken as the variable itself.

2. In evaluating expressions, the standard rules of algebra are applied.

3. Three functions have been included to specify the numerical values and evaluate the numerical algebraic values of expressions. Suppose x, y, z are symbolic variables, y has been assigned to an algebraic expression in terms of x, whereas z has been assigned to a numerical constant value, then

   - `x.set(const T num)` assigns the numeric constant num to the variable x.

   - `y.value()` returns the evaluated algebraic value of the expression y.

   - `z.nvalue()` returns the numerical value of the variable z.

Note that assigning a numerical constant to a variable by `x.set(num)` is different to using the assignment operator x=num. The `x.set()` operator merely gives the numerical counterpart of the variable x. This value could be modified and the symbolic properties of the variable x still remains. However, the assignment operator replaces x by the number num. x is now semantically equivalent to num, and after the operation, it loses its symbolic properties in an expression and these cannot be retrieved. Their different properties are demonstrated in the following program:

```
// setvalue.cxx

#include <iostream.h>
#include "MSymbol.h"

void main()
{
    Sum<double> x("x",0), y("y",0), z("z",0);
    double      c1 = 0.5, c2 = 1.2;
    y = x*x + z/2.0;
    cout << "y = " << y << endl;
    cout << "The value of y is " << y.value() << endl;
    cout << endl;

    x.set(c1); z.set(c2);
    cout << "Put x = " << c1 << ", z = " << c2 << endl;
```

```
        cout << "The value of y = " << y << " is " << y.value() << endl;
        cout << endl;

        cout << "Substitute x with " << c1 << " and z with " << c2
             << ", then" << endl;
        x = c1;
        z = c2;

        cout << "y = " << y << endl; cout << endl;

        x.clear(); z.clear();

        y = x*x*z + 0.7*z - x*z;
        cout << "y = " << y << endl;

        x.set(c1);
        cout << "Put x = " << c1 << endl;
        cout << "The value of y is " << y.value() << endl;
        cout << endl;

        z.set(c2);
        cout << "Put z = " << c2 << endl;
        cout << "The value of y is " << y.value() << endl;
    }
/*
Result
======
y = x^(2)+0.5*z
The value of y is x^(2)+0.5*z

Put x = 0.5, z = 1.2
The value of y = x^(2)+0.5*z is 0.85

Substitute x with 0.5 and z with 1.2, then
y = 0.85

y = x^(2)*z+0.7*z-x*z
Put x = 0.5
The value of y is 0.45*z

Put z = 1.2
The value of y is 0.54
*/
```

7.9   Summary

This chapter introduced a new computer algebra system which is built using the concept of object-oriented programming. We pointed out that the mathematical hierarchy is parallel with the philosophy of object-oriented programming, which served as the motivation for this development using C++. In this chapter we described the following:

- *Object-oriented design*: We explained how mathematical expressions may be defined using the production rules and how it is related to our symbolic system.

- *Data fields and types of symbol*: We outlined the data fields defined in the classes and the basic types of symbol used in the system.

- *Operators*: The functionalities and properties of the arithmetic operators and functions were described. Differentiation and integration are explained.

- *Functions*: Special mathematical functions like sine, cosine, etc. were explained.

- *Simplification*: It is an important part of a computer algebra system, whereby complicated expressions are reduced to simpler forms.

- *Commutativity*: We described how commutativity plays an important role in the system.

- *Symbolic and numeric interface*: A computer algebra system is never complete without the interface between the symbolic and numerical aspects. Here, the rules and member functions that provide such an interface were described.

In the next chapter, we apply the classes implemented in Chapter 6 and this chapter to applications in different areas, such as number theory, chaotic dynamics, spherical harmonics, curvature and so on.

Chapter 8

Applications

The functionalities of a computer algebra system are best illustrated by applications. In this chapter, we present the applications of the classes introduced in Chapters 6 and 7. The problems presented here relate to number theory, nonlinear dynamics, special functions in mathematics and physics, etc. In general, applications are categorized under different classes. In each application, the mathematical background is first described, followed by the proposed solution to the problem and then the excerpt of a program that solves the problem symbolically. There are twenty-eight applications presented in this chapter:

(1) Prime Numbers
(2) Big Prime Numbers
(3) Inverse Map and Denumerable Set
(4) Gödel Numbering
(5) Logistic Map
(6) Contracting Mapping Theorem
(7) Ghost Solutions
(8) Iterated Function Systems
(9) Logistic Map and Ljapunov Exponent
(10) Mandelbrot Set
(11) Polynomials
(12) Cumulant Expansion
(13) Exterior Product
(14) First Integrals
(15) Spherical Harmonics
(16) Nambu Mechanics
(17) Taylor Expansion of Differential Equations
(18) Commutator of Two Vector Fields
(19) Lie Derivative and Killing Vector Field
(20) Hilbert-Schmidt Norm
(21) Lax Pair and Hamilton System
(22) Padé Approximant

(23) Pseudospherical Surfaces
       and Soliton Equations
(24) Picard's Method
(25) Lie Series Techniques
(26) Spectra of Small Spin Clusters
(27) Systems and Nonlinear Maps
       with Chaotic Behaviour
(28) Numerical-Symbolic Application

8.1   Bit Vector Class

8.1.1   Prime Numbers

In this section, we apply the `BitVector` class to number theory — the *prime numbers*. A prime number is a positive integer $p > 1$ such that no other integer divides p except 1 and p. The sequence formed by the prime numbers is perhaps the most famous sequence in number theory. We generate the prime number sequence using the "sieve of Eratosthenes" described in Section 2.2.

To implement the sieve algorithm, one must decide how numbers are stored and crossed out in the sequence. We could maintain an array of flags with each entry corresponding to a number in the sequence. The array is initialized by setting all the flags *true*. The crossing out of multiples is equivalent to setting the respective flag to *false*. When we reach a prime p with $p^2 > N$, all the primes below N have been found, then the array entries with flags equal to *true* will correspond to all the prime numbers $\leq N$.

Usually, an array of flags is represented by a bit field. It is simply a vector of 0/1 values. The field is maintained by a vector of `unsigned char`. Since each character is composed of 8 bits, it can store 8 flags. This data structure is very effective in terms of storage space.

Actually, we only need to store flags for the odd numbers, because we know that all even numbers besides 2 are not prime. In this way, we need $1000000/(8 \times 2) = 62500$ bytes to store the table of primes less than one million. The space complexity of the algorithm is $O(N)$.

In the following we list the class `Prime`, which generates the prime number sequence. Note that the header file `"Bitvec.h"` is included.

```
// Prime.h

#ifndef PRIME_H
#define PRIME_H

#include <assert.h>
#include <iostream.h>
#include "Bitvec.h"

class Prime
{
 private:
   // Data Fields
   unsigned int max_num, max_index, index, p, q, current;
```

```
    BitVector bvec;

  public:
    // Constructors
    Prime();
    Prime(unsigned int num);
    void reset(unsigned int);
    int  step();
    void run();
    int  is_prime(unsigned int) const;
    unsigned int current_prime() const;

    unsigned int operator () (unsigned int) const;
};

Prime::Prime()
    : max_num(0), max_index(0), index(0),
      p(0), q(0), current(0), bvec() {}

Prime::Prime(unsigned int num)
    : max_num(num), max_index((num+1)/2 - 1), index(0),
      p(3), q(3), current(2), bvec(max_index, 0xFF) {}

void Prime::reset(unsigned int num)
{
    assert(num > 1);
    max_num = num;
    max_index = (num+1)/2 - 1;
    index = 0; p = 3; q = 3; current = 2;
    bvec.reset(max_index, 0xFF);
}

int Prime::step()
{
    if(index < max_index)
    {
      while(!bvec.test(index))
      {
          ++index;
          p += 2;
          if(q < max_index) q += p+p-2;
          if(index > max_index) return 0;
      }
      current = p;
```

```
        if(q < max_index)
        {
            // cross out all odd multiples of p, starting with p^2
            // k = index of p^2
            unsigned int k = q;

            while(k < max_index) { bvec.clear(k); k += p; }

            ++index;
            p += 2;
            q += p+p - 2;
            return 1;
        }
        // p^2 > n, so bvec[] has all primes <= n recorded
        else // to next odd number
        {
            p += 2;
            ++index;
            return 2;
        }
    }
    else return 0;
}

void Prime::run()
{  while(step() == 1) ; }

int Prime::is_prime(unsigned int num) const
{
    if(!(num%2)) return 0;
    if(bvec.test((num-3)/2)) return 1;
    return 0;
}

unsigned int Prime::current_prime() const
{  return current; }

unsigned int Prime::operator () (unsigned int idx) const
{
    if(idx == 0) return 2;

    for(unsigned int i=0; idx && i<max_index; i++)
        if(bvec.test(i)) --idx;
```

```
        return i+i + 1;
    }
#endif
```

The constructor `Prime(unsigned int N)` specifies the upper limit N of the prime number sequence $\{p_k\}$ where $1 < p_k \leq N$, sets up the bit vector and turns all the bits on.

Suppose p is an instance of the class `Prime`, the six member functions work as follows:

- `step()` finds out the next prime number, step by step, starting from 3 and crosses out all the multiples of that prime number.

- `run()` repeatedly executes the function `step()` until all the prime numbers less than N are found.

- `reset(int M)` re-specifies the upper limit of the prime number sequence.

- `is_prime(int num)` checks if num is a prime number. It works provided the bit table has already been built.

- `current_prime()` returns the current prime number being iterated by the function `step()`.

- The subscript operator `p(int M)` returns the Mth prime number on the sequence.

Let us look at a simple program that calculates the total number of primes below 100, 1000, 10000, 100000, 1000000 and displays the first 20 prime numbers. It also demonstrates the ability to access the ith prime number.

```
// sprime.cxx

#include <iostream.h>
#include "Prime.h"

void main()
{
    unsigned int i, j, max0 = 1000000, count = 0;

    Prime p(max0);    // specifies the upper limit of the sequence
    p.run();          // generates the prime number sequence

    // for all odd numbers greater than 3
    // check if they are prime
```

```
for(i=3, j=100; i<=max0; i+=2)
{
    if(p.is_prime(i)) count++;

    // sum the number of primes below 100, 1000, ..., 1000000
    if(i == j-1)
    {
        j *= 10;
        cout << "There are " << count + 1
             << " primes below " << i+1 << endl;
    }
}
cout << endl;

// print the first 20 primes
for(i=0; i<20; i++) cout << p(i) << " ";
cout << endl; cout << endl;

// randomly pick some primes
cout << "The   100th prime is " << p(99)   << endl;
cout << "The   200th prime is " << p(199)  << endl;
cout << "The  3000th prime is " << p(2999) << endl;
cout << "The 10000th prime is " << p(9999) << endl;
cout << "The 12500th prime is " << p(12499) << endl;
}
```

Result
======
There are 25 primes below 100
There are 168 primes below 1000
There are 1229 primes below 10000
There are 9592 primes below 100000
There are 78498 primes below 1000000

2 3 5 7 11 13 17 19 23 29 31 37 41 43 47 53 59 61 67 71

The   100th prime is 541
The   200th prime is 1223
The 3000th prime is 27449
The 10000th prime is 104729
The 12500th prime is 134053

8.2 Verylong Class

8.2.1 Big Prime Numbers

In this section, we make use of the `Verylong` class to test whether a large positive integer number is a *prime number*. A prime number is a positive integer $p > 1$ such that no other integer divides p except 1 and p.

The function `is_prime(unsigned int num)` which we described in Section 8.1.1 can be used for checking whether `num` is a prime. It works provided the bit table has already been built. For a large number greater than 2^{32}, the built-in type `unsigned int` is too small. The `Verylong` integer can be used to overcome this problem. However, the main problem with this algorithm is the huge space requirement for the bit table. For large numbers such as a 20-digit number, it becomes highly impractical.

We describe another algorithm that tests the primality for a positive integer with little memory requirement. It is an obvious algorithm for a primality test. For a given positive integer N, we divide N by successive primes $p = 2, 3, 5, \ldots$ until a smallest p for which $N \bmod p = 0$, then N is not prime and has a factor p. If at any stage we find that $N \bmod p \neq 0$ but $\lfloor N/p \rfloor \leq p$, we conclude that N is prime.

Let us restate the arguments above:

Given a positive integer N, we divide N by a sequence of "trial divisors"

$$2 = d_0 < d_1 < d_2 < d_3 < \cdots < d_n$$

which includes all prime numbers $\leq \sqrt{N}$ (the sequence may also include some non-prime numbers if it is convenient). The last divisor in the sequence d_n is the smallest number $\geq \sqrt{N}$. If, at any stage, $N \bmod d_i = 0$ for $0 \leq i \leq n$, then N is not prime. Otherwise, it is prime.

The trial divisors sequence $d_0, d_1, d_2, \ldots, d_n$ works best when it contains only prime numbers. Sometimes it is convenient to include some non-prime numbers as well because such a sequence is usually easier to generate. One good sequence can be generated using the following recurrence relation:

$$d_0 = 2, \quad d_1 = 3, \quad d_2 = 5,$$
$$d_k = d_{k-1} + 2 \quad \text{for odd } k > 2$$
$$d_k = d_{k-1} + 4 \quad \text{for even } k > 2$$

The first few trial divisors generated using the relation above are as follows:

$$2 \ 3 \ 5 \ 7 \ 11 \ 13 \ 17 \ 19 \ 23 \ 25 \ 29 \ 31 \ 35 \ 37 \ 41 \ 43 \ \cdots$$

This sequence contains no multiples of 2 or 3 and the first 9 numbers are all prime. However, it also includes some non-prime numbers such as 25, 35, 49, etc.

```
// isprime.cxx

#include <iostream.h>
#include "Verylong.h"

template <class T> int is_prime(T p)
{
    T j(2), zero(0), one(1), two(2), four(4);
    T limit = T(sqrt(p)) + one;

    if(j < limit && p%j == zero) return 0;
    j++;
    if(j < limit && p%j == zero) return 0;
    j += two;
    while(j < limit)
    {
        if(p%j == zero) return 0;
        j += two;
        if(p%j == zero) return 0;
        j += four;
    }
    return 1;
}

void main()
{
    Verylong x;

    cout << "Please enter a positive integer number : ";
    cin >> x;

    if(is_prime(x)) cout << "The number " << x << " is prime " << endl;
    else cout << "The number " << x << " is not prime " << endl;
}
```

Result
======
Please enter a positive integer number : 20
The number 20 is not prime

Please enter a positive integer number : 624682384467
The number 624682384467 is not prime

Please enter a positive integer number : 48957934859
The number 48957934859 is prime

Although this is not the best way to check for primality, it illustrates how the Verylong class could be incorporated into number theory. Sometimes, it is good to have the prime numbers table (see Section 8.1.1) as part of the program. For example, if the table contains all the 78498 prime numbers less than one million, we could test the primality of N less than 10^{12}. Such a table could be built easily by the class Prime which we have discussed in Section 8.1.1.

So far, there are no known efficiently methods to test for the primality of large numbers. However, there are some algorithms [27] that will perform the testing in a reasonable amount of time.

An interesting extension of the program would be to find all the prime twins, for example $21, 23$. Why are there no prime triplets other than $2, 3, 5$?

The numbers $2^k - 1$ ($k \in \mathbf{N}$) are called *Mersenne numbers*. For $k = 2$, 3, 4, 7, 13, 19, 31, 61, 89, 107, 127, 521, 607, 1279, 2203, 2281, 3217, 4253, 4423, 9689, 9941, 11213, ..., we obtain prime numbers. Big prime numbers also play an important role in data encryption and network security.

The prime number $2^k - 1$ ($k = 11213$) can be found using the following small program. It is a large number consisting of 2816 digits.

```
// Mersenne.cxx

#include <iostream.h>
#include "Verylong.h"

int main()
{
    Verylong p,
             one("1"),
             two("2"),
             mersenne("11213");

    p = pow(two,mersenne) - one;
    cout << p << endl;

    return 0;
}
```

8.2.2   Inverse Map and Denumerable Set

The set $N \times N$ is *denumerable* because it is equipotent to the natural numbers N. In other words, there exists a 1-1 map between N and $N \times N$.

Let us write the elements of $N \times N$ in the form of an array as follows:

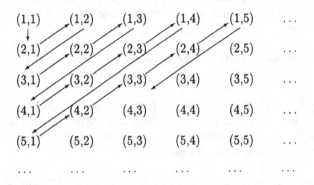

From the figure we see that we could arrange the elements of $N \times N$ into a linear sequence as indicated by the arrows, i.e.

$$(1,1), \; (2,1), \; (1,2), \; (3,1), \; (2,2), \; (1,3), \; (4,1), \; (3,2), \; \cdots$$

Therefore a 1-1 map between N and $N \times N$ exists. If $(m,n) \in N \times N$ we have

$$f(m,n) = \frac{1}{2}(m+n-1)(m+n-2) + n. \tag{8.1}$$

The above relation can be obtained as follows. The pair (m,n) lies in the $(m+n-1)^{\text{th}}$ diagonal stripe in the above figure and it is the nth pair counting from the left of the stripe. The first stripe contains only one pair. In the second, there are two pairs and so on. Thus (m,n) is located at the position numbered by

$$n + \sum_{k=1}^{m+n-2} k$$

in the counting procedure. Thus (8.1) follows. How could we find the inverse of the problem? This means to find m and n if $f(m,n)$ is given.

To find the inverse map, we first have to find out in which diagonal stripe $f(m,n)$ lies. Note that $m = 1$ corresponds to the last element on the diagonal stripe. Let $k = m + n - 1$, thus for $m = 1$, $k = 1 + n - 1 = n$. By (8.1),

$$f(1,n) = \frac{1}{2}(n)(-1+n) + n = \frac{n^2 + n}{2} = \frac{k^2 + k}{2}.$$

The last element in the kth diagonal is given by

$$f(m,n) = \frac{k^2 + k}{2}$$

whereas the last element in the $(k-1)^{\text{th}}$ diagonal is given by

$$f(m,n) = \frac{(k-1)^2 + (k-1)}{2} = \frac{k^2 - k}{2}.$$

Thus, if $f(m,n)$ lies in the kth diagonal stripe, then

$$\frac{k^2 - k}{2} < f(m,n) \leq \frac{k^2 + k}{2}. \tag{8.2}$$

Let $x \in \mathbf{R}$ such that

$$\frac{x^2 + x}{2} = f(m,n) \tag{8.3}$$

then

$$x^2 + x - 2f(m,n) = 0 \quad \text{and} \quad x = \frac{-1 + \sqrt{1 + 8f(m,n)}}{2}$$

where we obviously choose the positive root. Using (8.2) and (8.3), we get $k - 1 < x \leq k$. Thus, k is the smallest positive integer number greater than or equal to x.

After we have obtained k, the values of m and n can readily be calculated by (8.1):

$$n = f(m,n) - \frac{k(k-1)}{2}, \quad m = k - n + 1.$$

We implemented the inverse map using the algorithm described above. The program makes use of the **Verylong** class to handle very long integers. The **sqrt()** function used here is implemented in the **Verylong** class. It evaluates the integer square root of a **Verylong** number. An interesting extension would be to find a 1-1 map between N and N × N × N.

```
// inverse.cxx

#include <iostream.h>
#include "Verylong.h"

void InverseMap(Verylong f, Verylong &m, Verylong &n)
{
    Verylong zero(0), one(1), two(2), eight(8), k;
    k = (sqrt(one+eight*f)-one)/two + one;
    n = f - k*(k-one)/two;
    if (n==zero) {--k; n += k;}
    m = k-n+one;
}

void main()
{
    Verylong one(1), two(2);
    Verylong f, m, n, result;

    cout << "Enter a number : ";
    cin >> f;
    InverseMap(f,m,n);
    cout << "The corresponding m is " << m << endl;
    cout << "                and n is " << n << endl;
}
```
```
Result
======
Enter a number : 123456789012345
The corresponding m is 8553526
                and n is 7159959
```

8.2.3 Gödel Numbering

We can work with an alphabet which contains only a single letter, e.g. the letter |. The words constructed from this alphabet (apart from the empty word) are: |, ||, |||, etc. These words can, in a trivial way, be identified with the natural numbers 0, 1, 2, ... Such an extreme standardization of the "material" is advisable for some considerations. On the other hand, it is often convenient to disperse the diversity of an alphabet consisting of several elements.

The use of an alphabet consisting of *one element* does not imply any essential limitation. We can associate the words W over an alphabet $\mathbf{A}$ consisting of N elements with natural numbers $G(W)$, in such a way that each natural number is associated with at most one word. Similar arguments apply to words of an alphabet consisting of *one element*. Such a representation of G is called a *Gödel numbering* [14] (also called *arithmetization*) and $G(W)$ is the *Gödel number* of the the word W with respect to G. The following are the requirements for an arithmetization of W:

1. If $W_1 \neq W_2$ then $G(W_1) \neq G(W_2)$.

2. There exists an algorithm such that for any given word W, the corresponding natural number $G(W)$ can be computed in a finite number of steps.

3. For any natural number n, it can be decided whether n is the Gödel number of a word W over $\mathbf{A}$ in a finite number of steps.

4. There exists an algorithm such that if n is the Gödel number of a word W over $\mathbf{A}$, then this word W (which is unique by argument (1)) can be constructed in a finite number of steps.

Here is an example of a Gödel numbering. Consider the alphabet with the letters a, b, c. A word is constructed by any finite concatenation of these – that is, a placement of these letters side by side in a line. For example, $abcbba$ is a word. We can then number the words as follows:

Given a word $x_1 x_2 \ldots x_n$ where each x_i is a, b or c, we assign to it the number

$$2^{d_0} * 3^{d_1} * \ldots * p_n^{d_n}$$

where p_i is the ith prime number (and 2 is the 0th prime) and

$$d_i := \begin{cases} 1 & \text{if } x_i \text{ is } a \\ 2 & \text{if } x_i \text{ is } b \\ 3 & \text{if } x_i \text{ is } c \end{cases}$$

The empty word is given the number 0.

For example, the word *acbc* has number $2^1 * 3^3 * 5^2 * 7^3 = 463050$, and *abc* has the number $2^1 * 3^2 * 5^3 = 2250$. The number 7350 represents *aabb* because $7350 = 2^1 * 3^1 * 5^2 * 7^2$.

To show that this numbering satisfies the criteria given above, we use the *fundamental theorem of arithmetic*:

> *Any natural number ≥ 2 can be represented as a product of primes, and that product is, except for the order of the primes, unique.*

We may number all kinds of objects, not just alphabets. In general, the criteria for a numbering to be useful are:

1. No two objects have the same number.

2. Given any object, we can "effectively" find the number that corresponds to it.

3. Given any number, we can "effectively" find if it is assigned to an object and, if so, to which object.

In the following, we list the `Goedel` class as described above. It uses `Prime.h` implemented in Section 8.1.1:

```
// Goedel.h

#ifndef GOEDEL_H
#define GOEDEL_H

#include <assert.h>
#include <iostream.h>
#include "Prime.h"
#include "MString.h"
#include "Verylong.h"

class Goedel
{
  private:
    // Data Fields
    Verylong nvalue;
    String   wvalue;
    int      is_G;
  public:
    // Constructors
    Goedel();
    Goedel(String);
    Goedel(Verylong);
```

```
    // Member Functions
    Verylong number() const;
    String word() const;
    int is_goedel() const;
    void rename(String);
    void resize(Verylong);
};

Goedel::Goedel()
    : nvalue(Verylong("0")), wvalue(String("")), is_G(1) {}

Goedel::Goedel(String s)
    : nvalue(Verylong("1")), wvalue(s), is_G(1)
{
    static Prime p(100);
    Verylong temp, prim, zero("0"), one("1");
    p.run();

    for(int i=0; i<s.length(); i++)
    {
        if(s[i]>='a' && s[i]<='c')
        {
            temp = one;
            prim = Verylong(p(i));
            for(char j='a'; j<=s[i]; j++) temp *= prim;
            nvalue *= temp;
        }
        else
        {
            nvalue = zero;
            break;
        }
    }
}

Goedel::Goedel(Verylong num) : nvalue(num), wvalue(String(""))
{
    static Prime p(100);
    static Verylong zero("0"), one("1");
    Verylong prim;
    int i=0, factor;
    p.run();

    while(num>one)
```

```
    {
        factor = 0; prim = Verylong(p(i));
        while(num % prim == zero) { num /= prim; ++factor; }
        switch(factor)
        {
            case 1: wvalue = wvalue + String("a"); break;
            case 2: wvalue = wvalue + String("b"); break;
            case 3: wvalue = wvalue + String("c"); break;
            default : is_G = 0; return;
        }
        ++i;
    }
    is_G = 1;
}


Verylong Goedel::number() const  { return nvalue; }

String Goedel::word() const
{
    if(is_G) return wvalue;
    return String("");
}

int Goedel::is_goedel() const  { return is_G; }

void Goedel::rename(String s)
{
    static Prime p(100);
    static Verylong zero("0"), one("1");
    Verylong temp, prim;
    p.run();
    nvalue = one;
    wvalue = s;
    is_G = 1;

    for(int i=0; i<s.length(); i++)
    {
        if(s[i]>='a' && s[i]<='c')
        {
            temp = one;
            prim = Verylong(p(i));
            for(char j='a'; j<=s[i]; j++) temp *= prim;
            nvalue *= temp;
        }
```

```
        else
        {
            nvalue = zero;
            break;
        }
    }
}

void Goedel::resize(Verylong num)
{
    static Verylong zero("0"), one("1");
    assert(num >= zero);
    if(num == one) { is_G = 0; return; }

    static Prime p(100);
    int i=0, factor;
    Verylong prim;
    p.run();
    nvalue = num; wvalue = String("");

    while(num > one)
    {
        factor = 0; prim = Verylong(p(i));
        while(num % prim == zero) { num /= prim; ++factor; }
        switch(factor)
        {
            case 1: wvalue = wvalue + String("a"); break;
            case 2: wvalue = wvalue + String("b"); break;
            case 3: wvalue = wvalue + String("c"); break;
            default : is_G = 0; return;
        }
        ++i;
    }
    is_G = 1;
}
#endif
```

Next let us look at a simple program which calculates the corresponding Gödel number of acbc and aabb. It also lists all the Gödel numbers below 1500:

```cpp
// sgoedel1.cxx

#include <iostream.h>
#include "Goedel.h"

void main()
{
   Goedel g(String("acbc"));

   cout << "The word is " << g.word();
   cout << " and its corresponding Goedel number is "
        << g.number() << endl;
   cout << endl;

   g.rename("aabb");

   cout << "The word is " << g.word();
   cout << " and its corresponding Goedel number is "
        << g.number() << endl;
   cout << endl;

   g.rename("abcbc");

   cout << "The word is " << g.word();
   cout << " and its corresponding Goedel number is "
        << g.number() << endl;
   cout << endl;

   Goedel h;

   // List all the Goedel Numbers below 1500
   for (int i=0; i<1500; i++)
   {
      h.resize(i);
      if (h.is_goedel())
         cout << i << " => " << h.word()
              << " is a Goedel number" << endl;
   }
}
```

Result
======

The word is acbc and its corresponding Goedel number is 463050

The word is aabb and its corresponding Goedel number is 7350

The word is abcbc and its corresponding Goedel number is 146742750


```
0 =>  is a Goedel number
2 => a is a Goedel number
4 => b is a Goedel number
6 => aa is a Goedel number
8 => c is a Goedel number
12 => ba is a Goedel number
18 => ab is a Goedel number
24 => ca is a Goedel number
30 => aaa is a Goedel number
36 => bb is a Goedel number
54 => ac is a Goedel number
60 => baa is a Goedel number
72 => cb is a Goedel number
90 => aba is a Goedel number
108 => bc is a Goedel number
120 => caa is a Goedel number
150 => aab is a Goedel number
180 => bba is a Goedel number
210 => aaaa is a Goedel number
216 => cc is a Goedel number
270 => aca is a Goedel number
300 => bab is a Goedel number
360 => cba is a Goedel number
420 => baaa is a Goedel number
450 => abb is a Goedel number
540 => bca is a Goedel number
600 => cab is a Goedel number
630 => abaa is a Goedel number
750 => aac is a Goedel number
840 => caaa is a Goedel number
900 => bbb is a Goedel number
1050 => aaba is a Goedel number
1080 => cca is a Goedel number
1260 => bbaa is a Goedel number
1350 => acb is a Goedel number
1470 => aaab is a Goedel number
```

Let us consider another example. It calculates the Gödel numbers for the strings: a, aa, aaa, aaaa, ... Note that some of the Gödel numbers exceed the limit of the built-in unsigned long type.

```
// sgoedel2.cxx

#include <iostream.h>
#include "MString.h"
#include "Goedel.h"

void main()
{
    String s;
    Goedel g;
    for(int i=1; i<=20; i++)
    {
        s = s + String('a');
        g.rename(s);
        cout << i << "  " << g.word() << " => " << g.number() << endl;
    }
}
```

Result
======
```
1   a => 2
2   aa => 6
3   aaa => 30
4   aaaa => 210
5   aaaaa => 2310
6   aaaaaa => 30030
7   aaaaaaa => 510510
8   aaaaaaaa => 9699690
9   aaaaaaaaa => 223092870
10  aaaaaaaaaa => 6469693230
11  aaaaaaaaaaa => 200560490130
12  aaaaaaaaaaaa => 7420738134810
13  aaaaaaaaaaaaa => 304250263527210
14  aaaaaaaaaaaaaa => 13082761331670030
15  aaaaaaaaaaaaaaa => 614889782588491410
16  aaaaaaaaaaaaaaaa => 32589158477190044730
17  aaaaaaaaaaaaaaaaa => 1922760350154212639070
18  aaaaaaaaaaaaaaaaaa => 117288381359406970983270
19  aaaaaaaaaaaaaaaaaaa => 7858321551080267055879090
20  aaaaaaaaaaaaaaaaaaaa => 557940830126698960967415390
```

8.3 Verylong and Rational Classes

8.3.1 Logistic Map

The *logistic map* $f : [0,1] \to [0,1]$ is given by

$$f(x) = 4x(1-x).$$

It can also be written as the difference equation

$$x_{t+1} = 4x_t(1-x_t), \qquad t = 0,1,2,\ldots, \quad x_0 \in [0,1]$$

where $x_t \in [0,1]$ for all $t \in \mathbf{N} \cup \{0\}$. Let $x_0 = 1/3$ be the initial value. Then

$$x_1 = \frac{8}{9}, \quad x_2 = \frac{32}{81}, \quad x_3 = \frac{6272}{6561}, \quad x_4 = \frac{7250432}{43046721}, \quad \cdots$$

This is a so-called *chaotic orbit* [51]. The exact solution of the logistic map is given by

$$x_t = \frac{1}{2} - \frac{1}{2}\cos(2^t \arccos(1-2x_0)).$$

Depending on the data type of x_0, we may get an approximate or exact orbit of the map. For an approximate orbit, we use the built-in data type double for x_0, whereas for the exact orbit of the logistic map, we use the Rational class that we have developed in Chapter 6.

We demonstrate the features of the Rational class as well as the coupling with the Verylong class. We compare Rational<long> and Rational<Verylong> in the iteration of the map.

From the result obtained we notice that the lengths of the numerator and denominator increase so fast that Rational<long> could only hold the result up to the fourth iteration. We conclude that the data type Rational<Verylong> is very useful for a calculation like this.

Quite often long fractions do not give a good idea of how large the value is; the floating point representation may be better. The conversion operator to double in the class provides such a function.

```
// logistic.cxx

#include <iostream.h>
#include "Rational.h"
#include "Verylong.h"

void main()
{
   Rational<long> a(4,1), b(1,1), x0(1,3); // initial value x0 = 1/3
   int i;
   cout << "x[0] = " << x0 << " or " << double(x0) << endl;

   for(i=1; i<=4; i++)      // cannot use higher values than 4
   {                        // out of range for data type int
   x0 = a*x0*(b - x0);
   cout << "x[" << i << "] = " << x0 << " or " << double(x0) << endl;
   }
   cout << endl;

   Rational<Verylong> c1("1","1"), c2("4","1"),
                      y0("1","3");          // initial value y0 = 1/3
   cout << "y[0] = " << y0 << " or " << double(y0) << endl;

   for(i=1; i<=6; i++)
   {
   y0 = c2*y0*(c1 - y0);
   cout << "y[" << i << "] = " << y0 << " or " << double(y0) << endl;
   }
}
```

Result
======
x[0] = 1/3 or 0.333333
x[1] = 8/9 or 0.888889
x[2] = 32/81 or 0.395062
x[3] = 6272/6561 or 0.955952
x[4] = 7250432/43046721 or 0.168432
y[0] = 1/3 or 0.333333
y[1] = 8/9 or 0.888889
y[2] = 32/81 or 0.395062
y[3] = 6272/6561 or 0.955952
y[4] = 7250432/43046721 or 0.168432
y[5] = 1038154236987392/1853020188851841 or 0.56025
y[6] = 3383826162019367796397224108032/3433683820292512484657849089281
       or 0.98548

8.3.2 Contracting Mapping Theorem

Let S be a closed set on a complete metric space X. A *contracting mapping* is a mapping $f : S \to S$ such that

$$d(f(x), f(y)) \leq kd(x, y), \quad 0 \leq k < 1, \quad d \text{ is the distance in } X.$$

One also says that "f is *lipschitzian* of order $k < 1$".

Contracting mapping theorem. A contracting mapping f has strictly one *fixed point*; i.e. there is one and only one point x^* such that

$$x^* = f(x^*).$$

The proof is by successive iteration. Let $x^* \in S$, then

$$f(x_0) \in S, \ldots, f^{(n)}(x_0) = f(f^{(n-1)}(x_0)) \in S$$

and

$$d(f^{(n)}(x_0), f^{(n-1)}(x_0)) \leq kd(f^{(n-1)}(x_0), f^{(n-2)}(x_0)) \leq \ldots \leq k^{n-1}d(f(x_0), x_0).$$

Since $k < 1$ the sequence $f^{(n)}(x_0)$ is a *Cauchy sequence* and it tends to a limit $x^* \in S$ when n tends to infinity

$$x^* = \lim_{n \to \infty} f^{(n)}(x_0) = \lim_{n \to \infty} f(f^{(n-1)}(x_0)) = f(x^*).$$

The uniqueness of x^* results from the defining property of contracting mappings. Assume that there is another point y^* such that $y^* = f(y^*)$, then

$$d(f(y^*), f(x^*)) = d(y^*, x^*)$$

On the other hand $d(f(y^*), f(x^*)) \leq kd(y^*, x^*)$, where $k < 1$. Hence, $d(y^*, x^*) = 0$ and $y^* = x^*$. $\square$

In the following program, we consider the map $f : (0, 1] \to (0, 1]$

$$f(x) = \frac{5}{2}x(1 - x).$$

Obviously the map f has a stable fixed point $x^* = 3/5$. We apply the contracting mapping theorem with the initial value $x_0 = 9/10$.

A brief description of the program is given below:

- The constructor of the program reads in two arguments. The first argument is the function name of the mapping while the second argument is the initial value.

- The increment operator (++) applies the mapping on the current value to obtain the next iteration value.

- The function Is_FP() checks if the values have converged to a fixed point by considering the relative difference between the previous and current values. If the relative difference is less than 10^{-5}, we claim that a fixed point has been obtained.

From the result obtained, we notice that the numerical value of the mapping converges to 0.6, which agrees with our discussion. However, the rational values (exact values) of the iteration grow very quickly.

```
// contract.cxx

#include <iostream.h>
#include "Rational.h"
#include "Verylong.h"

class Map
{
private:
   Rational<Verylong> (*function)(Rational<Verylong>);
   Rational<Verylong> value;          // current iterated value
public:
   Map(Rational<Verylong> (*f)(Rational<Verylong>),Rational<Verylong>);
   void operator ++ ();               // next iteration
   void Is_FP();                      // Is there a fixed point?
};

Map::Map(Rational<Verylong> (*f)(Rational<Verylong>),
         Rational<Verylong> x0) : function(f), value(x0) {}

void Map::operator ++ ()
{  value = (*function)(value); }

// Is there a fixed point?
void Map::Is_FP()
{
    Rational<Verylong> temp,   // the value in previous step
                       dist;   // the relative difference
```

```
                         // between previous and current step
   do
   {
   temp = value;
   ++(*this);
   dist = abs((value-temp)/temp); // the relative difference
   cout << "Value : " << value << " or " << double(value) << endl;
   cout << endl;
   } while(double(dist) > 1e-5);
}

// f(x) = 5/2 x(1-x)
Rational<Verylong> mapping(Rational<Verylong> x)
{
return(Rational<Verylong>("5","2")*x*(Rational<Verylong>("1")-x));
}

void main()
{
// initial value = 9/10
Map M(mapping, Rational<Verylong>("9","10"));
M.Is_FP();
}
```

Result
======

Value: 9/40 or 0.225

Value: 279/640 or 0.435938

Value: 100719/163840 or 0.61474

Value: 6357483999/10737418240 or 0.592087

Value: 27845361853829709759/46116860184273879040 or 0.6038

Value: 50877648262286330031975357175093713279/
       8507059173023461586584365185794205286640 or 0.598064

Value: 17396565550814513824057897676647375331886675751088414
       9486312212745209801068787197/
       28948022309329048855892746252171976963317496166410141010
       9864396001978282409984040 or 0.600959

....

8.3.3   Ghost Solutions

Consider the ordinary differential equation

$$\frac{du}{dt} = u(1-u)$$

with initial condition $u(0) = u_0 > 0$. The fixed points are given by

$$u^* = 0, \quad u^* = 1.$$

The fixed point $u^* = 1$ is asymptotically stable. The exact solution of the differential equation is given by

$$u(t) = \frac{u_0 e^t}{1 - u_0 + u_0 e^t} \qquad \text{for } 0 < u_0 < 1.$$

The exact solution starting from initial value $u_0 = 99/100 = 0.99$ is monotonously increasing and it converges to 1 as t tends to ∞. For $t = \ln 99 \approx 4.5951$, we find that

$$u(t) = 0.9999 .$$

So, $u(t)$ is already quite near the asymptotically stable fixed point $u^* = 1$. In order to integrate this equation by a finite difference scheme, we apply the *central difference scheme*:

$$\frac{u_{n+1} - u_{n-1}}{2h} = u_n(1 - u_n) \qquad n = 1, 2, 3, \ldots \tag{8.4}$$

with initial conditions $u_0 = x_0$, $u_1 = x_0 + hx_0(1 - x_0)$. This second order difference equation can be rewritten as a system of two first order difference equations:

$$
\begin{aligned}
v_{n+1} &= u_n \\
u_{n+1} &= v_n + 2hu_n(1 - u_n) \qquad \text{for } n = 1, 2, 3, \ldots
\end{aligned}
$$

with $v_1 = u_0 = x_0$ and $u_1 = x_0 + hx_0(1 - x_0)$. We iterate the system of nonlinear difference equations using the `Rational` and `Verylong` class with initial value $x_0 = 99/100$ and time-mesh length $h = 0.1$. This system does not converge to the fixed point $u^* = 1$ and we find oscillating behaviour. Such a phenomenon is called the *ghost solution* (also called *spurious solution*).

```
// ghost.cpp

#include <fstream.h>
#include <math.h>
#include "Rational.h"
#include "Verylong.h"
const Rational<Verylong> h("1","10");     // time-mesh length
const Rational<Verylong> x0("99","100"); // initial value

void main()
{
   Rational<Verylong> u, v, u1, twoh, t,
                     zero("0"), one("1"), two("2"), fifty("50");
   ofstream sout("ghosts.dat"); // contains rational number values
   // initial values
   u = x0; v = u + h*u*(one-u);
   t = zero; twoh = two*h;

   while(t <= fifty)
   {
      u1 = u; u = v;
      v = u1 + twoh*u*(one-u);
      t += h;
      sout << t << " " << v << endl;
   }
}
```

For $0 < t < 1.4$, the numerical solution gives a good approximation to the true solution. The solution u_n increases monotonously approaching 1. After $t = 1.5$, the numerical solution is no longer monotonous. At $t = 3.0$, the value u_n becomes greater than 1 for the first time and the solution begins to oscillate thereafter. The amplitude of the oscillation grows larger and larger. The growth of this amplitude is geometric and the rate of growth is such that the amplitude is multiplied by about $e \approx 2.71828$ for each unit t increment. When $t = 10$, the oscillation loses its symmetry with respect to $u^* = 1$. The repetition of such cycles seems to be nearly periodic. The ghost solutions also appear even if h is quite small. One of the reasons for this phenomenon is that the central difference scheme is a second order difference scheme and that the instability enters at $u^* = 1$ and $u^* = 0$. The global behaviour of the solution computed is very sensitive to the initial condition and the time-mesh length. To show that the behaviour of (8.4) is not caused by the finite precision used in a digital computer we iterate the equations using exact arithmetic (ghost.cxx) to avoid rounding-off error. The Rational and Verylong class have been employed. From the result obtained, we see that the oscillating behaviour exists even for the exact arithmetics. The behaviour is an inherited property of the difference equation.

8.3.4   Iterated Function Systems

A *hyperbolic iterated function* [5] system consists of a complete metric space $(\mathbf{X}, d)$ together with a finite set of contraction mappings $w_n : \mathbf{X} \to \mathbf{X}$, with *contractivity factors* s_n, for $n = 1, 2, \ldots, N$. The notation for the iterated function system is $\{\mathbf{X}; w_n, n = 1, 2, \ldots, N\}$ and its contractivity factor is

$$s = \max\{s_n : n = 1, 2, \ldots, N\}.$$

Let $(\mathbf{X}, d)$ be a complete metric space and $(\mathcal{H}(\mathbf{X}), h(d))$ denotes the corresponding space of non-empty compact subsets, with the Hausdorff metric $h(d)$. The following theorem summarizes the facts about a hyperbolic iterated function system:

Theorem. Let
$$\{\mathbf{X}; w_n, n = 1, 2, \ldots, N\}$$

be a hyperbolic iterated function system with contractivity factor s, then the transformation $W : \mathcal{H}(\mathbf{X}) \to \mathcal{H}(\mathbf{X})$ defined by

$$W(B) := \bigcup_{n=1}^{N} w_n(B)$$

for all $B \in \mathcal{H}(\mathbf{X})$, is a contraction mapping on the complete metric space $(\mathcal{H}(\mathbf{X}), h(d))$ with contractivity factor s, i.e.

$$h(W(B), W(C)) \leq s \cdot h(B, C)$$

for all $B, C \in \mathcal{H}(\mathbf{X})$. Its unique fixed point, $A \in \mathcal{H}(\mathbf{X})$, obeys

$$A := W(A) = \bigcup_{n=1}^{N} w_n(A)$$

and is given by
$$A := \lim_{n \to \infty} W^{(n)}(B)$$

for any $B \in \mathcal{H}(\mathbf{X})$. The fixed point $A \in \mathcal{H}(\mathbf{X})$ is called the attractor of the iterated function system. For the proof we refer to the excellent book of Barnsley [5].

Next, we present an interesting example in fractals – the standard *Cantor set*.

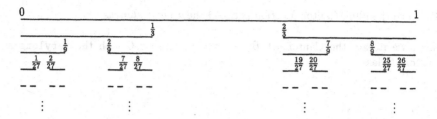

Figure 8.1: *Construction of the Cantor set*

The standard Cantor set is described by the following iterated function system

$$\{\,\mathbf{R}:w_1,w_2\,\} \qquad \text{where} \quad w_1(x) := \frac{1}{3}x \quad \text{and} \quad w_2(x) := \frac{1}{3}x + \frac{2}{3}\,.$$

This is an iterated function system with contractivity factor $s = 1/3$. Suppose

$$B_0 = [0,1] \quad \text{and} \quad B_n = W^{(n)}(B_0)$$

Then $B = \lim_{n \to \infty} B_n$ is the standard *Cantor set*. It is also called the Cantor middle third set or ternary Cantor set. The sets $B_0, B_1, B_2, \ldots$ are given by

$$B_0 \;=\; [0,1]$$

$$B_1 \;=\; \left[0,\frac{1}{3}\right] \cup \left[\frac{2}{3},1\right]$$

$$B_2 \;=\; \left[0,\frac{1}{9}\right] \cup \left[\frac{2}{9},\frac{1}{3}\right] \cup \left[\frac{2}{3},\frac{7}{9}\right] \cup \left[\frac{8}{9},1\right]$$

$$\vdots \qquad \vdots$$

This means we remove the open middle third, i.e. the interval $(\frac{1}{3},\frac{2}{3})$ for the first step and remove the pair of intervals $(\frac{1}{9},\frac{2}{9})$ and $(\frac{7}{9},\frac{8}{9})$ in the second step. Continuing to remove the middle thirds in this fashion, we arrive at the Cantor set as $n \to \infty$ (Figure 8.1).

The standard Cantor set has cardinal c, is perfect, nowhere dense and has Lebesgue measure zero. Every $x \in B$ can be written as

$$x = a_1 3^{-1} + a_2 3^{-2} + a_3 3^{-3} + \ldots \qquad \text{where} \quad a_j \in \{0,2\}.$$

The corresponding Cantor function is called the *Devil's staircase*.

Next, we iterate the Cantor set B_n for $n = 1, 2, 3, 4, 5, 6$ with the `Verylong` and Rational class:

```
// Cantor.cxx

#include <iostream.h>
#include "Rational.h"
#include "Verylong.h"
#include "Vector.h"

const Rational<Verylong> a = Rational<Verylong>(0); // lower limit
const Rational<Verylong> b = Rational<Verylong>(1); // upper limit

class Cantor
{
private:
   Vector<Rational<Verylong> > CS;
   int currentSize;
public:
   Cantor(int);
   Cantor(const Cantor&);
   int step();
   void run();
   friend ostream& operator << (ostream&,const Cantor&);
};

Cantor::Cantor(int numStep)
   : CS(power(2,numStep+1)), currentSize(2)
{  CS[0] = a; CS[1] = b; }

Cantor::Cantor(const Cantor& s)
   : CS(s.CS), currentSize(s.currentSize) {}

int Cantor::step()
{
   int i, newSize;
   static Rational<Verylong> three(3), tt(2,3);
   static int maxSize = CS.length();

   if(currentSize < maxSize)
   {
```

```
        for(i=0; i<currentSize; i++) CS[i] /= three;
        newSize = currentSize + currentSize;
        for(i=currentSize; i<newSize; i++)
            CS[i] = CS[i-currentSize] + tt;
        currentSize = newSize;
        return 1;
    }
    return 0;
}

void Cantor::run()  {  while (step()) ; }

ostream& operator << (ostream& s,const Cantor& c)
{
    for(int i=0; i<c.currentSize; i+=2)
    {
        s << "[" << c.CS[i] << " ";
        s << c.CS[i+1] << "] ";
    }
    return s;
}

void main()
{
    const int N = 6;
    Cantor C(N);
    cout << C << endl;

    for(int i=0; i<N; i++)
    {
        C.step();
        cout << C << endl;
    }
}
```

Result
======
[0 1]
[0 1/3] [2/3 1]
[0 1/9] [2/9 1/3] [2/3 7/9] [8/9 1]
. . . .

8.4   Verylong, Rational and Derive Classes

8.4.1   Logistic Map and Ljapunov Exponent

In this section, we calculate the *Ljapunov exponent* of the *logistic map* [51] which is given by

$$x_{t+1} = 4x_t(1 - x_t)$$

where $t = 0, 1, 2, \ldots$ and $x_0 \in [0, 1]$. The *variational equation* of the logistic map (also called the *linearized equation*) is defined by

$$y_{t+1} = \frac{df(x_t)}{dx} y_t$$

where

$$f(x) = 4x(1 - x).$$

Since

$$\frac{df}{dx} = 4 - 8x$$

it follows that

$$y_{t+1} = (4 - 8x_t)y_t$$

with $y_0 \neq 0$. The *Ljapunov exponent* λ is defined as

$$\lambda(x_0, y_0) := \lim_{T \to \infty} \frac{1}{T} \ln \left| \frac{y_T}{y_0} \right|, \qquad y_0 \neq 0.$$

For almost all initial values we find that the Ljapunov exponent is given by

$$\lambda = \ln 2.$$

We iterate the logistic map and the variational equation to find the Ljapunov exponent. The following program makes use of the Verylong, Rational and Derive classes to approximate the Ljapunov exponent. The Derive class will do the differentiation. Thus the variational equation is obtained via exact differentiation.

```
// Ljapunov.cpp

#include <iostream.h>
#include <math.h>
#include "Verylong.h"
#include "Rational.h"
#include "Derive.h"

void main()
{
   int N = 100;
   double x1, x = 1.0/3.0, y = 1.0;
   Derive<double> C1(1.0);  // constant 1.0
   Derive<double> C4(4.0);  // constant 4.0
   Derive<double> X;

   cout << "i = 0   x = " << x << "    " << "y = " << y << endl;
   for(int i=1; i<=N; i++)
   {
      x1 = x;
      x = 4.0*x1*(1.0 - x1);
      X.set(x1);
      Derive<double> Y = C4*X*(C1 - X);
      y = df(Y)*y;
      cout << "i = " << i << "    "
           << "x = " << x << "    " << "y = " << y << endl;
   }
   double lam = log(fabs(y))/N;
   cout << "Approximation value for lambda = " << lam << endl;
   cout << endl;

   int M = 9;
   Rational<Verylong> u1;
   Rational<Verylong> u("1","3"), v("1");
   Rational<Verylong> K1("1");
   Rational<Verylong> K2("4");
   Derive<Rational<Verylong> > D1(K1); // constant 1
   Derive<Rational<Verylong> > D4(K2); // constant 4
   Derive<Rational<Verylong> > U;

   cout << "j = 0   u = " << u << "    " << "v = " << v << endl;
   for(int j=1; j<=M; j++)
   {
      u1 = u;
```

```
        u = K2*u1*(K1 - u1);
        U.set(Rational<Verylong>(u1));
        Derive<Rational<Verylong> > V = D4*U*(D1 - U);
        v = df(V)*v;
        cout << "j = " << j << "   "
             << "u = " << u << "   " << "v = " << v << endl;
    }
    lam = log(fabs(double(v)))/M;
    cout << "Approximation value for lambda = " << lam << endl;


}
Result
======
i = 0    x = 0.333333    y = 1
i = 1    x = 0.888889    y = 1.33333
i = 2    x = 0.395062    y = -4.14815
i = 3    x = 0.955952    y = -3.4824
i = 4    x = 0.168432    y = 12.7024
              . . .
i = 96   x = 0.298676    y = -7.69211e+28
i = 97   x = 0.837875    y = -1.23888e+29
i = 98   x = 0.543363    y = 3.3487e+29
i = 99   x = 0.992479    y = -1.16167e+29
i = 100   x = 0.0298588   y = 4.57677e+29
Approximation value for lambda = 0.68296


j = 0    u = 1/3    v = 1
j = 1    u = 8/9    v = 4/3
j = 2    u = 32/81     v = -112/27
j = 3    u = 6272/6561    v = -7616/2187
j = 4    u = 7250432/43046721    v = 182266112/14348907
              . . .
Approximation value for lambda = 0.69225
```

8.5 Verylong, Rational and Complex Classes

8.5.1 Mandelbrot Set

Suppose $\mathbf{C}$ is the complex plane, then the *Mandelbrot set* M is defined by

$$M := \left\{ c \in \mathbf{C} \; : \; c, \; c^2 + c, \; (c^2 + c)^2 + c, \; \dots \not\to \infty \right\}.$$

To find the Mandelbrot set we study the recursion relation

$$z_{t+1} = z_t^2 + c, \qquad t = 0, 1, 2, \dots$$

with initial value $z_0 = 0$. Since $z := x + iy$ and $c := c_1 + ic_2$ with x, y, c_1, $c_2 \in \mathbf{R}$, we can write the recursion relation as

$$\begin{aligned}
x_{t+1} &= x_t^2 - y_t^2 + c_1 \\
y_{t+1} &= 2x_t y_t + c_2
\end{aligned}$$

with the initial value $(x_0, y_0) = (0, 0)$. Let us consider some points in the complex plane $\mathbf{C}$ and determine whether they belong to the Mandelbrot set M:

- $(c_1, c_2) = (0, 0)$ belongs to M

- $(c_1, c_2) = (\frac{1}{4}, \frac{1}{4})$ belongs to M

- $(c_1, c_2) = (\frac{1}{2}, 0)$ does not belong to M, because

$$z_0 = 0, \; z_1 = \tfrac{1}{2}, \; z_2 = \tfrac{3}{4}, \; z_3 = \tfrac{17}{16}, \; \dots \to \infty$$

We iterate the recursion relation with the starting point

$$z = \frac{1}{8} + i\frac{1}{3}.$$

Here, we have coupled three classes: the `Verylong`, `Rational` and `Complex` classes. We wrap up this section by comparing the virtues of the two data types:

`Complex<Rational<long> >` and `Complex<Rational<Verylong> >`.

It is obvious that `Rational<long>` has limited application since it is restricted by the precision of the built-in data type `long`. However, `Rational<Verylong>` removes this restriction and it can be applied to problems that require multi-precision.

```cpp
// Mandel.cpp

#include <iostream.h>
#include "MComplex.h"
#include "Rational.h"
#include "Verylong.h"

void main()
{
  Complex<Rational<long> >
     c(Rational<long>(1,8),Rational<long>(1,3)),
     z0(Rational<long>(0,1),Rational<long>(0,1));

  cout << "Using data type Rational<long>" << endl;
  for(int i=1; i<=3; i++)
  {
     z0 = z0*z0 + c;
     cout << "z[" << i << "] = " << z0 << endl;
  }
  cout << endl;
  Complex<Rational<Verylong> >
     d(Rational<Verylong>(1,8),Rational<Verylong>(1,3)),
     w0(Rational<Verylong>(0,1),Rational<Verylong>(0,1));

   cout << "Using data type Rational<Verylong>" << endl;
   for(int j=1; j<=5; j++)
   {
     w0 = w0*w0 + d;
     cout << "w[" << j << "] = " << w0 << endl;
   }
}
```

```
Result
======
Using data type Rational<long>
z[1] = (1/8,1/3i)
z[2] = (17/576,5/12i)
z[3] = (-15839/331776,1237/3456i)
Using data type Rational<Verylong>
w[1] = (1/8,1/3i)
w[2] = (17/576,5/12i)
w[3] = (-15839/331776,1237/3456i)
w[4] = (-91749311/110075314176,171510133/5733089281)
w[5] = (43019895380028096729/1211657479094510655897,1050212412505202552525/
31553580184752881664i)
```

8.6 Symbolic Class

8.6.1 Polynomials

Legendre Polynomials

The *Legendre differential equation* is given by

$$\left((1 - x^2)\frac{d^2}{dx^2} - 2x\frac{d}{dx} + n(n+1) \right) P_n(x) = 0 \qquad \text{where } n \text{ is a constant.}$$

This equation arises in the solution of problems in mechanics, quantum mechanics, etc. The solution to this differential equation is called the *Legendre polynomials* which can be represented by *Rodrigue's formula*

$$P_0(x) := 1, \qquad P_n(x) := \frac{1}{2^n n!}\frac{d^n}{dx^n}(x^2 - 1)^n, \qquad n = 1, 2, \ldots$$

The above equation can be rewritten as

$$P_n(x) = \frac{2^n}{n!}\frac{d^n}{dx^n}\left(\left(\frac{x-1}{2}\right)^n \left(\frac{x+1}{2}\right)^n \right).$$

Now, if we apply the well-known *Leibniz rule* for differentiating products, we get

$$P_n(x) = \sum_{k=0}^{n} \binom{n}{k}^2 \left(\frac{x-1}{2}\right)^{n-k} \left(\frac{x+1}{2}\right)^{k}.$$

The Legendre polynomials form a set of orthonormal functions on $(-1, 1)$, that is

$$\int_{-1}^{+1} P_m(x)P_n(x)dx = \frac{2}{2n+1}\delta_{mn}.$$

Furthermore, we have the recursion relation

$$(n+1)P_{n+1}(x) = (2n+1)xP_n(x) - nP_{n-1}(x), \qquad n = 1, 2, \ldots$$

with $P_0(x) = 1$ and $P_1(x) = x$. The *generating function* for the Legendre polynomials is given by

$$\frac{1}{\sqrt{1 - 2tx + t^2}} = \sum_{n=0}^{\infty} P_n(x)t^n.$$

In the following program, we make use of the recursion relation to generate the first few Legendre polynomials

$$P_0(x) = 1, \qquad P_1(x) = x, \qquad P_2(x) = \frac{1}{2}(3x^2 - 1),$$

$$P_3(x) = \frac{1}{2}(5x^3 - 3x), \qquad P_4(x) = \frac{1}{8}(35x^4 - 30x^2 + 3).$$

We also show that the Legendre differential equation holds for $n = 4$.

```
// Legendre.h

#include <iostream.h>
#include <assert.h>
#include "MSymbol.h"
#include "Rational.h"

class Legendre
{
private:
    int maxTerm, currentStep;
    Sum<Rational<int> > P, Q;
    const Sum<Rational<int> >& x;

public:
    Legendre(int,const Sum<Rational<int> >&);

    int step();
    void run();
    void reset();
    Sum<Rational<int> > current() const;
    Sum<Rational<int> > operator () (int);

    friend ostream& operator << (ostream&,const Legendre&);
```

```
};

Legendre::Legendre(int n, const Sum<Rational<int> > &kernal)
    : maxTerm(n), currentStep(0), P(1), x(kernal) {}

int Legendre::step()
{
    int prev = currentStep;
    Sum<Rational<int> > R, one(1);
    ++currentStep;

    if(currentStep==1) { Q = one; P = x; return 1; }

    if(currentStep <= maxTerm)
    {
        R = Rational<int>(prev+currentStep,currentStep)*x*P
            - Rational<int>(prev,currentStep)*Q;
        Q = P; P = R;

        return 1;
    }
    return 0;
}

void Legendre::run()
{ while (step()) ; }

void Legendre::reset()
{
    currentStep = 0;
    P = Rational<int>(1);
}

Sum<Rational<int> > Legendre::current() const
{ return P; }

Sum<Rational<int> > Legendre::operator () (int m)
{
    assert(m <= maxTerm);

    reset();
    for(int i=0; i<m; i++) step();

    return P;
```

```
}

ostream& operator << (ostream& s,const Legendre& L)
{ return s << L.P; }

// Legendre.cxx

#include <iostream.h>
#include "MSymbol.h"
#include "Rational.h"
#include "Legendre.h"

void main()
{
   int n=4;
   Sum<Rational<int> > x("x",0), one(1);
   Legendre P(n,x);

   // Calculate the first few Legendre polynomials
   cout << "P(0) = " << P << endl;
   for(int i=1; i<=n; i++)
   {
      P.step();
      cout << "P("<< i << ") = " << P << endl;
   }
   cout << endl;

   // Another way to access the Legendre polynomial
   cout << "P(1) = " << P(1) << endl;
   cout << "P(2) = " << P(2) << endl;
   cout << "P(3) = " << P(3) << endl;
   cout << "P(4) = " << P(4) << endl; cout << endl;

   // To show that for n=4,
   //    the Legendre differential equation is satisfied
   Sum<Rational<int> > result;
   result = df((one-x*x)*df(P.current(),x),x)
            + Sum<Rational<int> >(n*(n+1))*P.current();

   cout << result << endl;   // ==> 0
}
```

Associated Legendre Functions

Having solved the Legendre differential equation, we now obtain the solution of

$$\left((1-x^2)\frac{d^2}{dx^2} - 2x\frac{d}{dx} + n(n+1) - \frac{m^2}{1-x^2}\right) P_n^m(x) = 0$$

where m, n are constants and m is not necessarily equal to zero. The solution to this equation $P_n^{|m|}(x)$ is called the *associated Legendre functions* of degree n ($n = 0, 1, 2, \ldots$) and order $|m| \le n$. It is defined by the relation

$$P_n^{|m|}(x) := (1-x^2)^{|m|/2}\frac{d^{|m|}}{dx^{|m|}}P_n(x), \quad |m| = 0, 1, 2, \ldots, \le n. \qquad (8.5)$$

Note that for $m = 0$, we have $P_n^0(x) = P_n(x)$ which are the Legendre polynomials. The functions $P_n^{|m|}$ satisfy the recurrence relation

$$(2n+1)xP_n^{|m|}(x) = (n - |m| + 1)P_{n+1}^{|m|}(x) + (n + |m|)P_{n-1}^{|m|}(x)$$

and the orthogonality relation

$$\int_{-1}^{+1} P_n^{|m|}(x)P_{n'}^{|m|}(x)dx = \frac{2}{2n+1}\frac{(n+|m|)!}{(n-|m|)!}\delta_{nn'}.$$

Finally, it can be shown that the $P_n^{|m|}$ form a complete set in the Hilbert space $L_2(-1, 1)$. The first few associated Legendre functions are given by

$$P_1^1(x) = (1-x^2)^{1/2}, \quad P_2^1(x) = 3x(1-x^2)^{1/2}, \quad P_2^2(x) = 3(1-x^2),$$

$$P_3^1(x) = \frac{3}{2}(1-x^2)^{1/2}(5x^2 - 1), \quad P_3^2(x) = 15x(1-x^2), \quad P_3^3(x) = 15(1-x^2)^{3/2}.$$

In the following program, we make use of (8.5) and the Legendre polynomials developed in the previous section to construct the associated Legendre polynomials.

```
// AssLegendre.h
#include <iostream.h>
#include "MSymbol.h"
#include "Rational.h"
#include "Legendre.h"

class AssLegendre
{
private:
   Sum<Rational<int> > P;
   const Sum<Rational<int> >& x;
public:
   AssLegendre(const Sum<Rational<int> >&);
   AssLegendre(int,int,const Sum<Rational<int> >&);
   void redefine(int,int);
   Sum<Rational<int> > current() const;
   friend ostream& operator << (ostream&,const AssLegendre&);
};

AssLegendre::AssLegendre(const Sum<Rational<int> >& kernal)
   : x(kernal) {}

AssLegendre::AssLegendre(int l,int m,const Sum<Rational<int> > &kernal)
   : x(kernal)
{
   int i, absm = abs(m);
   Rational<int> one(1);
   Legendre L(l,x);
   P = L(l);
   for(i=0; i<absm; i++) P = df(P,x);
   P *= power(one-x*x,Rational<int>(absm,2));
}

void AssLegendre::redefine(int l,int m)
{
   int i, absm = abs(m);
   Rational<int> one(1);
   Legendre L(l,x);
   P = L(l);
   for(i=0; i<absm; i++) P = df(P,x);
   P *= power(one-x*x,Rational<int>(absm,2));
}

Sum<Rational<int> > AssLegendre::current() const {   return P; }
```

```
    ostream& operator << (ostream& s,const AssLegendre& L)
    {   return s << L.P; }

    // AssLegendre.cxx

    #include <iostream.h>
    #include "MSymbol.h"
    #include "Rational.h"
    #include "AssLegendre.h"

    void main()
    {
        int i, j, n=5;
        Sum<Rational<int> > x("x",0), Y;
        AssLegendre P(x);
        for(i=0; i<n; i++)
        {
            for(j=0; j<=i; j++)
            {
                P.redefine(i,j);
                Y = P.current();
                cout << "P(" << i << "," << j << ") = " << Y << endl;
            }
            cout << endl;
        }
    }
Result
======
P(0,0) = 1
P(1,0) = x
P(1,1) = (1-x^(2))^(1/2)
P(2,0) = 3/2*x^(2)-1/2
P(2,1) = 3*x*(1-x^(2))^(1/2)
P(2,2) = 3-3*x^(2)
P(3,0) = 5/2*x^(3)-3/2*x
P(3,1) = 15/2*x^(2)*(1-x^(2))^(1/2)-3/2*(1-x^(2))^(1/2)
P(3,2) = 15*x-15*x^(3)
P(3,3) = 15*(1-x^(2))^(3/2)
P(4,0) = 35/8*x^(4)-15/4*x^(2)+3/8
P(4,1) = 35/2*x^(3)*(1-x^(2))^(1/2)-15/2*x*(1-x^(2))^(1/2)
P(4,2) = 60*x^(2)-105/2*x^(4)-15/2
P(4,3) = 105*x*(1-x^(2))^(3/2)
P(4,4) = 105-210*x^(2)+105*x^(4)
```

Laguerre Polynomials

The *Laguerre polynomials* may be defined by *Rodrigue's formula*:

$$L_n(x) := e^x \frac{d^n}{dx^n}(x^n e^{-x}), \qquad n = 0, 1, \ldots$$

where $L_0(x) = 1$. Using the *Leibniz rule*, we obtain

$$L_n(x) = n! \sum_{m=0}^{n} (-1)^m \left(\begin{array}{c} n \\ m \end{array} \right) \frac{x^m}{m!}.$$

The Laguerre polynomials L_n are the solutions of the linear second order differential equation

$$x \frac{d^2 L_n}{dx^2} + (1 - x) \frac{dL_n}{dx} + n L_n = 0. \tag{8.6}$$

They obey the recursion relation

$$L_{n+1}(x) = (2n + 1 - x) L_n(x) - n^2 L_{n-1}(x)$$

where $L_0(x) = 1$ and $L_1(x) = 1 - x$. The Laguerre polynomials can also be defined by the generating function

$$\frac{1}{1 - t} \exp \left(\frac{-xt}{1 - t} \right) = \sum_{n=0}^{\infty} \frac{L_n(x) t^n}{n!}.$$

The orthogonality relation of the Laguerre polynomials is given by

$$\int_0^{\infty} e^{-x} L_m(x) L_n(x) dx = (n!)^2 \delta_{mn}.$$

Furthermore we have

$$\int_0^x L_n(t) dt = L_n(x) - \frac{L_{n+1}(x)}{n + 1}.$$

In the following program, we make use of the recursion relation to generate the first few Laguerre polynomials. We also show that (8.6) holds for $n = 4$. The first few Laguerre polynomials are given by

$$L_0(x) = 1, \quad L_1(x) = 1 - x, \quad L_2(x) = 2 - 4x + x^2,$$

$$L_3(x) = 6 - 18x + 9x^2 - x^3, \quad L_4(x) = 24 - 96x + 72x^2 - 16x^3 + x^4.$$

```
// Laguerre.cxx

#include <iostream.h>
#include "MSymbol.h"

class Laguerre
{
private:
   int maxTerm, currentStep;
   Sum<int> P, Q;
   const Sum<int> &x;
public:
   Laguerre(int,const Sum<int>&);
   int step();
   void run();
   Sum<int> current() const;
   friend ostream& operator << (ostream&,const Laguerre&);
};

Laguerre::Laguerre(int n,const Sum<int> &kernal)
   : maxTerm(n), currentStep(0), P(1), x(kernal) {}

int Laguerre::step()
{
   int prev = currentStep;
   Sum<int> R;
   ++currentStep;
   if(currentStep==1) { Q = 1; P = 1-x; return 1; }
   if(currentStep <= maxTerm)
   {
      R = (2*prev+1-x)*P-prev*prev*Q;
      Q = P; P = R;
      return 1;
   }
   return 0;
```

```
    }

    void Laguerre::run()  {  while (step()) ; }

    Sum<int> Laguerre::current() const  {  return P; }

    ostream & operator << (ostream& s,const Laguerre& L)
    {  return s << L.P; }

    void main()
    {
        int n=4;
        Sum<int> x("x",0);
        Laguerre L(n,x);

        // Calculate the first few Laguerre polynomials
        cout << "L(0) = " << L << endl;
        for (int i=1; i<=n; i++)
        {
            L.step();
            cout << "L("<< i << ") = " << L << endl;
        }
        cout << endl;

        // To show that for n=4,
        // the Laguerre differential equation is satisfied.
        Sum<int> result;
        result = x*df(L.current(),x,2) + (1-x)*df(L.current(),x)
                 + n*L.current();
        cout << result << endl;  // ==> 0
    }
Result
======
L(0) = 1
L(1) = 1-x
L(2) = 2-4*x+x^(2)
L(3) = 6-18*x+9*x^(2)-x^(3)
L(4) = 24-96*x+72*x^(2)-16*x^(3)+x^(4)
0
```

Hermite Polynomials

The *Hermite polynomials* are defined by

$$H_n(x) := (-1)^n e^{x^2} \frac{d^n}{dx^n} e^{-x^2}$$

where $n = 0, 1, 2, \ldots$ It can be proved that they satisfy the linear differential equation

$$\frac{d^2 H_n}{dx^2} - 2x \frac{dH_n}{dx} + 2n H_n = 0. \tag{8.7}$$

Using the differential equation, we can prove that the Hermite polynomials are orthogonal on $(-\infty, \infty)$ with respect to the weight function e^{-x^2}. We have

$$\int_{-\infty}^{\infty} e^{-x^2} H_n(x) H_m(x) dx = \sqrt{\pi} 2^n n! \delta_{nm}$$

where $n, m = 0, 1, 2, \ldots$ and δ_{nm} is the *Kronecker delta*. The recursion formula takes the form

$$H_{n+1}(x) = 2x H_n(x) - 2n H_{n-1}(x)$$

where $H_0(x) = 1$ and $H_1(x) = 2x$. The Hermite polynomials can also be defined by the generating function:

$$e^{2tx - t^2} = \sum_{n=0}^{\infty} \frac{H_n(x) t^n}{n!}.$$

Furthermore we have

$$\int_0^x H_n(t) dt = \frac{H_{n+1}(x)}{2(n+1)} - \frac{H_{n+1}(0)}{2(n+1)}.$$

In the following program, we make use of the recursion relation to generate the first few Hermite polynomials, and show that (8.7) holds for $n = 4$:

$$H_0(x) = 1, \quad H_1(x) = 2x, \quad H_2(x) = 4x^2 - 2, \quad H_3(x) = 8x^3 - 12x,$$
$$H_4(x) = 16x^4 - 48x^2 + 12, \quad H_5(x) = 32x^5 - 160x^3 + 120x.$$

```
// Hermite.cxx

#include <iostream.h>
#include "MSymbol.h"

class Hermite
{
 private:
   int maxTerm, currentStep;
   Sum<int> P, Q;
   const Sum<int>& x;
 public:
   Hermite(int,const Sum<int>&);
   int step();
   void run();
   Sum<int> current() const;
   friend ostream& operator << (ostream&,const Hermite&);
};

Hermite::Hermite(int n,const Sum<int>& kernal)
   : maxTerm(n), currentStep(0), P(1), x(kernal) {}

int Hermite::step()
{
   int prev = currentStep;
   Sum<int> R;
   ++currentStep;
   if(currentStep==1)
   {
      Q = 1;
      P = 2*x;
      return 1;
   }

   if(currentStep <= maxTerm)
   {
      R = 2*x*P-2*prev*Q;
      Q = P; P = R;
      return 1;
   }
   return 0;
}

void Hermite::run()
```

```
{ while (step()); }

Sum<int> Hermite::current() const
{ return P; }

ostream& operator << (ostream& s,const Hermite& H)
{ return s << H.P; }

void main()
{
    int n=4;
    Sum<int> x("x",0);
    Hermite H(n,x);

    // Calculate the first few Hermite polynomials
    cout << "H(0) = " << H << endl;
    for (int i=1; i<=n; i++)
    {
        H.step();
        cout << "H("<< i << ") = " << H << endl;
    }
    cout << endl;

    // To show that for n=4,
    //    the Hermite differential equation is satisfied.
    Sum<int> result;

    result = df(H.current(),x,2) - 2*x*df(H.current(),x)+2*n*H.current();

    cout << result << endl;  // ==> 0
}
```
Result
======
H(0) = 1
H(1) = 2*x
H(2) = 4*x^(2)-2
H(3) = 8*x^(3)-12*x
H(4) = 16*x^(4)-48*x^(2)+12

0

Chebyshev Polynomials

The *Chebyshev polynomials* are defined by the relation

$$T_n(x) := \cos(n \arccos x), \quad n = 0, 1, 2, \ldots$$

Note that $T_{-n}(x) = T_n(x)$ and from the trigonometric formulae we get

$$T_{n+m}(x) + T_{n-m}(x) = 2T_n(x)T_m(x).$$

For $m = 1$, we obtain the recursion relation

$$T_{n+1}(x) = 2xT_n(x) - T_{n-1}(x)$$

where $T_0(x) = 1$ and $T_1(x) = x$. Thus we can successively compute all $T_n(x)$. Let $x := \cos\theta$, then

$$y = T_n(x) = \cos(n\theta)$$

$$\frac{dy}{dx} = \frac{n\sin(n\theta)}{\sin\theta}$$

$$\frac{d^2y}{dx^2} = \frac{-n^2\cos(n\theta) + n\sin(n\theta)\cot\theta}{\sin^2\theta}$$

$$= -\frac{n^2 y}{1 - x^2} + \frac{x}{1 - x^2}\frac{dy}{dx}.$$

Thus the polynomials $T_n(x)$ satisfy the second order linear ordinary differential equation

$$(1 - x^2)\frac{d^2y}{dx^2} - x\frac{dy}{dx} + n^2 y = 0.$$

The first few polynomials are

$$T_0(x) = 1, \quad T_1(x) = x, \quad T_2(x) = 2x^2 - 1,$$

$$T_3(x) = 4x^3 - 3x, \quad T_4(x) = 8x^4 - 8x^2 + 1.$$

```
// Chebyshev.cxx

#include <iostream.h>
#include "MSymbol.h"

class Chebyshev
{
private:
    int maxTerm, currentStep;
    Sum<int> P, Q;
    const Sum<int>& x;
public:
    Chebyshev(int,const Sum<int>&);
    int step();
    void run();
    Sum<int> current() const;
    friend ostream& operator << (ostream&,const Chebyshev&);
};

Chebyshev::Chebyshev(int n, const Sum<int> &kernal)
    : maxTerm(n), currentStep(0), P(1), x(kernal) {}

int Chebyshev::step()
{
    Sum<int> R;
    ++currentStep;

    if(currentStep==1)
    {
        Q = 1;
        P = x;
        return 1;
    }

    if(currentStep <= maxTerm)
    {
        R = 2*x*P-Q;
        Q = P; P = R;
        return 1;
    }
    return 0;
}

void Chebyshev::run()
```

```
    { while (step()); }

    Sum<int> Chebyshev::current() const
    { return P; }

    ostream & operator << (ostream &s, const Chebyshev &T)
    { return s << T.P; }

    void main()
    {
        int n=4;
        Sum<int> x("x",0);
        Chebyshev T(n,x);

        // Calculate the first few Chebyshev polynomials
        cout << "T(0) = " << T << endl;
        for(int i=1; i<=n; i++)
        {
            T.step();
            cout << "T("<< i << ") = " << T << endl;
        }
        cout << endl;

        // To show that for n=4,
        //     the Chebyshev differential equation is satisfied.
        Sum<int> result;

        result = (1-x*x)*df(T.current(),x,2) - x*df(T.current(),x)
                 + n*n*T.current();

        cout << result << endl;    // ==> 0
    }
Result
======
T(0) = 1
T(1) = x
T(2) = 2*x^(2)-1
T(3) = 4*x^(3)-3*x
T(4) = 8*x^(4)-8*x^(2)+1

0
```

8.6.2 Cumulant Expansion

Suppose $x, a_n, b_n \in \mathbf{R}$ and

$$\exp\left[\sum_{n=1}^{\infty} \frac{b_n x^n}{n!}\right] = \sum_{n=0}^{\infty} \frac{a_n x^n}{n!},$$

with $a_0 = 1$. We determine the relation between the coefficients a_n and b_n. The kth term of the exponential function on the left-hand side is given by

$$\frac{1}{k!}\left(\sum_{n=1}^{\infty} \frac{b_n x^n}{n!}\right)^k = \frac{1}{k!}\left(\sum_{n_1=1}^{\infty} \frac{b_{n_1} x^{n_1}}{n_1!}\right) \cdots \left(\sum_{n_k=1}^{\infty} \frac{b_{n_k} x^{n_k}}{n_k!}\right)$$

$$= \frac{1}{k!}\sum_{n_1=1}^{\infty}\sum_{n_2=1}^{\infty} \cdots \sum_{n_k=1}^{\infty} \frac{b_{n_1} b_{n_2} \cdots b_{n_k} x^{n_1+n_2+\cdots+n_k}}{n_1!n_2!\cdots n_k!}.$$

Therefore,

$$\exp\left[\sum_{n=1}^{\infty} \frac{b_n x^n}{n!}\right] = 1 + \sum_{n=1}^{\infty} \frac{b_n x^n}{n!} + \frac{1}{2!}\sum_{n_1=1}^{\infty}\sum_{n_2=1}^{\infty} \frac{b_{n_1} b_{n_2} x^{n_1+n_2}}{n_1!n_2!} + \cdots$$

$$+ \frac{1}{k!}\sum_{n_1=1}^{\infty}\sum_{n_2=1}^{\infty} \cdots \sum_{n_k=1}^{\infty} \frac{b_{n_1} b_{n_2} \cdots b_{n_k} x^{n_1+n_2+\cdots+n_k}}{n_1!n_2!\cdots n_k!} + \cdots$$

$$= \sum_{n=0}^{\infty} \frac{a_n x^n}{n!}.$$

Equating terms of the same order in x, we can obtain the relation between a_n and b_n for all positive integers n. The first three terms are

$$x^1 : a_1 = b_1, \qquad x^2 : a_2 = b_2 + b_1^2, \qquad x^3 : a_3 = b_3 + 3b_2 b_1 + b_1^3.$$

It follows that

$$b_1 = a_1, \qquad b_2 = a_2 - a_1^2, \qquad b_3 = a_3 - 3a_2 a_1 + 2a_1^3.$$

In the program, we repeat the process of deriving the coefficients. The process shows that a computer algebra system can obtain the coefficients efficiently.

```cpp
// cumu.cpp

#include <iostream.h>
#include "MSymbol.h"
#include "Vector.h"
#include "Rational.h"

Sum<Rational<int> >
Taylor(Sum<Rational<int> > u,Sum<Rational<int> >& x,int n)
{
   Rational<int> zero(0);
   x.set(zero);
   Sum<Rational<int> > series(u.value());
   int j, fac = 1;
   for(j=1; j<=n; j++)
   {
      u = df(u,x); fac = fac * j;
      series += Rational<int>(1,fac) * u.value() * power(x,j);
   }
   return series;
}

void main()
{
   int i, fac, n = 5;
   Vector<Sum<Rational<int> > > a(n), b(n);
   Sum<Rational<int> > x("x",0), y, P, Q,
               a0("a0",0), a1("a1",0), a2("a2",0),
               a3("a3",0), a4("a4",0),
               b0("b0",0), b1("b1",0), b2("b2",0),
               b3("b3",0), b4("b4",0);
   a[0] = a0; a[1] = a1; a[2] = a2; a[3] = a3; a[4] = a4;
   b[0] = b0; b[1] = b1; b[2] = b2; b[3] = b3; b[4] = b4;

   fac = 1; P = a[0];
   for(i=1; i<n; i++)
   {
      fac *= i;
      P += Rational<int>(1,fac) * a[i] * power(x,i);
      Q += Rational<int>(1,fac) * b[i] * power(x,i);
   }

   cout << "P = " << P << endl;
   cout << "Q = " << Q << endl; cout << endl;
```

```
        y = Taylor(exp(Q),x,5);

        cout << "Taylor series expansion of exp(Q) = " << endl;
        cout << y << endl; cout << endl;
        cout << "Coefficient of x :"<< endl;
        cout << "exp(Q) => " << y.coeff(x,1) << endl;
        cout << "  P    => " << P.coeff(x,1) << endl; cout << endl;
        cout << "Coefficient of x^2 :"<< endl;
        cout << "exp(Q) => " << y.coeff(x,2) << endl;
        cout << "  P    => " << P.coeff(x,2) << endl; cout << endl;
        cout << "Coefficient of x^3 :"<< endl;
        cout << "exp(Q) => " << y.coeff(x,3) << endl;
        cout << "  P    => " << P.coeff(x,3) << endl; cout << endl;
        cout << "Coefficient of x^4 :"<< endl;
        cout << "exp(Q) => " << y.coeff(x,4) << endl;
        cout << "  P    => " << P.coeff(x,4) << endl;
}
Result
======
P = a0+a1*x+1/2*a2*x^(2)+1/6*a3*x^(3)+1/24*a4*x^(4)
Q = b1*x+1/2*b2*x^(2)+1/6*b3*x^(3)+1/24*b4*x^(4)

Taylor series expansion of exp(Q) =
1+b1*x+1/2*b1^(2)*x^(2)+1/2*b2*x^(2)+1/6*b1^(3)*x^(3)+1/2*b1*b2*x^(3)
+1/6*b3*x^(3)+1/24*b1^(4)*x^(4)+1/4*b1^(2)*b2*x^(4)+1/6*b1*b3*x^(4)
+1/8*b2^(2)*x^(4)+1/24*b4*x^(4)+1/120*b1^(5)*x^(5)+1/12*b1^(3)*b2*x^(5)
+1/12*b1^(2)*b3*x^(5)+1/8*b1*b2^(2)*x^(5)+1/24*b1*b4*x^(5)
+1/12*b2*b3*x^(5)


Coefficient of x :
exp(Q) => b1
  P    => a1


Coefficient of x^2 :
exp(Q) => 1/2*b1^(2)+1/2*b2
  P    => 1/2*a2


Coefficient of x^3 :
exp(Q) => 1/6*b1^(3)+1/2*b1*b2+1/6*b3
  P    => 1/6*a3


Coefficient of x^4 :
exp(Q) => 1/24*b1^(4)+1/4*b1^(2)*b2+1/6*b1*b3+1/8*b2^(2)+1/24*b4
  P    => 1/24*a4
```

8.6.3   Exterior Product

In this section, we give an implementation of the *exterior product* (also called wedge product or Grassmann product) described in Section 2.13. In the program the exterior product $\wedge$ is written as $*$. We evaluate the determinant of the 4×4 matrix:

$$A := \begin{pmatrix} 1 & 2 & 5 & 2 \\ 0 & 1 & 2 & 3 \\ 1 & 0 & 1 & 0 \\ 0 & 3 & 0 & 7 \end{pmatrix}$$

by calculating the exterior product

$$\begin{pmatrix} 1 \\ 0 \\ 1 \\ 0 \end{pmatrix} \wedge \begin{pmatrix} 2 \\ 1 \\ 0 \\ 3 \end{pmatrix} \wedge \begin{pmatrix} 5 \\ 2 \\ 1 \\ 0 \end{pmatrix} \wedge \begin{pmatrix} 2 \\ 3 \\ 0 \\ 7 \end{pmatrix}.$$

The exterior product gives $24 e_1 \wedge e_2 \wedge e_3 \wedge e_4$, where e_1, e_2, e_3, e_4 is the standard basis in $\mathbf{R}^4$. Thus, we find that $\det A = 24$.

```
// Grass.cpp

#include <iostream.h>
#include "Vector.h"
#include "Matrix.h"
#include "MSymbol.h"

void main()
{
    int i,j,n=4;
    Sum<int> e0("e0",0), e1("e1",0), e2("e2",0), e3("e3",0),
             y, result;
    Matrix<Sum<int> > A(n,n);
    Vector<Sum<int> > e(n);

    Commutative(int(),0);

    A[0][0] = 1; A[0][1] = 2; A[0][2] = 5; A[0][3] = 2;
    A[1][0] = 0; A[1][1] = 1; A[1][2] = 2; A[1][3] = 3;
    A[2][0] = 1; A[2][1] = 0; A[2][2] = 1; A[2][3] = 0;
```

```
    A[3][0] = 0; A[3][1] = 3; A[3][2] = 0; A[3][3] = 7;

    e[0] = e0; e[1] = e1; e[2] = e2; e[3] = e3;

    result = 1;
    for(i=0; i<n; i++) result *= (A[i]|e);

    cout << result << endl; cout << endl;

    // for all i>j, put e[i]*e[j] into -e[j]*e[i]
    int flag;

    do
    {
       flag=0;
       for(i=0; i<n; i++)
          for(j=0; j<i; j++)
             flag += result.put(e[i]*e[j], -e[j]*e[i]);
    } while(flag);

    // put e[i]*e[i] into 0
    for(i=0; i<n; i++) result.put(e[i]*e[i],0);
    for(i=0; i<n; i++) result.put(e[i]*e[i]*e[i],0);

    cout << "result = " << result << endl;
  }
```

```
Result
======
3*e0*e1*e0*e1+7*e0*e1*e0*e3+3*e0*e1*e2*e1+7*e0*e1*e2*e3+6*e0*e2*e0*e1
+14*e0*e2*e0*e3+6*e0*e2^(2)*e1+14*e0*e2^(2)*e3+9*e0*e3*e0*e1+21*e0*e3
*e0*e3+9*e0*e3*e2*e1+21*e0*e3*e2*e3+6*e1^(2)*e0*e1+14*e1^(2)*e0*e3+6*
e1^(2)*e2*e1+14*e1^(2)*e2*e3+12*e1*e2*e0*e1+28*e1*e2*e0*e3+12*e1*e2^(2)
*e1+28*e1*e2^(2)*e3+18*e1*e3*e0*e1+42*e1*e3*e0*e3+18*e1*e3*e2*e1+42*e1
*e3*e2*e3+15*e2*e1*e0*e1+35*e2*e1*e0*e3+15*e2*e1*e2*e1+35*e2*e1*e2*e3
+30*e2^(2)*e0*e1+70*e2^(2)*e0*e3+30*e2^(3)*e1+70*e2^(3)*e3+45*e2*e3*e0
*e1+105*e2*e3*e0*e3+45*e2*e3*e2*e1+105*e2*e3*e2*e3+6*e3*e1*e0*e1+14*e3
*e1*e0*e3+6*e3*e1*e2*e1+14*e3*e1*e2*e3+12*e3*e2*e0*e1+28*e3*e2*e0*e3+12
*e3*e2^(2)*e1+28*e3*e2^(2)*e3+18*e3^(2)*e0*e1+42*e3^(2)*e0*e3+18*e3^(2)
*e2*e1+42*e3^(2)*e2*e3

result = 24*e0*e1*e2*e3
```

8.7 Symbolic Class and Symbolic Differentiation

8.7.1 First Integrals

Consider an autonomous system of first order ordinary differential equations

$$\frac{d\mathbf{u}}{dt} = V(\mathbf{u})$$

where $\mathbf{u} = (u_1, u_2, \ldots, u_n)$ and V_j are smooth functions of $u_1, u_2, \ldots, u_n$. A smooth function $I(\mathbf{u})$ is called a *first integral* of the differential equations if

$$\frac{d}{dt}I(\mathbf{u}(t)) \equiv \sum_{j=1}^{n} \frac{\partial I}{\partial u_j} \frac{du_j}{dt} \equiv \sum_{j=1}^{n} \frac{\partial I}{\partial u_j} V_j(\mathbf{u}) = 0.$$

A smooth function $I(\mathbf{u}(t), t)$ is called an explicitly *time-dependent first integral* of the differential equations if

$$\frac{d}{dt}I(\mathbf{u}(t), t) = \frac{\partial I}{\partial t} + \sum_{j=1}^{n} \frac{\partial I}{\partial u_j} V_j(\mathbf{u}(t), t) = 0.$$

Example 1. Consider the following differential equations

$$\frac{du_1}{dt} = u_2 u_3, \qquad \frac{du_2}{dt} = u_1 u_3, \qquad \frac{du_3}{dt} = u_1 u_2.$$

We show that I_1, I_2 and I_3 where

$$I_1(\mathbf{u}) = u_1^2 - u_2^2, \qquad I_2(\mathbf{u}) = u_3^2 - u_1^2, \qquad I_3(\mathbf{u}) = u_2^2 - u_3^2$$

are the first integrals of the system.

Example 2. Consider the *Lorenz model*

$$\frac{du_1}{dt} = -\sigma u_1 + \sigma u_2, \qquad \frac{du_2}{dt} = -u_1 u_3 + r u_1 - u_2, \qquad \frac{du_3}{dt} = u_1 u_2 - b u_3.$$

For certain values of the bifurcation parameters σ, b and r the system admits explicitly the time-dependent first integrals. We insert the ansatz

$$I(\mathbf{u}(t), t) = (u_2^2 + u_3^2) \exp(2t)$$

into the system in order to find the conditions on the coefficients σ, r, b such that $I(\mathbf{u}(t), t)$ is a first integral.

Example 1

```cpp
// first1.cpp

#include <iostream.h>
#include "MSymbol.h"
#include "Vector.h"

void main()
{
    int i, j, N=3;
    Sum<int> u1("u1",0), u2("u2",0), u3("u3",0),
             term, sum=0;
    Vector<Sum<int> > u(N), v(N), I(N);

    u[0] = u1;          u[1] = u2;          u[2] = u3;
    v[0] = u2*u3;       v[1] = u1*u3;       v[2] = u1*u2;
    I[0] = u1*u1-u2*u2; I[1] = u3*u3-u1*u1; I[2] = u2*u2-u3*u3;

    for(i=0; i<N; i++)
    {
        for(j=0; j<N; j++)
        {
            term = df(I[i],u[j]) * v[j];
            sum += term;
            cout << "Partial Term " << j << " = " << term << endl;
        }
        cout << "Sum = " << sum << endl;    // output : 0
        cout << endl;
    }
}
```

Example 2

```
// first2.cpp

#include <iostream.h>
#include "MSymbol.h"
#include "Vector.h"

void main()
{
    int i;
    Vector<Sum<int> > u(3), v(4);
    Sum<int> term, sum, I, R1, R2,
             u1("u1",0), u2("u2",0), u3("u3",0), t("t",0),
             s("s",0), b("b",0), r("r",0);
    u[0] = u1; u[1] = u2; u[2] = u3;

    // Lorenz Model
    v[0] = s*u2- s*u1;      v[1] = -u2 - u1*u3 + r*u1;
    v[2] = u1*u2 - b*u3;    v[3] = 1;
    // The ansatz
    I = (u2*u2 + u3*u3)*exp(2*t);

    sum = 0;
    for(i=0; i<3; i++) sum += v[i]*df(I,u[i]);
    sum += v[3]*df(I,t);
    cout << "sum = " << sum << endl; cout << endl;
    R1 = sum.coeff(u3,2);
    R1 = R1/(exp(2*t));
    cout << "R1 = " << R1 << endl;
    R2 = sum.coeff(u1,1); R2 = R2.coeff(u2,1);
    R2 = R2/(exp(2*t));
    cout << "R2 = " << R2 << endl;
}
```

```
Result
======
sum = 2*r*u1*u2*exp(2*t)-2*b*u3^(2)*exp(2*t)+2*u3^(2)*exp(2*t)
R1 = -2*b+2
R2 = 2*r
```

From the results obtained, it is obvious that for R1, R2 to be equal to zero, we need $b = 1$ and $r = 0$.

8.7.2 Spherical Harmonics

Spherical harmonics are complex, single-valued functions on the surface of the unit sphere, i.e. functions of two real-valued arguments $0 \leq \theta \leq \pi$, $0 \leq \phi < 2\pi$. They are defined by

$$Y_{lm}(\theta, \phi) := (-1)^m \left[\frac{(2l+1)(l-m)!}{4\pi(l+m)!} \right]^{\frac{1}{2}} P_l^m(\cos\theta) e^{im\phi}, \quad m \geq 0$$

where $P_l^m(w)$ is the associated Legendre functions defined as

$$P_l^m(w) := (1-w^2)^{|m|/2} \frac{d^{|m|}}{dw^{|m|}} P_l(w),$$

with $l = 0, 1, 2, \ldots$ and $m = -l, -l+1, \ldots, l$. For negative values of m we have

$$Y_{lm}(\theta, \phi) = (-1)^m Y_{l,-m}^*(\theta, \phi)$$

where the asterisk (*) denotes the complex conjugate. The spherical harmonics satisfy the orthonormality relations

$$\int Y_{l'm'}^*(\theta, \phi) Y_{lm}(\theta, \phi) d\Omega = \int_0^{2\pi} d\phi \int_0^\pi d\theta \sin\theta \, Y_{l'm'}^*(\theta, \phi) Y_{lm}(\theta, \phi)$$

$$= \delta_{ll'} \delta_{mm'}$$

where we have written $d\Omega \equiv \sin\theta d\theta d\phi$ and $\int \Omega$ means that we integrate over the full range of the angular variables (θ, ϕ), namely

$$\int \Omega \equiv \int_0^{2\pi} d\phi \int_0^\pi d\theta \sin\theta.$$

The spherical harmonics form a basis in the Hilbert space $L_2(S^2)$ where

$$S^2 := \left\{ (x_1, x_2, x_3) : x_1^2 + x_2^2 + x_3^2 = 1 \right\}.$$

The first few spherical harmonics are given by

$$Y_{00}(\theta, \phi) = \frac{1}{\sqrt{4\pi}}, \quad Y_{11}(\theta, \phi) = -\sqrt{\frac{3}{8\pi}} \sin\theta \, e^{i\phi},$$

$$Y_{10}(\theta, \phi) = \sqrt{\frac{3}{4\pi}} \cos\theta, \quad Y_{1-1}(\theta, \phi) = \sqrt{\frac{3}{8\pi}} \sin\theta \, e^{-i\phi}.$$

Any function in the Hilbert space $L_2(S^2)$ can be written as a linear combination of the spherical harmonics.

```
// spheric.cpp

#include <iostream.h>
#include "MSymbol.h"
#include "Rational.h"
#include "AssLegendre.h"

Sum<Rational<int> > PI("PI",0), I("I",0);

int factorial(int n)
{
   int i, result=1;
   for(i=2; i<=n; i++) result *= i;
   return result;
}

Sum<Rational<int> > Y(int l,int m,const Sum<Rational<int> >& phi,
                      const Sum<Rational<int> >& w)
{
   Sum<Rational<int> > a, b, u("u",0), result;
   int absm = abs(m);
   AssLegendre A(l,m,u);
   a = A.current(); a.put(u,cos(w));

   if(m>0 && m%2) a = -a;
   b = sqrt(Rational<int>((2*l+1)*factorial(l-absm),
            4*factorial(l+absm))/PI);
   result = a*b*exp(I*Rational<int>(m)*phi);
   return result;
}
```

```
    void main()
    {
        int i, j, n=3;
        Sum<Rational<int> > phi("phi",0), w("w",0), result, one(1);

        for(i=0; i<=n; i++)
        {
            for(j=-i; j<=i; j++)
            {
                result = Y(i,j,phi,w);
                result.put(cos(w)*cos(w),one-sin(w)*sin(w));
                cout << "Y(" << i << "," << j << ") = " << result << endl;
            }
            cout << endl;
        }
    }
```

Result
======

Y(0,0) = sqrt(1/4*PI^(-1))

Y(1,-1) = sin(w)*sqrt(3/8*PI^(-1))*exp(-I*phi)
Y(1,0) = cos(w)*sqrt(3/4*PI^(-1))
Y(1,1) = -sin(w)*sqrt(3/8*PI^(-1))*exp(I*phi)

Y(2,-2) = 3*sin(w)^(2)*sqrt(5/96*PI^(-1))*exp(-2*I*phi)
Y(2,-1) = 3*cos(w)*sin(w)*sqrt(5/24*PI^(-1))*exp(-I*phi)
Y(2,0) = sqrt(5/4*PI^(-1))-3/2*sin(w)^(2)*sqrt(5/4*PI^(-1))
Y(2,1) = -3*cos(w)*sin(w)*sqrt(5/24*PI^(-1))*exp(I*phi)
Y(2,2) = 3*sin(w)^(2)*sqrt(5/96*PI^(-1))*exp(2*I*phi)

Y(3,-3) = 15*sin(w)^(3)*sqrt(7/2880*PI^(-1))*exp(-3*I*phi)
Y(3,-2) = 15*cos(w)*sin(w)^(2)*sqrt(7/480*PI^(-1))*exp(-2*I*phi)
Y(3,-1) = 6*sin(w)*sqrt(7/48*PI^(-1))*exp(-I*phi)
          -15/2*sin(w)^(3)*sqrt(7/48*PI^(-1))*exp(-I*phi)
Y(3,0) = cos(w)*sqrt(7/4*PI^(-1))
          -5/2*cos(w)*sin(w)^(2)*sqrt(7/4*PI^(-1))
Y(3,1) = -6*sin(w)*sqrt(7/48*PI^(-1))*exp(I*phi)
          +15/2*sin(w)^(3)*sqrt(7/48*PI^(-1))*exp(I*phi)
Y(3,2) = 15*cos(w)*sin(w)^(2)*sqrt(7/480*PI^(-1))*exp(2*I*phi)
Y(3,3) = -15*sin(w)^(3)*sqrt(7/2880*PI^(-1))*exp(3*I*phi)
```
```

8.7.3   Nambu Mechanics

In *Nambu mechanics* the phase space is spanned by an n-tuple of dynamical variables u_i, for $i = 1, \ldots, n$. The equations of motion of Nambu mechanics (i.e., the autonomous system of first order ordinary differential equations) can be constructed as follows. Let $I_k : \mathbf{R}^n \to \mathbf{R}$, for $k = 1, \ldots, n-1$ be smooth functions, then

$$\frac{du_i}{dt} = \frac{\partial(u_i, I_1, \ldots, I_{n-1})}{\partial(u_1, u_2, \ldots, u_n)} \tag{8.8}$$

where $\partial(u_i, I_1, \ldots, I_{n-1})/\partial(u_1, u_2, \ldots, u_n)$ denotes the *Jacobian*. Consequently, the equations of motion can also be written as (summation convention)

$$\frac{du_i}{dt} = \epsilon_{ijk\ldots l} \partial_j I_1 \ldots \partial_l I_{n-1}$$

where $\epsilon_{ijk\ldots l}$ is the generalized *Levi-Cevita symbol* and $\partial_j \equiv \partial/\partial u_j$. The proof that $I_1, \ldots, I_{n-1}$ are first integrals of system (8.8) is as follows. Since (summation convention)

$$\frac{dI_i}{dt} = \frac{\partial I_i}{\partial u_j} \frac{du_j}{dt}$$

we have

$$\frac{dI_i}{dt} = (\partial_j I_i)\epsilon_{jl_1\ldots l_{n-1}}(\partial_{l_1} I_1) \ldots (\partial_{l_{n-1}} I_{n-1}) = \frac{\partial(I_i, I_1, \ldots, I_{n-1})}{\partial(u_1, \ldots, u_n)} = 0 \, .$$

Since the Jacobian matrix has two equal rows, it is singular. If the first integrals are polynomials, then the dynamical system (8.8) is algebraically completely integrable. For the case $n = 3$, we obtain the equations of motion

$$\frac{du_1}{dt} = \frac{\partial I_1}{\partial u_2} \frac{\partial I_2}{\partial u_3} - \frac{\partial I_1}{\partial u_3} \frac{\partial I_2}{\partial u_2}, \quad \frac{du_2}{dt} = \frac{\partial I_1}{\partial u_3} \frac{\partial I_2}{\partial u_1} - \frac{\partial I_1}{\partial u_1} \frac{\partial I_2}{\partial u_3}, \quad \frac{du_3}{dt} = \frac{\partial I_1}{\partial u_1} \frac{\partial I_2}{\partial u_2} - \frac{\partial I_1}{\partial u_2} \frac{\partial I_2}{\partial u_1} \, .$$

As an example, we consider the Nambu machanics of the following first integrals:

$$I_1 = u_1 + u_2 + u_3 \quad \text{and} \quad I_2 = u_1 u_2 u_3 \, .$$

The program uses (8.8) to construct the 3×3 matrices and calculate their determinants:

$$\frac{du_i}{dt} = \det \begin{pmatrix} \dfrac{\partial u_i}{\partial u_1} & \dfrac{\partial I_1}{\partial u_1} & \dfrac{\partial I_2}{\partial u_1} \\[2mm] \dfrac{\partial u_i}{\partial u_2} & \dfrac{\partial I_1}{\partial u_2} & \dfrac{\partial I_2}{\partial u_2} \\[2mm] \dfrac{\partial u_i}{\partial u_3} & \dfrac{\partial I_1}{\partial u_3} & \dfrac{\partial I_2}{\partial u_3} \end{pmatrix} \qquad \text{for } i = 1, 2, 3.$$

```
// nambu.cpp
#include <iostream.h>
#include "MSymbol.h"
#include "Vector.h"
#include "Matrix.h"

void nambu(Vector<Sum<double> > I,Vector<Sum<double> > u,int n)
{
    int i, j;
    Matrix<Sum<double> > J(n,n);
    for(i=0; i<n; i++)
        for(j=1; j<n; j++) J[i][j] = df(I[j-1], u[i]);

    for(i=0; i<n; i++)
    {
        for (j=0; j<n; j++) J[j][0] = df(u[i], u[j]);
        cout << "du(" << i << ")/dt = " << J.determinant() << endl;
    }
}

void main()
{
    Sum<double> u1("u1",0), u2("u2",0), u3("u3",0);
    Vector<Sum<double> > I(2), u(3);
    u[0] = u1; u[1] = u2; u[2] = u3;
    I[0] = u1 + u2 + u3; I[1] = u1 * u2 * u3;
    nambu(I,u,3);
}
Result
======
du(0)/dt = u1*u2-u1*u3
du(1)/dt = -u1*u2+u2*u3
du(2)/dt = u1*u3-u2*u3
```

8.7.4   Taylor Expansion of Differential Equations

Consider the first order ordinary differential equation

$$\frac{du}{dx} = f(x, u), \quad u(x_0) = u_0$$

where f is an analytic function of x and u. The *Taylor series expansion* gives an approximate solution about $x = x_0$ and the solution of the differential equation is given by

$$u(x_0 + h) = u(x_0) + h\left(\frac{du}{dx}\right)_{x=x_0} + \frac{h^2}{2!}\left(\frac{d^2 u}{dx^2}\right)_{x=x_0} + \ldots$$

The derivatives can be obtained by the successive differentiation of the differential equation,

$$\frac{d^2 u}{dx^2} = \frac{\partial f}{\partial x} + \frac{\partial f}{\partial u}\frac{du}{dx} = \frac{\partial f}{\partial x} + \frac{\partial f}{\partial u}f$$

$$\frac{d^3 u}{dx^3} = \frac{\partial^2 f}{\partial x^2} + 2\frac{\partial^2 f}{\partial x \partial u}f + \frac{\partial^2 f}{\partial u^2}f^2 + \frac{\partial f}{\partial u}\left(\frac{\partial f}{\partial x} + \frac{\partial f}{\partial u}f\right)$$

$$\vdots \qquad\qquad \vdots$$

The formulae for higher derivatives become very complicated. As an example consider the *Riccati differential equation*

$$\frac{du}{dx} = u^2 + x, \quad u(0) = 1.$$

This equation has no solution in terms of elementary functions. *Bessel functions* are needed to solve it. Suppose the solution may be represented by a Taylor series expansion,

$$u(x) = 1 + \sum_{j=1}^{\infty}\frac{x^j}{j!}\left(\frac{d^j u}{dx^j}\right)_{x=x_0}.$$

We compute all the successive derivatives

$$\frac{d^2u}{dx^2} = 2u\frac{du}{dx} + 1, \quad \frac{d^3u}{dx^3} = 2\left(\frac{du}{dx}\right)^2 + 2u\frac{d^2u}{dx^2}, \quad \ldots$$

These derivatives can be found using computer algebra. The initial value is then inserted and we obtain the Taylor series expansion up to order n. In Example 1, we solve the Riccati differential equation using the Taylor series method up to order $n = 4$, with initial value $u(0) = 1$. The solution is:

$$u(x) = 1 + x + \frac{3}{2}x^2 + \frac{4}{3}x^3 + \frac{17}{12}x^4 + \cdots$$

Example 1

```
// Taylor1.cpp

#include <iostream.h>
#include "MSymbol.h"
#include "Vector.h"
#include "Rational.h"

int factorial(int N)
{
   int result=1;

   for(int i=2; i<=N; i++) result *= i;

   return result;
}

void main()
{
   int i, j, n=4;
   Rational<int> one(1);
   Sum<Rational<int> > u("u",0), x("x",0), result;
   Vector<Sum<Rational<int> > > u0(n), y(n);

   u.depend(x);
```

```
    u0[0] = u*u+x;
    for(j=1; j<n; j++) u0[j] = df(u0[j-1],x);

    cout << u0 << endl;

    // initial condition u(0)=1
    x.set(0); u.set(1);
    u0[0] = u0[0].value();

    y[0] = u;
    for(i=1; i<n; i++) y[i] = df(y[i-1],x);

    // substitution of initial conditions
    for(i=1; i<n; i++)
    {
        for(j=i; j>0; j--) u0[i].put(y[j],u0[j-1]);
        u0[i].expand();
    }
    cout << u0 << endl;

    u = one;
    cout << u0 << endl;

    // Taylor series expansion
    result = one;
    for(i=0; i<n; i++)
        result += Rational<int>(1,factorial(i+1))*u0[i]*power(x,i+1);
    cout << "u(x) = " << result << endl;
}
```

```
Result
======
[u^(2)+x]
[2*u*df(u,x)+1]
[2*df(u,x)^(2)+2*u*df(df(u,x),x)]
[6*df(u,x)*df(df(u,x),x)+2*u*df(df(df(u,x),x),x)]

[1]
[2*u+1]
[2+4*u^(2)+2*u]
[16*u+6+8*u^(3)+4*u^(2)]

[1]
[3]
```

[8]
[34]

```
u(x) = 1+x+3/2*x^(2)+4/3*x^(3)+17/12*x^(4)
```

As another example, let us consider the nonlinear initial value problem

$$\frac{du}{dx} = x^2 + xu - u^2, \quad \text{where} \quad u(0) = 1.$$

It is straightforward, though tedious, to compute all the derivatives:

$$\frac{d^2u}{dx^2} = 2x + u + x\frac{du}{dx} - 2u\frac{du}{dx}$$

$$\frac{d^3u}{dx^3} = 2 + 2\frac{du}{dx} + x\frac{d^2u}{dx^2} - 2\left(\frac{du}{dx}\right)^2 - 2u\frac{d^2u}{dx^2}$$

$$\frac{d^4u}{dx^4} = 3\frac{d^2u}{dx^2} + x\frac{d^3u}{dx^3} - 6\frac{du}{dx}\frac{d^2u}{dx^2} - 2u\frac{d^3u}{dx^3}$$

$$\vdots \qquad \qquad \vdots$$

At the point $x = 0$, $u(x = 0) = 1$,

$$\left(\frac{du}{dx}\right)_{x=0} = -1, \quad \left(\frac{d^2u}{dx^2}\right)_{x=0} = 3, \quad \left(\frac{d^3u}{dx^3}\right)_{x=0} = -8, \quad \left(\frac{d^4u}{dx^4}\right)_{x=0} = 43, \quad \dots$$

Finally, the solution $u(x)$ of the initial value problem near $x = 0$ is given by

$$u(x) = 1 - x + \frac{3}{2}x^2 - \frac{4}{3}x^3 + \frac{43}{24}x^4 + \dots$$

In the following, we reproduce the procedure of the calculation using the algebra system.

Example 2

```cpp
// Taylor2.cpp

#include <iostream.h>
#include "MSymbol.h"
#include "Vector.h"
#include "Rational.h"

int factorial(int N)
{
    int result=1;

    for(int i=2; i<=N; i++) result *= i;

    return result;
}

void main()
{
    int i, j, n=4;
    Rational<int> zero(0), one(1);
    Sum<Rational<int> > x("x",0), u("u",0), result;
    Vector<Sum<Rational<int> > > u0(n), w(n);

    u.depend(x);
    u0[0] = x*x + x*u - u*u;

    for(i=1; i<n; i++) u0[i] = df(u0[i-1],x);

    cout << u0 << endl;

    // initial condition u(0)=1
    x.set(0); u.set(1);
    u0[0] = u0[0].value();

    w[0] = u;
    for(i=1; i<n; i++) w[i] = df(w[i-1],x);

    // substitution of initial conditions
    for(i=1; i<n; i++)
    {
        for(j=i; j>0; j--) u0[i].put(w[j],u0[j-1]);
        u0[i].expand();
```

```
        }
        cout << u0 << endl;

        x = zero; u = one;
        cout << u0 << endl;

        // Taylor series expansion
        x.clear();
        result = one;
        for(i=0; i<n; i++)
            result += Rational<int>(1,factorial(i+1))*u0[i]*power(x,i+1);
        cout << "u = " << result << endl;
    }
```

```
Result
======
[x^(2)+x*u-u^(2)]
[2*x+u+x*df(u,x)-2*u*df(u,x)]
[2+2*df(u,x)+x*df(df(u,x),x)-2*df(u,x)^(2)-2*u*df(df(u,x),x)]
[3*df(df(u,x),x)+x*df(df(df(u,x),x),x)-6*df(u,x)*df(df(u,x),x)
-2*u*df(df(df(u,x),x),x)]

[-1]
[x+3*u]
[x^(2)+x*u-2-6*u^(2)]
[7*x+31*u+x^(3)-x^(2)*u-8*x*u^(2)+12*u^(3)]

[-1]
[3]
[-8]
[43]

u = 1-x+3/2*x^(2)-4/3*x^(3)+43/24*x^(4)
```

8.7.5   Commutator of Two Vector Fields

Consider an autonomous system of first order ordinary differential equations

$$\frac{d\mathbf{x}}{dt} = V(\mathbf{x})$$

where $V_j : \mathbf{R}^n \to \mathbf{R}$ and $j = 1, 2, \ldots, n$. We assume that the functions V_j are analytic. We can associate the vector field

$$V = \sum_{j=1}^{n} V_j(\mathbf{x}) \frac{\partial}{\partial x_j}$$

with the differential equation. Let W be another vector field

$$W = \sum_{k=1}^{n} W_k(\mathbf{x}) \frac{\partial}{\partial x_k}.$$

We define the *commutator* of V and W as follows

$$[V, W] := \sum_{k=1}^{n} \sum_{j=1}^{n} \left(V_j \frac{\partial W_k}{\partial x_j} - W_j \frac{\partial V_k}{\partial x_j} \right) \frac{\partial}{\partial x_k}.$$

We find that the commutator satisfies the following properties:

- $[V, W] = -[W, V]$

- $[V, U + W] = [V, U] + [V, W]$

- the *Jacobi identity* $[[V, W], U] + [[U, V], W] + [[W, U], V] = 0$

where U is analytic vector field on $\mathbf{R}^n$. The analytic vector fields define a *Lie algebra*.

In the following program, we consider three vector fields V, W, U and show that they satisfy the three properties listed above.

```cpp
// comm.cpp

#include <iostream.h>
#include "MSymbol.h"
#include "Vector.h"

const int n = 3;
Vector<Sum<int> > x(n);

Vector<Sum<int> > commutator(Vector<Sum<int> > V,Vector<Sum<int> > W)
{
   int j,k;
   Vector<Sum<int> > U(V.length(),0);

   for(k=0; k<n; k++)
      for(j=0; j<n; j++)
         U[k] += V[j]*df(W[k],x[j]) - W[j]*df(V[k],x[j]);
   return U;
}

void main()
{
   Vector<Sum<int> > V(n), W(n), U(n), Y(n);
   Sum<int>          x1("x1",0), x2("x2",0), x3("x3",0);
   x[0] = x1;     x[1] = x2;     x[2] = x3;

   V[0] = x1*x1; V[1] = x2*x3; V[2] = x3*x3;
   W[0] = x3;     W[1] = x1-x2; W[2] = x2*x3;
   U[0] = x2*x1; U[1] = x2;     U[2] = x1-x3;

   // [V,W] = -[W,V]
   Y = commutator(V,W) + commutator(W,V);
   cout << Y << endl;  // output : [0 0 0]

   // [V,U+W] = [V,U] + [V,W]
   Y = commutator(V,U+W) - (commutator(V,U)+commutator(V,W));
   cout << Y << endl;  // output : [0 0 0]

   // Jacobian Identity [[V,W],U] + [[U,V],W] + [[W,U],V] = 0
   Y =    commutator(commutator(V,W),U)
       + commutator(commutator(U,V),W)
       + commutator(commutator(W,U),V);
   cout << Y << endl;  // output : [0 0 0]
}
```

8.7.6   Lie Derivative and Killing Vector Field

Let M be a Riemannian manifold with *metric tensor field*

$$g = \sum_{j=1}^{4}\sum_{k=1}^{4} g_{jk}(\mathbf{x})dx_j \otimes dx_k$$

and V be a smooth vector field defined on M. The vector field V is called a *Killing vector field* if

$$L_V g = 0$$

where $L_V g$ denotes the *Lie derivative* of g with respect to V. The Lie derivative is linear, i.e.

$$L_V(T + S) = L_V T + L_V S$$

where T and S are (r, s) tensor fields. Furthermore the Lie derivative obeys the product rule

$$L_V(S \otimes T) = (L_V S) \otimes T + S \otimes (L_V T)$$

where S is an (r, s) tensor field and T is a (p, q) tensor field. Finally, we have in local coordinates

$$L_V dx_j = d(V \lrcorner \, dx_j) = dV_j = \sum_{k=1}^{4} \frac{\partial V_j}{\partial x_k} dx_k$$

where V is given by

$$V = \sum_{j=1}^{4} V_j(\mathbf{x}) \frac{\partial}{\partial x_j}.$$

Consequently, we obtain

$$L_V g = \sum_{j=1}^{4}\sum_{k=1}^{4}\sum_{l=1}^{4} \left(V_l \frac{\partial g_{jk}}{\partial x_l} + g_{lk}\frac{\partial V_l}{\partial x_j} + g_{jl}\frac{\partial V_l}{\partial x_k} \right) dx_j \otimes dx_k.$$

In the program, we consider the *Gödel metric tensor field*

$$g = dx_1 \otimes dx_1 - \frac{1}{2} \exp(2x_1) \, dx_2 \otimes dx_2 + dx_3 \otimes dx_3 - dx_4 \otimes dx_4$$

$$- \exp(x_1) \, dx_2 \otimes dx_4 - \exp(x_1) \, dx_4 \otimes dx_2.$$

The metric tensor field can also be written in matrix form:

$$(g_{jk}) = \begin{pmatrix} 1 & 0 & 0 & 0 \\ 0 & -\exp(2x_1)/2 & 0 & -\exp(x_1) \\ 0 & 0 & 1 & 0 \\ 0 & -\exp(x_1) & 0 & -1 \end{pmatrix}.$$

We show that the vector field

$$V = x_2 \frac{\partial}{\partial x_1} + \left(\exp(-2x_1) - \frac{x_2^2}{2} \right) \frac{\partial}{\partial x_2} - 2 \exp(-x_1) \frac{\partial}{\partial x_4}$$

is the Killing vector field of g.

```
// kill.cpp

#include <iostream.h>
#include "Vector.h"
#include "Matrix.h"
#include "Rational.h"
#include "MSymbol.h"

void main()
{
   int j, k, l, N=4;
   Rational<int> zero(0), one(1), half(1,2), two(2);
   Matrix<Sum<Rational<int> > > g(N,N), Lg(N,N,zero);
   Vector<Sum<Rational<int> > > V(N), x(N);
   Sum<Rational<int> > x1("x1",0),x2("x2",0),x3("x3",0),x4("x4",0);

   x[0] = x1; x[1] = x2; x[2] = x3; x[3] = x4;
```

```
// The Goedel metric
g[0][0] = one;  g[0][1] = zero;     g[0][2] = zero; g[0][3] = zero;
g[1][0] = zero; g[1][1] = -half*exp(two*x1);
g[1][2] = zero; g[1][3] = -exp(x1);
g[2][0] = zero; g[2][1] = zero;     g[2][2] = one;  g[2][3] = zero;
g[3][0] = zero; g[3][1] = -exp(x1); g[3][2] = zero; g[3][3] = -one;

// The Killing vector field of the Goedel metric
V[0] = x2;    V[1] = exp(-two*x1) - half*x2*x2;
V[2] = zero; V[3] = -two*exp(-x1);

// The Lie derivative
for(j=0; j<N; j++)
   for(k=0; k<N; k++)
      for(l=0; l<N; l++)
         Lg[j][k] += V[l]*df(g[j][k],x[l]) + g[l][k]*df(V[l],x[j])
                    + g[j][l]*df(V[l],x[k]);

cout << "The Goedel Metric, g\n" << g << endl;
cout << "The Killing vector field of the Goedel metric, V\n"
     << V << endl; cout << endl;
cout << "The Lie derivative of g with respect to V, Lg\n"
     << Lg << endl;
}
```

Result
======

The Goedel Metric, g
[1 0 0 0]
[0 -1/2*exp(2*x1) 0 -exp(x1)]
[0 0 1 0]
[0 -exp(x1) 0 -1]

The Killing vector field of the Goedel metric,
[x2]
[exp(-2*x1)-1/2*x2^(2)]
[0]
[-2*exp(-x1)]

The Lie derivative of g with respect to V, Lg
[0 0 0 0]
[0 0 0 0]
[0 0 0 0]
[0 0 0 0]

8.8 Matrix Class

8.8.1 Hilbert-Schmidt Norm

Let A and B be two arbitrary $n \times n$ matrices over $\mathbf{R}$. Define

$$(A, B) := \text{tr}(AB^T)$$

where B^T denotes the transpose of B and $\text{tr}()$ denotes the trace. We find that

$$
\begin{aligned}
(A, A) &\geq 0 \\
(A, B) &= (B, A) \\
(cA, B) &= c(A, B) \\
(A_1 + A_2, B) &= (A_1, B) + (A_2, B)
\end{aligned}
$$

where $c \in \mathbf{R}$. Thus (A, B) defines a scalar product for $n \times n$ matrices over $\mathbf{R}$. The scalar product induces a norm, which is given by

$$||A|| := \sqrt{(A, A)} = \sqrt{\sum_{i=1}^{n} \sum_{j=1}^{n} |a_{ij}|^2}.$$

The norm is called the *Hilbert-Schmidt norm*. The results can be extended to infinite dimensions when we impose the condition

$$\sum_{i=1}^{\infty} \sum_{j=1}^{\infty} |a_{ij}|^2 < \infty.$$

If an infinite dimensional matrix A satisfies this condition, we call A a *Hilbert-Schmidt operator*.

In the following program we consider a 2×2 symbolic matrix

$$A := \begin{pmatrix} a & b \\ b & a \end{pmatrix}.$$

The norm is implemented in the **Matrix** class which has been described in Chapter 6.

```
// Hilbert.cpp

#include <iostream.h>
#include "Matrix.h"
#include "MatNorm.h"
#include "MSymbol.h"

void main()
{
    int n = 2;
    Sum<double> a("a",0), b("b",0), y1;
    Matrix<Sum<double> > A(n,n);

    A[0][0] = a; A[0][1] = b;
    A[1][0] = b; A[1][1] = a;

    cout << "The " << n << "x" << n << " matrix A is \n" << A << endl;

    y1 = normH(A);
    cout << "The Hilbert-Schmidt norm of matrix A is " << y1 << endl;
    cout << endl;

    a = 2.0; b = 3.0;
    cout << "Put a = " << a << " and b = " << b << endl;
    cout << "The Hilbert-Schmidt norm of matrix A is "
         << y1 << " or " << y1.value() << endl;
}
```

Result
======
The 2x2 matrix A is
[a b]
[b a]

The Hilbert-Schmidt norm of matrix A is sqrt(2*a^(2)+2*b^(2))

Put a = 2 and b = 3
The Hilbert-Schmidt norm of matrix A is sqrt(26) or 5.09902

8.8.2 Lax Pair and Hamilton System

Consider the *Hamilton function* (*Toda lattice*)

$$H(\mathbf{p}, \mathbf{q}) = \frac{1}{2}(p_1^2 + p_2^2 + p_3^2) + \exp(q_1 - q_2) + \exp(q_2 - q_3) + \exp(q_3 - q_1)$$

where $\mathbf{p}$, $\mathbf{q}$ are the momentum and position vectors, respectively. The *Hamilton equations of motion* are given by

$$\frac{dp_j}{dt} = -\frac{\partial H}{\partial q_j}, \qquad \frac{dq_j}{dt} = \frac{\partial H}{\partial p_j}, \qquad \text{for } j = 1, 2, 3.$$

Introducing the quantities

$$a_j := \frac{1}{2} \exp\left(\frac{1}{2}(q_j - q_{j+1})\right), \qquad b_j := \frac{1}{2} p_j$$

and cyclic boundary conditions (i.e. $q_4 \equiv q_1$), we find that the Hamilton equations of motion take the form (with $a_3 = 0$)

$$\frac{da_j}{dt} = a_j(b_j - b_{j+1}), \quad \frac{db_1}{dt} = -2a_1^2, \quad \frac{db_2}{dt} = 2(a_1^2 - a_2^2), \quad \frac{db_3}{dt} = 2a_2^2 \qquad (8.9)$$

where $j = 1, 2$. Now, let us introduce the matrices

$$L := \begin{pmatrix} b_1 & a_1 & 0 \\ a_1 & b_2 & a_2 \\ 0 & a_2 & b_3 \end{pmatrix}, \qquad A := \begin{pmatrix} 0 & -a_1 & 0 \\ a_1 & 0 & -a_2 \\ 0 & a_2 & 0 \end{pmatrix}.$$

The eigenvalues of the matrix L are constants of motion and thus the coefficients of the characteristic polynomial are also constants of motion. The equations of motion (8.9) can be rewritten as the *Lax representation*:

$$\frac{dL}{dt} = [A, L](t). \qquad (8.10)$$

It can be shown that the first integrals of the system take the forms

$$\text{tr}(L^n) \quad \text{where } n = 1, 2, \ldots$$

In the following program, we demonstrate that equations (8.9) and (8.10) are equivalent. In the second part of the program, we show that

$$\text{tr} L = b_1 + b_2 + b_3 \quad \text{and} \quad \text{tr}(L^2) = b_1^2 + b_2^2 + b_3^2 + 2a_1^2 + 2a_2^2$$

are the first integrals. The determinant of L,

$$\det(L) = b_1 b_2 b_3 - b_1 a_2^2 - b_3 a_1^2$$

is also a first integral of the system. Can $\det(L)$ be expressed in the form $\text{tr}(L^n)$ for some $n > 0$? Notice that $\det(L)$ is the product of the eigenvalues of L and $\text{tr}(L)$ is the sum of the eigenvalues of L.

```
// Lax.cpp

#include <iostream.h>
#include "MSymbol.h"
#include "Matrix.h"
#include "Rational.h"

void main()
{
  Matrix<Sum<Rational<int> > > L(3,3), A(3,3), Lt(3,3); // Lt == dL/dt
  Sum<Rational<int> > a1("a1",0), a2("a2",0),
                      b1("b1",0), b2("b2",0), b3("b3",0),
                      a1t, a2t, b1t, b2t, b3t;
  Rational<int> zero(0);

  L[0][0] = b1;   L[0][1] = a1; L[0][2] = zero;
  L[1][0] = a1;   L[1][1] = b2; L[1][2] = a2;
  L[2][0] = zero; L[2][1] = a2; L[2][2] = b3;
  A[0][0] = zero; A[0][1] = -a1;  A[0][2] = zero;
  A[1][0] = a1;   A[1][1] = zero; A[1][2] = -a2;
  A[2][0] = zero; A[2][1] = a2;   A[2][2] = zero;

  Lt = A*L - L*A; cout << "Lt =\n" << Lt << endl;

  b1t = Lt[0][0]; b2t = Lt[1][1]; b3t = Lt[2][2];
  a1t = Lt[0][1]; a2t = Lt[1][2];
  cout << "b1t = " << b1t << ", b2t = " << b2t << ", b3t = "
       << b3t << endl;
  cout << "a1t = " << a1t << ", a2t = " << a2t << endl;
```

```
    cout << endl;

    // Show that I[0],I[1],I[2] are first integrals
    int i, n=3;
    Sum<Rational<int> > result;
    Vector<Sum<Rational<int> > > I(n);

    I[0] = L.trace();        cout << "I[0] = " << I[0] << endl;
    I[1] = (L*L).trace();    cout << "I[1] = " << I[1] << endl;
    I[2] = L.determinant(); cout << "I[2] = " << I[2] << endl;
    cout << endl;

    for(i=0; i<n; i++)
    {
        result = b1t*df(I[i],b1) + b2t*df(I[i],b2) + b3t*df(I[i],b3)
                + a1t*df(I[i],a1) + a2t*df(I[i],a2);
        cout << "result" << i+1 << " = " << result << endl;
    }
}
Result
======

Lt =
[-2*a1^(2) -a1*b2+b1*a1 0]
[a1*b1-b2*a1 2*a1^(2)-2*a2^(2) -a2*b3+b2*a2]
[0 a2*b2-b3*a2 2*a2^(2)]

b1t = -2*a1^(2), b2t = 2*a1^(2)-2*a2^(2), b3t = 2*a2^(2)
a1t = -a1*b2+b1*a1, a2t = -a2*b3+b2*a2

I[0] = b1+b2+b3
I[1] = b1^(2)+2*a1^(2)+b2^(2)+2*a2^(2)+b3^(2)
I[2] = b1*b2*b3-b1*a2^(2)-a1^(2)*b3

result1 = 0
result2 = 0
result3 = 0
```

In the program, Lt corresponds to dL/dt and b1t, b2t, b3t, a1t, a2t correspond to db_1/dt, db_2/dt, db_3/dt, da_1/dt, da_2/dt, respectively.

The output shows that (8.9) is equivalent to (8.10) by comparing db_1/dt to the $(0,0)$-th, db_2/dt to the $(1,1)$-th, ... entry of the matrix L. **result1** = 0, **result2** = 0, **result3** = 0 indicate that I[0], I[1] and I[2] are first integrals of the system.

8.8.3   Padé Approximant

When a power series of a function diverges, the function has singularities in a certain region. A *Padé approximant* is a ratio of polynomials that contains the same information as the power series over an interval, often with information about whether singularities exist. In the $[N, M]$ Padé approximant the numerator has degree M and the denominator has degree N. The coefficients are determined by equating like powers of x in the following equation

$$f(x)Q(x) - P(x) = Ax^{M+N+1} + Bx^{M+N+2} + \ldots \quad \text{with } Q(0) = 1$$

where $P(x)/Q(x)$ is the $[N, M]$ Padé approximant to $f(x)$.

Suppose the solution to a differential equation can be expressed by a kth order Taylor series

$$f(x) = a_0 + a_1 x + a_2 x^2 + \ldots + a_k x^k.$$

The $[N, M]$ Padé approximant $P_N^M(x)$ is given explicitly in terms of the coefficients a_j

$$[N, M](x) := \frac{\det \begin{vmatrix} a_{M-N+1} & a_{M-N+2} & \cdots & a_{M+1} \\ \vdots & \vdots & & \vdots \\ a_M & a_{M+1} & \cdots & a_{M+N} \\ \sum_{j=N}^{M} a_{j-N} x^j & \sum_{j=N-1}^{M} a_{j-N+1} x^j & \cdots & \sum_{j=0}^{M} a_j x^j \end{vmatrix}}{\det \begin{vmatrix} a_{M-N+1} & a_{M-N+2} & \cdots & a_{M+1} \\ \vdots & \vdots & & \vdots \\ a_M & a_{M+1} & \cdots & a_{M+N} \\ x^N & x^{N-1} & \cdots & 1 \end{vmatrix}}$$

with $N + M + 1 = k$. Note that $a_j \equiv 0$ if $j < 0$, and the sums for which the initial value is larger than the final value are taken to be zero. It often happens that $P_N^M(x)$ converges to the true solution of the differential equation as $N, M \to \infty$ even when the Taylor series solution diverges. Usually we only consider the convergence of the Padé sequence

$$\left\{ P_0^J(x), P_1^{J+1}(x), P_2^{J+2}(x), \ldots \right\}$$

having $M = N + J$ and J is held constant while $N \to \infty$. The special sequence with $J = 0$ is called the *diagonal sequence*.

In the following program we calculate the Padé approximant $[1,1]$, $[2,2]$ and $[3,3]$ for

$$f(x) = \sin(x) = \sum_{k=0}^{\infty} \frac{(-1)^k x^{2k+1}}{(2k+1)!} = x - \frac{1}{6} x^3 + \cdots$$

Specifically, $[2,2]$ can be calculated as follows:

$$[2,2] = \frac{\det \begin{vmatrix} a_1 & a_2 & a_3 \\ a_2 & a_3 & a_4 \\ a_0 x^2 & a_0 x + a_1 x^2 & a_0 + a_1 x + a_2 x^2 \end{vmatrix}}{\det \begin{vmatrix} a_1 & a_2 & a_3 \\ a_2 & a_3 & a_4 \\ x^2 & x & 1 \end{vmatrix}}$$

with $a_0 = 0, a_1 = 1, a_2 = 0, a_3 = -1/6$ and $a_4 = 0$. We find that

$$[2,2] = \frac{x}{1 + \dfrac{x^2}{6}} .$$

```
// Pade.cpp
#include <iostream.h>
#include "MSymbol.h"
#include "Rational.h"
#include "Vector.h"
#include "Matrix.h"

Sum<Rational<int> >
Taylor(Sum<Rational<int> > u, Sum<Rational<int> > &x, int n)
{
   x.set(Rational<int>(0));
   Sum<Rational<int> > series(u.value());
   int j, fac = 1;
   for(j=1; j<=n; j++)
   {
      u = df(u,x); fac = fac * j;
      series += Rational<int>(u.nvalue(),fac) * power(x,j);
   }
```

```
      return series;
}


Sum<Rational<int> >
Pade(const Sum<Rational<int> > &f,Sum<Rational<int> > &x,int N,int M)
{
   int i, j, k, N1 = N+1, M1 = M+1, n = M + N1;
   Rational<int> zero(0);
   Sum<Rational<int> > y, z;
   Vector<Sum<Rational<int> > > a(n);
   Matrix<Sum<Rational<int> > > P(N1,N1), Q(N1,N1);
   y = Taylor(f,x,n);
   for(i=0; i<n; i++) a[i] = y.coeff(x,i);
   for(i=0; i<N; i++)
      for(j=0; j<N1; j++)
      {
         k = M-N+i+j+1;
         if(k >= 0) P[i][j] = Q[i][j] = a[k];
         else       P[i][j] = Q[i][j] = zero;
      }

   for(i=0; i<N1; i++)
   {
      for(j=N-i; j<M1; j++)
      {
         k = j-N+i;
         if(k >= 0) P[N][i] += a[k]*power(x,j);
      }
      Q[N][i] = power(x,N-i);
   }
   y = P.determinant();
   z = Q.determinant();
   return y/z;
}


void main()
{
   Sum<Rational<int> > x("x",0), f, g;
   f = sin(x);
   cout << Pade(f,x,1,1) << endl; // => x
   cout << Pade(f,x,2,2) << endl; // => -1/6*x*(-1/6-1/36*x^(2))^(-1)
   cout << Pade(f,x,3,3) << endl;
   // => (7/2160*x-1589885/3306816*x^(3))*(7/2160+7/43200*x^(2))^(-1)
}
```

8.9 Array and Symbolic Classes

8.9.1 Pseudospherical Surfaces and Soliton Equations

In Section 3.2.2, we described how to find the *sine-Gordon equation* from a metric
tensor field. An implementation using REDUCE was given. Here we give an implementation using SymbolicC++. In this case u is declared as a variable dependent
on x1 and x2. The operator df() denotes differentiation. Since terms of the form
$\cos^2(x)$ and $\sin^2(x)$ result from our calculation, we have included the identity

$$\sin^2(u) + \cos^2(u) \equiv 1$$

to simplify the expressions. This is done using the put() function in MSymbol.h.

```
// tensor.cpp

#include <iostream.h>
#include "Matrix.h"
#include "Array.h"
#include "MSymbol.h"

void main()
{
    int a,b,m,n,c,K=2;
    Matrix<Sum<double> > g(K,K),
                         g1(K,K);    // inverse of g
    Array1<Sum<double> >  x(K);
    Array2<Sum<double> >  Ricci(K,K), Ricci1(K,K);
    Array3<Sum<double> >  gamma(K,K,K);
    Array4<Sum<double> >  R(K,K,K,K);
    Sum<double> u("u",0), sum, RR,
                x1("x1",0), x2("x2",0);

    g[0][0] = 1;      g[0][1] = cos(u);
    g[1][0] = cos(u); g[1][1] = 1;

    x[0] = x1; x[1] = x2;

    g1 = g.inverse();
    u.depend(x1); u.depend(x2);

    for(a=0; a<K; a++)
        for(m=0; m<K; m++)
            for(n=0; n<K; n++)
            {
```

```
            sum = 0;
            for(b=0; b<K; b++)
               sum += g1[a][b]*(df(g[b][m],x[n]) + df(g[b][n],x[m])
                        - df(g[m][n],x[b]));
            gamma[a][m][n] = 0.5*sum;

            cout << "gamma(" << a << "," << m << "," << n << ") = "
                  << gamma[a][m][n] << endl;
        }
   cout << endl;

   for(a=0; a<K; a++)
      for(m=0; m<K; m++)
         for(n=0; n<K; n++)
            for(b=0; b<K; b++)
            {
               R[a][m][n][b] = df(gamma[a][m][b],x[n])
                              -df(gamma[a][m][n],x[b]);

               for(c=0; c<K; c++)
               {
                  R[a][m][n][b] += gamma[a][c][n]*gamma[c][m][b]
                              - gamma[a][c][b]*gamma[c][m][n];
               }
               R[a][m][n][b].put(cos(u)*cos(u),1-sin(u)*sin(u));
               }

   for(m=0; m<K; m++)
      for(n=0; n<K; n++)
      {
         Ricci[m][n] = 0;
         for(b=0; b<K; b++) Ricci[m][n] += R[b][m][b][n];

         cout << "Ricci(" << m << "," << n << ") = "
               << Ricci[m][n] << endl;
      }
   cout << endl;

   for(m=0; m<K; m++)
      for(n=0; n<K; n++)
      {
         Ricci1[m][n] = 0;
         for(b=0; b<K; b++) Ricci1[m][n] += g1[m][b]*Ricci[n][b];
      }
```

```
        RR = 0;

        for(b=0; b<K; b++) RR += Ricci1[b][b];
        RR.put(cos(u)*cos(u),1-sin(u)*sin(u));
        RR.expand();
        RR.put(df(df(u,x2),x1),df(df(u,x1),x2));
        cout << "R = " << RR << endl;
}
```

```
Result
======
gamma(0,0,0) = -cos(u)*(-1+cos(u)^(2))^(-1)*df(u,x0)*sin(u)
gamma(0,0,1) = 0
gamma(0,1,0) = 0
gamma(0,1,1) = (-1+cos(u)^(2))^(-1)*df(u,x1)*sin(u)
gamma(1,0,0) = (-1+cos(u)^(2))^(-1)*df(u,x0)*sin(u)
gamma(1,0,1) = 0
gamma(1,1,0) = 0
gamma(1,1,1) = -cos(u)*(-1+cos(u)^(2))^(-1)*df(u,x1)*sin(u)

Ricci(0,0) = -sin(u)^(-1)*df(df(u,x0),x1)
Ricci(0,1) = -cos(u)*sin(u)^(-1)*df(df(u,x0),x1)
Ricci(1,0) = -cos(u)*sin(u)^(-1)*df(df(u,x1),x0)
Ricci(1,1) = -sin(u)^(-1)*df(df(u,x1),x0)

R = -2*sin(u)^(-1)*df(df(u,x1),x2)
```

8.10   Polynomial and Symbolic Classes

8.10.1   Picard's Method

We consider *Picard's method* to approximate a solution to the differential equation

$$\frac{dy}{dx} = f(x, y)$$

with initial condition $y(x_0) = y_0$, where f is an analytic function of x and y. Integrating both sides yields

$$y(x) = y_0 + \int_{x_0}^{x} f(s, y(s))ds.$$

Now starting with y_0 this formula can be used to approach the exact solution iteratively if the procedure converges. The next approximation is given by

$$y_{n+1}(x) = y_0 + \int_{x_0}^{x} f(s, y_n(s))ds.$$

The example approximates the solution of the linear differential equation

$$\frac{dy}{dx} = x + y$$

and the nonlinear differential equation

$$\frac{dy}{dx} = x + y^2$$

using five and four steps of Picard's method. The initial conditions are $y(x = 0) = 1$. We also give the value $y(x = 2)$ for these approximations. In the first program `picard.cpp` we use the `Polynomial` class and integration from this class. In the second program `picard1.cpp` we use integration from the `Symbolic` class.

```cpp
// picard.cpp

#include "poly.h"
#include "rational.h"

int main(void)
{
 Polyterm<Rational<Verylong> > x("x");
 Polynomial<Rational<Verylong> > pic(x);
 Rational<Verylong> zero("0"), one("1"), two("2");
 int i;

 cout << endl << "x+y up to fifth approximation :" << endl;
 pic = one;
 cout << pic << endl;
 for(i=1;i<=5;i++)
 {
  //integrate and evaluate at the boundaries x and zero
  pic = one + Int(x+pic)-(Int(x+pic))(zero);
  cout << pic << endl;
 }
 cout << "The approximation at x=2 gives " << pic(two) << endl;

 cout << endl << "x+y^2 up to fourth approximation :" << endl;
 pic = one;
 cout << pic << endl;
 for(i=1;i<=4;i++)
 {
  //integrate and evaluate at the boundaries x and zero
  pic = one+Int(x+(pic^2))-(Int(x+(pic^2)))(zero);
  cout << pic << endl;
 }
 cout << "The approximation at x=2 gives " << pic(two) << endl;

 return 0;
}
```

```
Results
=======
```

```
x+y up to fifth approximation :
(1)
(1/2)x^2+x+(1)
(1/6)x^3+x^2+x+(1)
(1/24)x^4+(1/3)x^3+x^2+x+(1)
(1/120)x^5+(1/12)x^4+(1/3)x^3+x^2+x+(1)
(1/720)x^6+(1/60)x^5+(1/12)x^4+(1/3)x^3+x^2+x+(1)
The approximation at x=2 gives 523/45
```

```
x+y^2 up to fourth approximation :
(1)
(1/2)x^2+x+(1)
(1/20)x^5+(1/4)x^4+(2/3)x^3+(3/2)x^2+x+(1)
(1/4400)x^11+(1/400)x^10+(31/2160)x^9+(29/480)x^8+(233/1260)x^7
  +(13/30)x^6+(49/60)x^5+(13/12)x^4+(4/3)x^3+(3/2)x^2+x+(1)
(1/445280000)x^23+(1/19360000)x^22+(607/997920000)x^21
+(943/190080000)x^20+(265897/8532216000)x^19
+(569963/3592512000)x^18+(5506583/8143027200)x^17
+(1963837/798336000)x^16+(1350761/174636000)x^15
+(2967707/139708800)x^14+(190159/3706560)x^13
+(1096099/9979200)x^12+(116621/554400)x^11
+(2195/6048)x^10+(6479/11340)x^9+(1657/2016)x^8
+(271/252)x^7+(13/10)x^6+(17/12)x^5+(17/12)x^4
+(4/3)x^3+(3/2)x^2+x+(1)
```

```
The approximation at x=2 gives 2956833855481454447/86964389386875
```

```cpp
// spicard.cpp

#include "MSymbol.h"
#include "Rational.h"

int main(void)
{
 Sum<double> x("x",0);
 Sum<double> pic;
 Sum<double> zero(0.0),one(1.0),two(2.0);
 int i;

 cout<<endl<<"x+y up to fifth approximation :"<<endl;
 x.set(0.0);
 pic=one;
 cout<<pic<<endl;
 for(i=1;i<=5;i++)
 {
  //integrate and evaluate at the boundaries x and zero
  pic=one+Int(x+pic,x)-Int(x+pic,x).value();
  cout<<pic<<endl;
 }
 x.set(2.0);
 cout<<"The approximation at x=2 gives "<<pic.value()<<endl;

 cout<<endl<<"x+y^2 up to fourth approximation :"<<endl;
 x:set(0.0);
 pic=one;
 cout<<pic<<endl;
 for(i=1;i<=4;i++)
 {
  //integrate and evaluate at the boundaries x and zero
  pic=one+Int(x+pic*pic,x)-Int(x+pic*pic,x).value();
  cout<<pic<<endl;
 }
 x.set(2.0);
 cout<<"The approximation at x=2 gives "<<pic.value()<<endl;

 return 0;
}
```

```
Results
=======


x+y up to fifth approximation :
1
1+0.5*x^(2)+x
1+x^(2)+x+0.166667*x^(3)
1+x^(2)+x+0.333333*x^(3)+0.0416667*x^(4)
1+x^(2)+x+0.333333*x^(3)+0.0833333*x^(4)+0.00833333*x^(5)
1+x^(2)+x+0.333333*x^(3)+0.0833333*x^(4)+0.0166667*x^(5)
+0.00138889*x^(6)
The approximation at x=2 gives 11.6222


x+y^2 up to fourth approximation :
1
1+0.5*x^(2)+x
1+1.5*x^(2)+x+0.666667*x^(3)+0.05*x^(5)+0.25*x^(4)
1+1.5*x^(2)+x+1.33333*x^(3)+1.08333*x^(4)+0.433333*x^(6)+0.816667*x^(5)
+0.0604167*x^(8)+0.184921*x^(7)+0.0143519*x^(9)+0.000227273*x^(11)
+0.0025*x^(10)
1+1.5*x^(2)+x+1.33333*x^(3)+1.41667*x^(4)+1.41667*x^(5)+1.0754*x^(7)
+1.3*x^(6)+0.57134*x^(9)+0.821925*x^(8)+0.36293*x^(10)+0.109838*x^(12)
+0.210355*x^(11)+0.0212421*x^(14)+0.0513034*x^(13)+0.00773472*x^(15)
+0.00245991*x^(16)+0.000158653*x^(18)+0.000676233*x^(17)
+4.96107e-006*x^(20)+3.11639e-005*x^(19)+6.08265e-007*x^(21)
+2.24578e-009*x^(23)+5.16529e-008*x^(22)
The approximation at x=2 gives 3400.05                                    •
```

8.11 Lie Series Techniques

Let us consider an autonomous system of ordinary differential equations

$$\frac{d\mathbf{x}}{dt} = \mathbf{f}(\mathbf{x}), \quad \mathbf{x}(t = 0) \equiv \mathbf{x}_0$$

where $\mathbf{x} = (x_1, \ldots, x_n)^T$ and $\mathbf{x}_0$ is the initial value at $t = 0$. Let f_j be analytic functions defined on $\mathbf{R}^n$. We consider the analytic vector field

$$V := \sum_{j=1}^{n} f_j(\mathbf{x}) \frac{\partial}{\partial x_j}.$$

Then the solution of the initial value problem, for sufficiently small t, can be written as

$$x_j(t) = \exp(tV) x_j|_{\mathbf{x}=\mathbf{x}(0)}$$

where $j = 1, 2, \ldots, n$. Expanding the exponential function yields

$$x_j(t) = x_j(0) + tV(x_j)|_{\mathbf{x}=\mathbf{x}(0)} + \frac{t^2}{2} V(V(x_j))|_{\mathbf{x}=\mathbf{x}(0)} + \ldots$$

This method is called the *Lie series technique*. Let us consider the *Lorenz model* which is given by

$$\frac{dx_1}{dt} = \sigma(x_2 - x_1)$$

$$\frac{dx_2}{dt} = -x_1 x_3 + r x_1 - x_2$$

$$\frac{dx_3}{dt} = x_1 x_2 - b x_3$$

where σ, b and r are positive constants. The vector field is given by

$$V = \sigma(x_2 - x_1) \frac{\partial}{\partial x_1} + (-x_1 x_3 + r x_1 - x_2) \frac{\partial}{\partial x_2} + (x_1 x_2 - b x_3) \frac{\partial}{\partial x_3}.$$

Hence the solution for the system is

$$
\mathbf{x}(t) = \begin{pmatrix} e^{tV} x_1 \\ e^{tV} x_2 \\ e^{tV} x_3 \end{pmatrix}_{\mathbf{x}=\mathbf{x}(0)}
$$

$$
= \begin{pmatrix} x_1(0) + tV(x_1)|_{\mathbf{x}=\mathbf{x}(0)} + \dfrac{t^2}{2}V(V(x_1))|_{\mathbf{x}=\mathbf{x}(0)} + \cdots \\[2ex] x_2(0) + tV(x_2)|_{\mathbf{x}=\mathbf{x}(0)} + \dfrac{t^2}{2}V(V(x_2))|_{\mathbf{x}=\mathbf{x}(0)} + \cdots \\[2ex] x_3(0) + tV(x_3)|_{\mathbf{x}=\mathbf{x}(0)} + \dfrac{t^2}{2}V(V(x_3))|_{\mathbf{x}=\mathbf{x}(0)} + \cdots \end{pmatrix}
$$

where

$$
\begin{aligned}
V(x_1) &= \sigma(x_2 - x_1) \\
V(x_2) &= -x_2 - x_1 x_3 + r x_1 \\
V(x_3) &= x_1 x_2 - b x_3
\end{aligned}
$$

and

$$
\begin{aligned}
V(V(x_1)) &= -\sigma^2(x_2 - x_1) + \sigma(-x_2 - x_1 x_3 + r x_1) \\
V(V(x_2)) &= \sigma(x_2 - x_1)(-x_3 + r) + x_2 + x_1 x_3 - r x_1 - x_1(x_1 x_2 - b x_3) \\
V(V(x_3)) &= \sigma x_2(x_2 - x_1) + x_1(-x_2 - x_1 x_3 + r x_1) - b(x_1 x_2 - b x_3).
\end{aligned}
$$

The Lorenz model possesses a number of interesting properties. The divergence is $-(\sigma + b + 1)$. Hence each small volume element shrinks to zero as $t \to \infty$, at a rate independent of x_1, x_2, x_3. However, this does not imply that each volume element shrinks to a point. For certain parameter values (for example $\sigma = 10$, $b = 8/3$ and $r = 28$) we find that nearby trajectories separate exponentially. The system shows chaotic behaviour.

In the following program, we apply the Lie series techniques to the Lorenz model. Here, we have expanded the exponential function up to second order. In the second part of the program, we iterate the Lie Series solution with $\sigma = 10$, $b = 8/3$ and $r = 28$.

```cpp
// Lie.cpp

#include <iostream.h>
#include "Vector.h"
#include "MSymbol.h"

const int N=3;

Vector<Sum<double> > x(N), xt(N);

// The vector field V
template <class T> T V(const T& ss)
{
   T sum(0);

   for(int i=0; i<N; i++) sum += xt[i]*df(ss,x[i]);

   return sum;
}

void main()
{
   int i,j;
   Sum<double> x1("x1",0), x2("x2",0), x3("x3",0), t("t",0),
               s("s",0), b("b",0), r("r",0),
               p1("p1",0), p2("p2",0), p3("p3",0);
   Vector<Sum<double> > xs(N), xg(N);
   double half = 0.5;
   x[0]  = x1; x[1]  = x2; x[2]  = x3;
   xg[0] = p1; xg[1] = p2; xg[2] = p3;

   // Lorenz model
   xt[0] = s*(x2-x1);
   xt[1] = -x2 - x1*x3 + r*x1;
   xt[2] = x1*x2 - b*x3;

   // Taylor series expansion up to order 2
   for(i=0; i<N; i++)
      xs[i] = x[i] + t*V(x[i]) + half*t*t*V(V(x[i]));

   cout << "xs =\n" << xs << endl;

   // Evolution of the approximate solution
   t.set(0.01); r.set(28.0); s.set(10.0); b.set(8.0/3.0);
```

```
    xg[0].set(0.8); xg[1].set(0.8); xg[2].set(0.8);

    for(j=0; j<100; j++)
    {
        for(i=0; i<N; i++) x[i].set(xg[i].nvalue());

        for(i=0; i<N; i++)
        {
            xg[i].set(xs[i].nvalue());
            cout << "x[" << i << "] = " << xg[i].value() << endl;
        }
        cout << endl;
    }
}
```

Result
======
xs =
[x1+t*s*x2-t*s*x1-0.5*t^(2)*s^(2)*x2+0.5*t^(2)*s^(2)*x1-0.5*t^(2)*x2*s
-0.5*t^(2)*x1*x3*s+0.5*t^(2)*r*x1*s]
[x2-t*x2-t*x1*x3+t*r*x1-0.5*t^(2)*s*x2*x3+0.5*t^(2)*s*x2*r
+0.5*t^(2)*s*x1*x3-0.5*t^(2)*s*x1*r+0.5*t^(2)*x2+0.5*t^(2)*x1*x3
-0.5*t^(2)*r*x1-0.5*t^(2)*x1^(2)*x2+0.5*t^(2)*b*x3*x1]
[x3+t*x1*x2-t*b*x3+0.5*t^(2)*s*x2^(2)-0.5*t^(2)*s*x1*x2
-0.5*t^(2)*x2*x1-0.5*t^(2)*x1^(2)*x3+0.5*t^(2)*r*x1^(2)
-0.5*t^(2)*x1*x2*b+0.5*t^(2)*b^(2)*x3]

x[0] = 0.81048
x[1] = 1.00861
x[2] = 0.786104

x[0] = 0.839826
x[1] = 1.22078
x[2] = 0.774439

x[0] = 0.886839
x[1] = 1.44137
x[2] = 0.76532

x[0] = 0.950876
x[1] = 1.67493
x[2] = 0.759203
....
```
```

8.12 Spectra of Small Spin Clusters

Consider the spin Hamilton operator

$$\hat{H} = a \sum_{j=1}^{3} \sigma_3(j)\sigma_3(j+1) + b \sum_{j=1}^{3} \sigma_1(j)$$

where a, b are real constants and

$$\sigma_i(1) = \sigma_i \otimes I \otimes I$$
$$\sigma_i(2) = I \otimes \sigma_i \otimes I$$
$$\sigma_i(3) = I \otimes I \otimes \sigma_i$$

where σ_i are the Pauli matrices given by

$$\sigma_1 := \begin{pmatrix} 0 & 1 \\ 1 & 0 \end{pmatrix}, \qquad \sigma_2 := \begin{pmatrix} 0 & -i \\ i & 0 \end{pmatrix}, \qquad \sigma_3 := \begin{pmatrix} 1 & 0 \\ 0 & -1 \end{pmatrix}.$$

Here, we adopt the cyclic boundary conditions, i.e.,

$$\sigma_3(4) \equiv \sigma_3(1).$$

The matrix representation of the first term on the right-hand side is a diagonal matrix

$$\sum_{j=1}^{3} \sigma_3(j)\sigma_3(j+1) = \mathrm{diag}(3a, -a, -a, -a, -a, -a, -a, 3a).$$

The second term leads to non-diagonal terms. The 8×8 symmetric matrix for $\hat{H}$ becomes

$$\begin{pmatrix}
3a & b & b & 0 & b & 0 & 0 & 0 \\
b & -a & 0 & b & 0 & b & 0 & 0 \\
b & 0 & -a & b & 0 & 0 & b & 0 \\
0 & b & b & -a & 0 & 0 & 0 & b \\
b & 0 & 0 & 0 & -a & b & b & 0 \\
0 & b & 0 & 0 & b & -a & 0 & b \\
0 & 0 & b & 0 & b & 0 & -a & b \\
0 & 0 & 0 & b & 0 & b & b & 3a
\end{pmatrix}.$$

In the following program, we calculate the spin Hamilton operator symbolically. We determine the trace and determinant of the matrix symbolically. Then we substitute

$$a = \frac{1}{2}, \qquad b = \frac{1}{3}$$

into the matrix and the determinant to get a matrix with numerical values.

```cpp
// spinS3.cpp

#include <iostream.h>
#include <fstream.h>
#include "MSymbol.h"
#include "Matrix.h"
#include "Rational.h"

ofstream fout("spinS3.dat");

Matrix<Sum<Rational<int> > > sigma(char coord,int index,int N)
{
   int i;
   Matrix<Sum<Rational<int> > > I(2,2);
   I.identity();
   Matrix<Sum<Rational<int> > > result, s(2,2);
   Sum<Rational<int> > zero(0), one(1);

   if(coord == 'x')
   {
      s[0][0] = zero;  s[0][1] = one;
      s[1][0] = one;   s[1][1] = zero;
   }
   else // 'z'
   {
      s[0][0] = one;   s[0][1] = zero;
      s[1][0] = zero;  s[1][1] = -one;
   }

   if(index == 0)
   {
      result = s;
      for(i=1; i<N; i++) result = kron(result,I);
   }
```

```
    else
    {
       result = I;
       for(i=1; i<index; i++)   result = kron(result,I);
       result = kron(result,s);
       for(i=index+1; i<N; i++) result = kron(result,I);
    }
    return result;
}


template <class T>
Matrix<T> H(int N,T a,T b)
{
    int i, size=pow(2,N);
    Matrix<T> result, part(size,size,T(0)), *sigmaX, *sigmaZ;
    sigmaX = new Matrix<T>[N];
    sigmaZ = new Matrix<T>[N];

    for(i=0; i<N; i++) sigmaX[i] = sigma('x',i,N);
    for(i=0; i<N; i++) sigmaZ[i] = sigma('z',i,N);
    for(i=1; i<=N; i++) part += sigmaZ[i-1]*sigmaZ[i%N];
    result = a * part;

    part.fill(T(0));

    for(i=0; i<N; i++) part += sigmaX[i];
    result += b * part;

    delete [] sigmaX; delete [] sigmaZ;
    return result;
}


void main()
{
    int N=3, size = pow(2,N);
    Rational<int> c(1,2), d(1,3);
    Sum<Rational<int> > a("a",0), b("b",0), p("p",0), det;
    Matrix<Sum<Rational<int> > > result, HI, I(size,size);
    I.identity();
    result = H(N,a,b);
    fout << result << endl;
    fout << "trace = " << result..trace() << endl;

    det = result.determinant();
```

```
    fout << "determinant = " << det << endl;
    fout << endl;

    // Assigning numerical values
    a = c; b = d;
    fout << "Put a = " << a << " and b = " << d << endl;
    fout << endl;

    fout << "The matrix becomes:" << endl;
    fout << result << endl;

    fout << "determinant = " << det << endl;
}
Result
======
[3*a b b 0 b 0 0 0]
[b -a 0 b 0 b 0 0]
[b 0 -a b 0 0 b 0]
[0 b b -a 0 0 0 b]
[b 0 0 0 -a b b 0]
[0 b 0 0 b -a 0 b]
[0 0 b 0 b 0 -a b]
[0 0 0 b 0 b b 3*a]

trace = 0
determinant = 9*a^(8)-36*a^(6)*b^(2)+54*a^(4)*b^(4)-36*a^(2)*b^(6)
              +9*b^(8)

Put a = 1/2 and b = 1/3

The matrix becomes:
[3/2 1/3 1/3 0 1/3 0 0 0]
[1/3 -1/2 0 1/3 0 1/3 0 0]
[1/3 0 -1/2 1/3 0 0 1/3 0]
[0 1/3 1/3 -1/2 0 0 0 1/3]
[1/3 0 0 0 -1/2 1/3 1/3 0]
[0 1/3 0 0 1/3 -1/2 0 1/3]
[0 0 1/3 0 1/3 0 -1/2 1/3]
[0 0 0 1/3 0 1/3 1/3 3/2]

determinant = 625/186624
```

8.13 Nonlinear Maps and Chaotic Behaviour

We consider the map

$$
\begin{aligned}
x_{t+1} &= r(3y_t + 1)x_t(1 - x_t) \\
y_{t+1} &= r(3x_t + 1)y_t(1 - y_t)
\end{aligned}
$$

which shows the Ruelle–Takens–Newhouse transition into chaos. The bifurcation parameter is r. We calculate the variational equation symbolically

$$
\begin{aligned}
u_{t+1} &= r(3y_t + 1)(1 - 2x_t)u_t + 3rx_t(1 - x_t)v_t \\
v_{t+1} &= 3ry_t(1 - y_t)u_t + r(3x_t + 1)(1 - 2y_t)v_t
\end{aligned}
$$

and then iterate these four equations using the data type **double**. The largest Ljapunov exponent is calculated approximately for $r = 1.0834$

$$
\lambda \approx \frac{1}{T} \ln \left(|u_T| + |v_T| \right)
$$

where T is large. The fixed points of the map are given as the solutions of

$$
r(3y^* + 1)x^*(1 - x^*) = x^*, \qquad r(3x^* + 1)y^*(1 - y^*) = y^*.
$$

We find

$$
x_1^* = \frac{1}{3r}\left(-\sqrt{4r^2 - 3r} + r\right), \qquad y_1^* = \frac{1}{3r}\left(-\sqrt{4r^2 - 3r} + r\right),
$$

$$
x_2^* = \frac{1}{3r}\left(\sqrt{4r^2 - 3r} + r\right), \qquad y_2^* = \frac{1}{3r}\left(\sqrt{4r^2 - 3r} + r\right),
$$

$$
x_3^* = \frac{r - 1}{r}, \qquad y_3^* = 0,
$$

$$
x_4^* = 0, \qquad y_4^* = 0,
$$

$$
x_5^* = 0, \qquad y_5^* = \frac{r - 1}{r}.
$$

The fixed points (x_1^*, y_1^*) and (x_2^*, y_2^*) exist only for $r \geq 3/4$.

In the following program, we calculate symbolically the variational equations of the
map described above. Then, we calculate the Ljapunov exponent with $r = 1.0834$
and the initial conditions

$$x_0 = 0.3, \qquad y_0 = 0.4, \qquad u_0 = 0.5, \qquad v_0 = 0.6.$$

From the result obtained, we see that the Ljapunov exponent for the map is approximately 0.22.

```
// var.cpp

#include <iostream.h>
#include <math.h>
#include "MSymbol.h"

template <class T> T f(T x,T y,T r)
{
   return r*(3*y+1)*x*(1-x);
}

template <class T> T g(T x,T y,T r)
{
   return r*(3*x+1)*y*(1-y);
}

void main()
{
   int T, N = 1000;
   double x2, y2, u2, v2;
   Sum<double> x("x",0), x1("x1",0), y("y",0), y1("y1",0), r("r",0),
               u("u",0), u1("u1",0), v("v",0), v1("v1",0);

   x1 = f(x,y,r);
   y1 = g(x,y,r);
   cout << "x1 = " << x1 << endl;
   cout << "y1 = " << y1 << endl;

   // Variational Equation
   u1 = df(x1,x)*u + df(x1,y)*v;
   v1 = df(y1,y)*v + df(y1,x)*u;
   cout << "u1 = " << u1 << endl;
   cout << "v1 = " << v1 << endl;
   cout << endl;
```

```
// Calculation of the Ljapunov exponent
// by iterating the four equations.

// Initial values
x.set(0.3); y.set(0.4); r.set(1.0834);
u.set(0.5); v.set(0.6);

for(T=1; T<N; T++)
{
    x2 = x1.nvalue();
    y2 = y1.nvalue();
    u2 = u1.nvalue();
    v2 = v1.nvalue();

    x.set(x2); y.set(y2); u.set(u2); v.set(v2);
    cout << "The Ljapunov exponent for T = " << T << " is "
         << log(fabs(u.nvalue())+fabs(v.nvalue()))/T << endl;
}
}
```

Result
======

```
x1 = 3*r*y*x-3*r*y*x^(2)+r*x-r*x^(2)
y1 = 3*r*x*y-3*r*x*y^(2)+r*y-r*y^(2)
u1 = 3*r*y*u-6*r*y*x*u+r*u-2*r*x*u+3*r*x*v-3*r*x^(2)*v
v1 = 3*r*x*v-6*r*x*y*v+r*v-2*r*y*v+3*r*y*u-3*r*y^(2)*u

The Ljapunov exponent for T = 1 is 0.420853
The Ljapunov exponent for T = 2 is 0.113849
The Ljapunov exponent for T = 3 is -0.2017
The Ljapunov exponent for T = 4 is -0.0178022
The Ljapunov exponent for T = 5 is 0.1242
The Ljapunov exponent for T = 6 is 0.224871
....


The Ljapunov exponent for T = 994 is 0.216136
The Ljapunov exponent for T = 995 is 0.216531
The Ljapunov exponent for T = 996 is 0.215979
The Ljapunov exponent for T = 997 is 0.215501
The Ljapunov exponent for T = 998 is 0.215698
The Ljapunov exponent for T = 999 is 0.215851
```

8.14   Numerical-Symbolic Application

Consider the equation

$$f(x) = 0$$

where it is assumed that f is at least twice differentiable. Let I be some interval containing a root of f. A method that approximates the root of f can be derived by taking the tangent line to the curve $y = f(x)$ at the point $(x_n, f(x_n))$ corresponding to the current estimate, x_n, of the root. The intersection of this line with the x-axis gives the next estimate to the root, x_{n+1}. The gradient of the curve $y = f(x)$ at the point $(x_n, f(x_n))$ is $f'(x_n)$. The tangent line at this point has the form $y = f'(x_n) * x + b$. Since this passes through $(x_n, f(x_n))$ we see that $b = f(x_n) - x_n * f'(x_n)$. Therefore the tangent line is

$$y = f'(x_n) * x + f(x_n) - x_n * f'(x_n).$$

To determine where this line cuts the x-axis we set $y = 0$. Taking the point of intersection as the next estimate, x_{n+1}, to the root, we have $0 = f'(x_n) * x_{n+1} + f(x_n) - x_n * f'(x_n)$. It follows that

$$x_{n+1} = x_n - \frac{f(x_n)}{f'(x_n)}.$$

This is the *Newton-Raphson scheme* [15], which has the following form

$$\text{next-estimate} = \text{current estimate} + \text{correction term} \quad .$$

The correction term is $-f(x_n)/f'(x_n)$ and this must be small when x_n is close to the root if convergence is to be achieved. This depends on the behaviour of $f'(x)$ near the root and, in particular, difficulty will be encountered when $f'(x)$ and $f(x)$ have roots close together. The Newton–Raphson method is of the form $x_{n+1} = g(x_n)$ with $g(x) := x - f(x)/f'(x)$. The order of the method can be examined. Differentiating this equation leads to $g'(x) = (f(x)f''(x))/((f'(x))^2)$. For convergence we require that

$$\left| \frac{f(x)f''(x)}{(f'(x))^2} \right| < 1$$

for all x in some interval I containing the root. Since $f(\alpha) = 0$, the above condition is satisfied at the root $x = \alpha$ provided that $f'(\alpha) \neq 0$. Then, provided that $g(x)$

is continuous, an interval I must exist in the neighbourhood of the root over which the condition is satisfied. Difficulty is sometimes encountered when the interval I is small because the initial guess must be taken within this interval. This usually arises when $f(x)$ and $f'(x)$ have roots close together since the correction term is inversely proportional to $f'(x)$.

In the following, we make use of the method described above to find the root of the function

$$f(x) = 4x - \cos(x).$$

In the program, we compare the built-in data type double and the symbolic iteration with numerical substitution at each step. The sequence converges to the value $x = 0.2426746806$.

```cpp
// Newton.cpp

#include <iostream.h>
#include <iomanip.h>    // for setprecision()
#include "MSymbol.h"

template <class T> T f(T x)  // f(x)
{
    return 4*x-cos(x);
}

void main()
{
    int i, N=7;
    double y, u0 = -1.0;
    Sum<double> x("x",0), ff, ff1, ff2, C;

    ff = f(x);
    ff1 = df(ff,x);
    ff2 = df(ff1,x);

    // Set numerical precision to 10 decimal places
    cout << setprecision(10);

    cout << "f(x) = " << ff << endl;
    cout << "f'(x) = " << ff1 << endl;
    cout << "f''(x) = " << ff2 << endl;
    cout << endl;
```

```
// ======= Condition for convergence =======
C = ff*ff2/ff1;
x.set(u0);
cout << "C = " << C << endl;
cout << "|C(x=u0)| = " << fabs(C.nvalue()) << endl;
cout << endl;

// ======= Symbolic computation =========
x.set(u0);
for(i=0; i<N; i++)
{
    y = (x - ff/ff1).nvalue();
    x.set(y);
    cout << "x = " << x.nvalue() << endl;
}
}
Result
======
f(x) = 4*x-cos(x)
f'(x) = 4+sin(x)
f''(x) = cos(x)

C = 4*x*cos(x)*(4+sin(x))^(-1)-cos(x)^(2)*(4+sin(x))^(-1)
|C(x=u0)| = 0.7766703403

x = 0.4374736734
x = 0.2466653145
x = 0.2426765006
x = 0.2426746806
x = 0.2426746806
x = 0.2426746806
x = 0.2426746806
```

8.15   Summary

In this chapter, we presented many useful applications in mathematics and physics, such as the Mandelbrot set, first integrals, Killing vector fields, Lie series technique and so on. We demonstrated the usefulness of the classes which we developed in previous chapters. However, the readers should take note that the applications presented here are far from being complete. Many possible fields of study will have applications in different aspects using computer algebra systems.

Chapter 9

Lisp and Computer Algebra

9.1 Introduction

Lisp is short for *List Processing*. It is a computer language that is used in many applications of artificial intelligence. One of its major qualities is that it can manipulate lists easily. Lisp was developed in the 1950s by John McCarthy [32]. It is one of the most commonly used languages for writing a computer algebra system. Reduce, Macsyma, Derive and Axiom are based on Lisp. The basic data types in Lisp are atoms and dotted pairs. Lists are built from dotted pairs. Most of the functions in Lisp operate on lists. Moreover arithmetic operations are also possible in Lisp.

In this chapter we show how a computer algebra system can be built using Lisp. This includes how simplification, differentiation, and polynomials are handled. We make use of functions and rely strongly on recursion. There are different dialects in Lisp. In this chapter we use *Portable Standard Lisp*. The programs listed here also run under *Common Lisp* with some small modifications. In Portable Standard Lisp, a function is indicated by (de), whereas in Common Lisp a function is indicated by (defun). A comment in Portable Standard Lisp is written as

```
% This is a single line comment in Portable Standard Lisp
```

whereas a comment line in Common Lisp is indicated by a semicolon (;).

```
; This is a single line comment in Common Lisp
```

The function (mapcan) is also implemented differently in Common Lisp and Portable Standard Lisp.

Lisp is described in a number of excellent textbooks [2], [60]. Some of them also discuss the implementations of symbolic manipulations [8], [20], [38], [58].

9.2   Basic Functions of Lisp

Values in Lisp are termed *S-expressions*, a contraction for symbolic expressions. An S-expression may be either an *atom*, which is written as a symbol, such as

```
a
apple
part2
```

or a *dotted pair*, written in the form

```
(s_1 . s_2)
```

where s_1 and s_2 stand for arbitrary S-expressions. Some examples of dotted pairs are

```
(a . b)
(a . (b1 . b2))
((u . v) . (x . (y . z)))
```

An important subset of the S-expressions is the *list*, which satisfies the following constraints

- An atom is a list if it is the atom nil.

- A dotted pair (s_1 . s_2) is a list if s_2 is a list.

The atom nil is regarded as the null list. Therefore, a list in LISP is a sequence whose components are S-expressions. The following S-expressions

```
nil
(apple . nil)
(a . (b . (c . nil)))
((apple . a2) . nil)
```

are all lists. Lists are more conveniently expressed in list notation

```
(s_1  s_2 ... s_n)
```

for n ≥ 0. This is an abbreviation of

```
(s_1 . (s_2 . ( ... (s_n . nil) ... )))
```

For example,

```
(a b c)
```

abbreviates

```
(a . (b . (c . nil)))
```

The list (a) abbreviates

    (a . nil)

and the list ( ) (the so-called empty list) abbreviates

    nil

Note the difference between the dotted pair (a . b), and the two-component list (a b) which may also be written as (a . (b . nil)). Obviously a and (a) are not equivalent, because the latter abbreviates the pair (a . nil).

S-expressions are the values manipulated by Lisp programs. However, Lisp programs are also S-expressions. For example, the notation for literals in Lisp is

    (quote S)

where S is an S-expression. Instead of (quote S) we can also write 'S. Syntactically, this is just a two-component list. The first component is the atom quote and the second is an S-expression S. Semantically, its value (relative to any state) is the S-expression S. For example, the value of

    (quote apple)

is the atom apple, and the value of

    (quote (a . b))

is the dotted pair (a . b). In Lisp, an unquoted atom used as an expression is an *identifier*, except when it is nil or t, which always denote themselves.

Lisp provides five primitive operations for constructing, selecting and testing S-expression values

          cons   car   cdr   atom   null

The function cons constructs a dotted pair from its two arguments. For example, the value of

    (cons (quote a) (quote b))

is (a . b). Of course, the actual parameters of an invocation need not be literals. For example, the value of

    (cons (quote a) (cons (quote b) nil))

is (a b). The functions car and cdr require a dotted pair (s_1 . s_2) as an argument and return the components s_1 and s_2, respectively. For example, the values of

```
(car (quote (a . b)))   and   (cdr (quote (a . b)))
```

are a and b, respectively. Next we consider the results of the functions cons, car and cdr on list arguments. The value of

```
(cons (quote a) (quote (b c d)))
```

is the list (a b c d). The result of applying car and cdr to a list argument is the first component of the list and the rest of the list, respectively. For example, the values of

```
(car (quote (a b c d)))   and   (cdr (quote (a b c d)))
```

are a and (b c d), respectively. Note also that the values of

```
(cons (quote a) nil),   (car (quote (a))),   (cdr (quote (a)))
```

are (a), a, and nil, respectively. The remaining two primitive functions in Lisp are predicates, for which the result is one of the atoms — nil (representing false) or t (representing true). The value of

```
(atom E)
```

is t if the S-expression E is an atom, and nil if it is a dotted pair. The function null takes one argument. It returns t if its argument evaluates to nil, and returns nil otherwise. An example is

```
(null nil)
```

where the return value is t. The function greaterp takes one or more arguments and returns t if they are ordered in decreasing order numerically. Otherwise, the function returns nil. For example,

```
(greaterp 9 4)    % return value : t
```

The function zerop takes a numeric argument and returns t if the argument evaluates to 0. Otherwise, the function returns nil. For example,

```
(zerop 1)         % return value : nil
```

The arithmetic function plus adds the arguments. For example,

```
(plus 2 4 -7)     % return value : -1
```

The arithmetic function times multiplies the arguments. For example,

```
(times 4 5 6 7)   % return value : 840
```

The function list accepts one or more arguments. It places all of its arguments in a list. This function can also be invoked with no arguments, in which case it returns nil.

The function cond is used for *conditional processing*. It takes zero or more arguments, called *cases*. Every case is a list, whose first element is a test and the remaining elements are actions. The cases in a cond are evaluated one at a time, from first to last. When one of the tests returns a non-nil value, the remaining cases are skipped. The actions in the case are evaluated and the result is returned. If none of the tests returns a non-nil value, cond returns nil. For example, the function length_list, defined as

```
(de length_list (L)
  (cond
      ((null L) 0)
      (t (plus 1 (length_list (cdr L))))))
```

can be applied as follows:

```
(setq L '(1 2 3 4 5 xx))   % return value : (1 2 3 4 5 xx)
(length_list L)            % return value : 6
```

The function mapcan takes two arguments: a function and a list. It maps the function over the elements in the list. The function used here must return a list, and the lists from the mapping are destructively spliced together. For example in Common Lisp we have

```
(mapcan 'cdr '((a b c) (d e)))   % return value : (b c e)
```

In Portable Standard Lisp this is expressed as

```
(mapcan '((a b c) (d e)) 'cdr)   % return value : (b c e)
```

Many functions in Lisp can be defined in terms of other functions. For example, caar can be defined in terms of car. It is, therefore, natural to ask whether there is a smallest set of primitives necessary to implement the language. In fact, there is no single "best" minimal set of primitives; it all depends on the implementation. One possible set of primitives might include car, cdr and cons for manipulation of S-expressions, quote, atom, read, print, eq for equality, cond for conditionals, setq for assignment, and defun for definitions.

9.3   Examples from Symbolic Computation

9.3.1   Polynomials

In this section we show how to add and multiply polynomials [12] of the form

$$p(x) = \sum_{j=0}^{n} a_j x^j.$$

First we have to give a representation for the polynomial p, using a list of dotted pairs. The first element of the dotted pair is the exponent and the second element is the factor. Thus the car of the dotted pair is the exponent and the cdr of the dotted pair is the coefficient, e.g.

```
(car '(3 . 5))   % return value : 3
(cdr '(3 . 5))   % return value : 5
```

Consider, for example, the polynomial

$$p(x) = 3x^2 + 7x + 1.$$

The representation as a list of dotted pairs is given by

```
((2 . 3) (1 . 7) (0 . 1))
```

Consider another example: the polynomial $3x^2 + 2$ is represented as

```
((2 .  3) (0 . 2))
```

Using this representation, the polynomial 0 can be represented by nil or ((0 . 0)), whereas the polynomial 1 is represented by ((0 . 1)).

In the following we give an implementation for the addition and multiplication of two polynomials [12]. *Recursion* is used here (i.e. the function calls itself). If the polynomial P has m terms and the polynomial Q has n terms, the calculation time (that is the number of Lisp operations) for add is bounded by $O(m + n)$, and for multiply is bounded by $O(m^2 n)$. Roughly speaking, we ought to sort the terms of the product so that they appear in decreasing order, and make use of the add function, corresponding to a sorting algorithm by insertion. Of course, the use of a better sorting method (such as quicksort) offers a more efficient multiplication algorithm, say $O(mn \log m)$. But most systems use an algorithm similar to the procedure described.

```
% poly.sl

(de add (P Q)
   (cond
      ((null P) Q)
      ((null Q) P)
      ((greaterp (caar P) (caar Q)) (cons (car P) (add (cdr P) Q)))
      ((greaterp (caar Q) (caar P)) (cons (car Q) (add P (cdr Q))))
      ((zerop (plus (cdar P) (cdar Q))) (add (cdr P) (cdr Q)))
      (t (cons (cons (caar P) (plus (cdar P) (cdar Q)))
         (add (cdr P) (cdr Q)))))))

(de multiply (P Q)
   (cond
      ((or (null P) (null Q)) nil)
      (t (cons (cons (plus (caar P) (caar Q)) (times (cdar P) (cdar Q)))
      (add (multiply (list (car P)) (cdr Q))
         (multiply (cdr P) Q)))))))

% Applications of the functions
% add and multiply are as follows:

(add '((2 . 3) (1 . 7) (0 . 1)) '((2 . 4) (0 . 2)))
(add '((0 . 0)) '((3 . 4) (1 . 2) (0 . 7)))
(multiply '((3 . 3) (0 . 2)) '((2 . 4) (1 . 1)))
(multiply '((3 . 3) (0 . 2)) '((0 . 0)))
```

In the first example, we add the two polynomials

$$(3x^2 + 7x + 1) + (4x^2 + 2) = 7x^2 + 7x + 3.$$

In the second example, we add the two polynomials

$$0 + (4x^3 + 2x + 7) = 4x^3 + 2x + 7.$$

In the third example, we multiply two polynomials

$$(3x^3 + 2) * (4x^2 + x) = 12x^5 + 3x^4 + 8x^2 + 2x.$$

In the fourth example, we multiply two polynomials

$$(3x^3 + 2) * 0 = 0.$$

The output of the last example is ((3 . 0) (0 . 0)) which is $0 * x^3 + 0 * x^0$. This simplifies to 0.

9.3.2   Simplifications

Here we show how simplification can be implemented in Lisp. We assume that the mathematical expression is given in *prefix notation*. This means an expression is arranged in the way that the operator appears before its operands.

In simp1.sl we implement the rules

```
+ x            =>  x
- 0            =>  0
exp(0)         =>  1
log(1)         =>  0
sin(0)         =>  0
cos(0)         =>  1
arcsin(0)      =>  0
arctan(0)      =>  0
sinh(0)        =>  0
cosh(0)        =>  1
```

Basically, the simplification is done by comparing the expression case by case. If there is a match, replace the expression by the simplified form. Notice that

```
(simp-unary '(plus (plus x)))
(simp-unary '(minus (plus x)))
(simp-unary '(plus (exp 0)))
(simp-unary '(exp (sin 0)))
(simp-unary '(log (exp (sin 0))))
```

are not simplified completely. This indicates that a recursion process is needed for the simplification to apply on different levels. Furthermore, we find that

```
(simp-unary '(plus x y))
```

gives a wrong answer, and

```
(simp-unary 'x)
(simp-unary '(x))
```

give error messages ! This has prompted us that extra attention has to be given to the number and types of the arguments for the function simp-unary. The next attempt (simp2.sl) tries to overcome these problems.

```
% simp1.sl

(de simp-unary (f)
   (prog (op opd)
   (setq op (car f))
   (setq opd (cadr f))
   (return (cond
      ((eq op 'plus)                      opd)  % + x        => x
      ((and (eq op 'minus) (zerop opd))    0)   % - 0        => 0
      ((and (eq op 'exp)   (zerop opd))    1)   % exp(0)     => 1
      ((and (eq op 'log)   (onep opd))     0)   % log(1)     => 0
      ((and (eq op 'sin)   (zerop opd))    0)   % sin(0)     => 0
      ((and (eq op 'cos)   (zerop opd))    1)   % cos(0)     => 1
      ((and (eq op 'arcsin) (zerop opd))   0)   % arcsin(0) => 0
      ((and (eq op 'arctan) (zerop opd))   0)   % arctan(0) => 0
      ((and (eq op 'sinh)  (zerop opd))    0)   % sinh(0)    => 0
      ((and (eq op 'cosh)  (zerop opd))    1)   % cosh(0)    => 1
      (t (list op opd)))))
)

% Applications of the simplification

(simp-unary '(plus x))   % => x
(simp-unary '(exp 0))    % => 1
(simp-unary '(log 1))    % => 0
(simp-unary '(cosh 0))   % => 1
(simp-unary '(exp 1))    % => (exp 1)
(simp-unary '(minus x))  % => (minus x)
(simp-unary '(minus 0))  % => 0

(simp-unary '(plus (plus x)))    % => (plus x)          <--- not simplified !
(simp-unary '(minus (plus x)))   % => (minus (plus x))  <--- not simplified !
(simp-unary '(plus (exp 0)))     % => (exp 0)           <--- not simplified !

(simp-unary '(exp (sin 0)))      % => (exp (sin 0))     <--- not simplified !
(simp-unary '(log (exp (sin 0))))
                                 % => (log (exp (sin 0))))  <--- not simplified !

(simp-unary 'x)                  % An attempt was made to do car on 'x',
                                 % which is not a pair <--- ERROR !

(simp-unary '(x))                % An attempt was made to do car on 'nil',
                                 % which is not a pair <--- ERROR !

(simp-unary '(plus x y))         % => x                 <--- WRONG !
```

In the program `simp2.sl`, we have successfully overcome all the problems existing in the previous version. In this version, simplification may be applied on

- atoms, for which the original expression is returned;

- a list with only one element, for which we apply the function `simp` again on the argument;

- a list with two elements, for which we apply the function `simp-unary`, with the first argument being the operator and the other being the operand;

- a list with more than two elements, for which the message `cannot_simplify` is printed, because this expression can only be simplified by a binary operator.

We have built one more level on top of the function `simp-unary`. The new function `simp` plays an important role in handling the arguments provided by the users. It checks for the types as well as the number of arguments. A different action is taken for each case.

In the function `simp-unary`, most of the statements are the same as the previous program `simp1.sl`. We have changed only one statement (line number 4) in this function to make it recursive:

```
(setq opd (simp (cadr f)))
```

This statement applies the simplification on the second argument of the function, enabling the simplification to apply on different levels of the operand.

Note that the statement

```
(simp '(minus (minus x)))
```

is not simplified because the program does not cater for the interaction between different levels of expressions, i.e. the `minus` operators do not interact and cancel out each other. Furthermore, we see that the statement

```
(simp '(plus x y))
```

prompts the message `cannot_simplify`. This is because we have not implemented the binary operators yet. In the next program `simp3.sl`, we consider the simplification of binary operators like `plus`, `minus`, `times`, `quotient` and `power`.

```
% simp2.sl

(de simp (f)
   (cond
      ((atom f) f)                        % f is an atom
      ((null (cdr f)) (simp (car f)))     % f has only one element
      ((null (cddr f)) (simp-unary f))    % f has two elements
      (t (quote cannot_simplify)) ))      % f has more than two elements

(de simp-unary (f)
   (prog (op opd)                         % local variables : op, opd
   (setq op (car f))
   (setq opd (simp (cadr f)))
   (return (cond
      ((eq op 'plus)                              opd)  % + x        => x
      ((and (eq op 'minus)  (zerop opd))  0)             % - 0        => 0
      ((and (eq op 'exp)    (zerop opd))  1)             % exp(0)     => 1
      ((and (eq op 'log)    (onep opd))   0)             % log(1)     => 0
      ((and (eq op 'sin)    (zerop opd))  0)             % sin(0)     => 0
      ((and (eq op 'cos)    (zerop opd))  1)             % cos(0)     => 1
      ((and (eq op 'arcsin) (zerop opd))  0)             % arcsin(0) => 0
      ((and (eq op 'arctan) (zerop opd))  0)             % arctan(0) => 0
      ((and (eq op 'sinh)   (zerop opd))  0)             % sinh(0)    => 0
      ((and (eq op 'cosh)   (zerop opd))  1)             % cosh(0)    => 1
      (t (list op opd)))))
)

% Applications of the simplification
(simp 'x)                  % => x
(simp '(x))                % => x

(simp '(plus x))           % => x
(simp '(plus (plus x)))    % => x
(simp '(plus (minus x)))   % => (minus x)
(simp '(plus (minus 0)))   % => 0
(simp '(minus (plus x)))   % => (minus x)

(simp '(minus (sin 0)))    % => 0
(simp '(plus (exp 0)))     % => 1
(simp '(minus (exp 0)))    % => (minus 1)
(simp '(plus (log 1)))     % => 0
(simp '(plus (cosh 0)))    % => 1
(simp '(exp (sin 0)))      % => 1
(simp '(log (exp (sin 0)))) % => 0

(simp '(minus (minus x)))  % => (minus (minus x))
(simp '(plus x y))         % => cannot_simplify
```

In the program `simp3.sl`, on top of the unary operators simplification, we implement the following binary operators:

plus   minus   times   quotient   power

In the function `simp`, we change the statement in line number 6 from

```
(t (quote cannot_simplify))
```

to

```
(t (simp-binary f))
```

This statement simply means if the argument contains more than two elements, apply the binary operator simplification.

In the new function `simp-binary`, we consider the following simplification:

- operation $x + y$:

  if $x = 0$ return y
  if $y = 0$ return x

- operation $x - y$:

  if $x = 0$,
      if $y = 0$ return 0
      else return $-y$
  if $y = 0$ return x
  if $x = y$ return 0

- operation $x * y$:

  if $x = 0$ or $y = 0$ return 0
  if $x = 1$ return y
  if $y = 1$ return x

- operation x/y:

  if $x = 0$ return 0
  if $y = 0$ return `infinity`
  if $y = 1$ return x

- operation x^y:

  if $x = 0$ return 0
  if $y = 0$ or $x = 1$ return 1
  if $y = 1$ return x

```
% simp3.sl

(de simp (f)
    (cond
        ((atom f) f)                          % f is an atom
        ((null (cdr f)) (simp (car f)))       % f has only one element
        ((null (cddr f)) (simp-unary f))      % f has two elements
        (t (simp-binary f)) ))                % f has more than two elements

(de simp-unary (f)
    (prog (op opd)                            % local variables: op, opd
    (setq op (car f))
    (setq opd (simp (cadr f)))
    (return (cond
        ((eq op 'plus)                opd)    % + x         => x
        ((and (eq op 'minus)  (zerop opd)) 0) % - 0         => 0
        ((and (eq op 'exp)    (zerop opd)) 1) % exp(0)      => 1
        ((and (eq op 'log)    (onep opd))  0) % log(1)      => 0
        ((and (eq op 'sin)    (zerop opd)) 0) % sin(0)      => 0
        ((and (eq op 'cos)    (zerop opd)) 1) % cos(0)      => 1
        ((and (eq op 'arcsin) (zerop opd)) 0) % arcsin(0) => 0
        ((and (eq op 'arctan) (zerop opd)) 0) % arctan(0) => 0
        ((and (eq op 'sinh)   (zerop opd)) 0) % sinh(0)    => 0
        ((and (eq op 'cosh)   (zerop opd)) 1) % cosh(0)    => 1
        (t (list op opd)))))
)

(de simp-binary (f)
    (prog (op opd1 opd2)
    (setq op (car f))
    (setq opd1 (simp (cadr f)))               % simplify first operand
    (setq opd2 (simp (caddr f)))              % simplify second operand

    (return (cond
        ((and (eq op 'plus)                   % operation: x + y
            (cond ((zerop opd1) opd2)         % if x=0 return y
                  ((zerop opd2) opd1))))      % if y=0 return x

        ((and (eq op 'minus)                  % operation: x - y
            (cond ((zerop opd1)               % if x=0,
                      (cond ((zerop opd2) 0)  %    if y=0 return 0
                            (t (list 'minus opd2)))) %    else return -y
                  ((zerop opd2) opd1)         % if y=0 return x
                  ((eq opd1 opd2) 0))))       % if x=y return 0

        ((and (eq op 'times)                  % operation: x * y
            (cond ((or (zerop opd1) (zerop opd2)) 0) % if x=0 or y=0 return 0
```

```
            ((onep opd1) opd2)                  % if x=1 return y
            ((onep opd2) opd1))))               % if y=1 return x

        ((and (eq op 'quotient)                 % operation: x / y
         (cond ((zerop opd1) 0)                 % if x=0 return 0
               ((zerop opd2) 'infinity)         % if y=0 return infinity
               ((onep opd2) opd1))))            % if y=1 return x

        ((and (eq op 'power)                    % operation: x^y
         (cond ((zerop opd1) 0)                 % if x=0 return 0
               ((or (zerop opd2) (onep opd1)) 1) % if y=0 or x=1 return 1
               ((onep opd2) opd1))))            % if y=1 return x

        (t (list op opd1 opd2)))))
)


% Applications of the simplification

(simp '(plus (plus x)))          % => x
(simp '(plus (minus x)))         % => (minus x)
(simp '(minus (minus x)))        % => (minus (minus x)) <- not simplified !

(simp '(plus (sin 0)))           % => 0
(simp '(plus (exp 0)))           % => 1
(simp '(plus (log 1)))           % => 0
(simp '(minus (cosh 0)))         % => (minus 1)
(simp '(minus (sin 0)))          % => 0
(simp '(minus (arctan 0)))       % => 0

(simp '(plus x 0))               % => x
(simp '(plus 0 x))               % => x
(simp '(minus x 0))              % => x
(simp '(times x 0))              % => 0
(simp '(quotient x 0))           % => infinity
(simp '(quotient 0 x))           % => 0
(simp '(minus (quotient x 0)))   % => (minus infinity)
(simp '(power x 0))              % => 1

(simp '(plus x y))                       % => (plus x y)
(simp '(plus (times x 0) (times x 1)))   % => x
(simp '(minus (plus 0 x) (times 1 x)))   % => 0

(simp '(minus (plus x y) (minus y x)))
           % => (minus (plus x y) (minus y x))  <- not simplified !
```

9.3.3 Differentiation

In this section, we consider the differentiation of algebraic (polynomial) expressions. The following rules are implemented

$$\frac{dc}{dx} = 0, \qquad c \text{ is a constant}$$

$$\frac{dx}{dx} = 1$$

$$\frac{d}{dx}(f(x) + g(x)) = \frac{df}{dx} + \frac{dg}{dx}$$

$$\frac{d}{dx}(f(x) - g(x)) = \frac{df}{dx} - \frac{dg}{dx}$$

$$\frac{d}{dx}(f(x) * g(x)) = f(x) * \frac{dg}{dx} + \frac{df}{dx} * g(x) \,.$$

In the program `differ.sl`, `ex` is the expression to be differentiated and `v` stands for the variable. The expression is given in *prefix notation*. A prefix notation is one in which the operator appears before its operands. For example, the mathematical expression

$$(3 + x) * (x - a),$$

when it is expressed in the prefix notation using Lisp, becomes

```
(times (plus 3 x) (minus x a))
```

Basically, the program proceeds by checking and matching the operators. When a match is found, it applies the corresponding differentiation rules on the expression. Sometimes, the expression becomes more complicated after the differentiation. Simplification of the expression becomes necessary. Therefore, we apply the simplification program `simp3.sl`, which we have just developed, when such a case arises.

The implementation of the function diff together with some applications is as follows.

```
% differ.sl

(de diff (ex v)                          % d(ex)/dv
   (cond
       ((atom ex)
          (cond ((eq ex v) 1) (t 0)))    % d(v)/dv = 1, d(constant)/dv = 0

       ((eq (car ex) 'plus)              % d(a+b)/dv = da/dv + db/dv
          (list 'plus (diff (cadr ex) v) (diff (caddr ex) v)))

       ((eq (car ex) 'times)             % d(a*b)/dv = da/dv * b + a * db/dv
          (list 'plus
              (list 'times (diff (cadr ex) v) (caddr ex))
              (list 'times (cadr ex) (diff (caddr ex) v))))

       ((eq (car ex) 'minus)             % d(a-b)/dv = da/dv - db/dv
          (list 'minus (diff (cadr ex) v) (diff (caddr ex) v))))
)

% Applications

(diff 'x 'x)  %  => 1
(diff '2 'x)  %  => 0
(diff 2 'x)   %  => 0
(diff 'x 'u)  %  => 0

(diff '(plus x x) 'x)              % => (plus 1 1)

(diff '(times x x) 'x)            % => (plus (times 1 x) (times x 1))
(simp (diff '(times x x) 'x))     % => (plus x x)

(diff '(times (plus 3 x) (minus a x)) 'x)
% => (plus (times (plus 0 1) (minus a x))
%     (times (plus 3 x) (minus 0 1)))

(simp (diff '(times (plus 3 x) (minus a x)) 'x))
% => (plus (minus a x) (times (plus 3 x) (minus 1)))

(diff '(times (times x x) x) 'x)
% => (plus (times (plus (times 1 x) (times 1 x)) x)
%     (times (times x x) 1))

(simp (diff '(times (times x x) x) 'x))
% => (plus (times (plus x x) x) (times x x))
```

9.4 Lisp System based on C++

In this section, we show how a Lisp system can be implemented using object-oriented programming with C++. In the program, we implement the following functions:

car   cdr   atom   cons   cond   append

The basic data structures of a Lisp system are the atom and dotted pair. Lists are built from dotted pairs. A list may consist of basic data types (e.g. int, double) or abstract data types (e.g. String, Verylong, Matrix).

Basically, the program consists of three classes related to each other as shown in Figure 9.1.

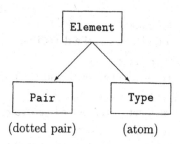

(dotted pair)   (atom)

Figure 9.1: Schematic diagram of the inheritance relationship

The inheritance hierarchy consists of an abstract base class Element and two derived classes: Pair and Type. The Pair class corresponds to the dotted pairs in Lisp, whereas the Type class corresponds to the atoms in Lisp.

The Element class, which is an abstract base class, specifies only the interface but not the implementation of the member functions available in the derived classes. All the functions in the class are declared as virtual. This allows the derived classes to override the definition of the functions. There are four member functions declared in the class:

- virtual Element* clone() const = 0;
  it declares a function that duplicates an Element.

- virtual void print(ostream&) const = 0;
  it declares a function that displays the content of an Element.

- virtual int atom() const = 0;
  it declares a function that checks if an Element is an atom.

- friend ostream& operator << (ostream&,const Element*);
  it declares an output stream function.

The structure and functionalities of the `Pair` class may be summarized as follows:

- Data field: The class contains only two data fields `_car` and `_cdr`. They represent the first and second element of the dotted pair, respectively.

- Constructors and destructor:
  - `Pair()`: The default constructor assigns NULL to both `_car` and `_cdr`.
  - `Pair(const Element *ee)`: it assigns `ee` to `_car` and NULL to `_cdr`.
  - `Pair(const Element *e1,const Element *e2)`: it assigns `e1` and `e2` to the first and second element of the dotted pair, respectively.
  - The copy constructor and destructor are overloaded as well.

- Member functions:
  - `Element* car() const;`
    it returns the first element of the dotted pair.
  - `Element* cdr() const;`
    it returns the second element of the dotted pair.
  - `void car(const Element *ee);`
    it sets the first element of the dotted pair to `ee`.
  - `void cdr(const Element *ee);`
    it sets the second element of the dotted pair to `ee`.
  - The virtual functions

        ```
        Element* clone() const;
        void print(ostream&) const;
        int atom() const;
        ```

    and the output stream operator

        ```
        ostream& operator << (ostream&,const Pair*);
        ```

    are overridden with new definitions for the class.

The `Type` class is implemented as a **template** class. Its structure and functionalities may be summarized as follows:

- Data field: The only data member in the class is `thing`. It represents the numerical value of an instance of the `Type` class.

- Constructors:
  - `Type()`: The default constructor sets the value of `thing` to zero.
  - `Type(T vv)`: It sets the value of `thing` to `vv`.
  - The copy constructor is overloaded.

- Member functions:

    - `const T& value() const;`
      it returns the value of thing.

    - The virtual functions

        ```
        Element* clone() const;
        void print(ostream&) const;
        int atom() const;
        ```

      and the output stream operator

        ```
        ostream& operator << (ostream&,const Type<T>*);
        ```

      are overridden with new definitions for the class.

A class has been derived from `Element` to provide the atoms `nil` and `t`. The atoms are passed as function parameters using the address of operator (`&`) for example `cons(&t,&nil)`; Notice that only the function `atom()` is declared as a member function for the class. The rest of the functions which we intend to implement are declared as global functions:

- `Element* car (const Element*);`
  It applies on a dotted pair and returns the first element of the pair. An error message is reported if the argument is an atom.

- `Element* cdr (const Element*);`
  It applies on a dotted pair and returns the second element of the pair. If the argument is an atom, an error message is reported.

- `Pair* cons(const Element *p1,const Element *p2);`
  It forms a dotted pair with `p1` and `p2` as the first and second element of the pair, respectively.

- `Pair* append(const Element *p1,const Element *p2);`
  It replaces the last element of the dotted pair `p1` with `p2`. Note that `p2` may be an atom or dotted pair(s). An error message is reported if `p1` is an atom.

- `Element* null(Element *e);`
  Returns `&t` if `e` points to the `nil` atom and `&nil` otherwise.

- `int is_lisp_list(Element *l);`
  Determines if `l` points to a valid Lisp list.

- `int is_nonempty_lisp_list(Element *l);`
  Determines if `l` points to a valid Lisp list which is not empty.

- `Element *cond(Element *c);`
  Implements the Lisp `cond` conditional statement. `c` must point to a non-empty
  list of lists where each list represents a condition to evaluate. `cond` returns `&nil`
  if no condition is satisfied.

In the following, we list the program of the implementation. The listing is followed
by an application program `lisprun.cpp`, which makes use of the functions. Notice
that the dotted pairs may contain elements of different types, such as `Verylong` and
`double`.

```
// Lisp.h

#ifndef __LISP_H
#define __LISP_H

#include <iostream.h>

// Abstract base class
class Element
{
public:
   virtual Element* clone() const = 0;
   virtual void print(ostream&) const = 0;
   virtual int atom() const = 0;
   friend ostream& operator << (ostream&,const Element*);
};

//unique class for atoms nil and t
class __nil_and_t_lisp_class : public Element
{
 private:
        int is_t;
 public:
        __nil_and_t_lisp_class(int i=0) : is_t(i) {}
        void print(ostream &s) const { s<<((is_t)?"t":"nil"); }
        Element *clone(void) const { return (Element*)this; }
        int atom(void) const {return 1;}
} nil(0), t(1);

class Pair : public Element  // Dotted pair
{
private:
   Element *_car;            // First element of the dotted pair
   Element *_cdr;            // Second element of the dotted pair
```

```
public:
   // Constructors
   Pair();
   Pair(const Element*);
   Pair(const Element*,const Element*);
   Pair(const Pair&);
   ~Pair();

   Pair& operator = (const Pair&);

   // Member functions
   Element* car() const;
   Element* cdr() const;
   void car(const Element*);
   void cdr(const Element*);
   Element* clone() const;
   void print(ostream&) const;
   int atom() const;

   // Friend function
   friend ostream& operator << (ostream&,const Pair*);
};

template <class T> class Type : public Element  // Atom
{
private:
   T thing;
public:
   // Constructors
   Type();
   Type(T);
   Type(const Type<T>&);

   // Member functions
   const T& value() const;

   Type<T>& operator = (const Type<T>&);
   Element* clone() const;
   void print(ostream&) const;
   int atom() const;

   // Friend function
   friend ostream& operator << (ostream&,const Type<T>*);
};
```

```
// Global functions
Pair* cons(const Element*,const Element *);
Element* append(const Element*,const Element*);
Element* car(const Element*);
Element* cdr(const Element*);

// Implementation

// class Pair
Pair::Pair() : _car(&nil), _cdr(&nil) {}

Pair::Pair(const Element *e) : _car(e->clone()), _cdr(&nil) {}

Pair::Pair(const Element *e1, const Element *e2) :
  _car(e1->clone()), _cdr(e2->clone()) {}

Pair::Pair(const Pair& p)
{
   _car = &nil;   _cdr = &nil;
   *this = p;
}

Pair::~Pair()
{
   // don't delete constants of the system
   if((_car != &nil)&&(_car != &t))
     delete _car; _car = &nil;
   if((_cdr != &nil)&&(_cdr != &t))
     delete _cdr; _cdr = &nil;
}

Pair& Pair::operator = (const Pair& p)
{
   if(this != &p)
   {
      if((_car != &nil)&&(_car != &t))
       delete _car;
      if((_car != &nil)&&(_car != &t))
       delete _cdr;

      _car = (p.car())->clone();
      _cdr = (p.cdr())->clone();
   }
```

```
     return *this;
}

Element* Pair::car() const
{ return _car; }

Element* Pair::cdr() const
{ return _cdr; }

void Pair::car(const Element *e)
{
   if(e != &nil) _car = e->clone();
}

void Pair::cdr(const Element *e)
{
   if(e != &nil) _cdr = e->clone();
}

Element* Pair::clone() const
{
   Pair* p = new Pair(*this);
   return p;
}

void Pair::print(ostream &os) const
{
  os << "(" << _car << " . " << _cdr << ")";
}

int Pair::atom() const
{ return 0; }

ostream& operator << (ostream&os,const Pair* p)
{
   if(p != NULL)
   {
     p->print(os);
     return os;
   }
   os << "nil";
   return os;
}
```

```
ostream& operator << (ostream& os,const Element* e)
{
   if(e != NULL)
   {
      e->print(os);
      return os;
   }
   os << "nil";
   return os;
}

template <class T> Type<T>::Type() : thing(T(0)) {}

template <class T> Type<T>::Type(T t) : thing(t) {}

template <class T> Type<T>::Type(const Type<T>& t)
{ thing = t.value(); }

template <class T>
Type<T> & Type<T>::operator = (const Type<T>& t)
{
   if(this != &t) thing = t.value();
   return *this;
}

template <class T> const T& Type<T>::value() const
{ return thing; }

template <class T> int Type<T>::atom() const
{  return 1; }

template <class T> Element* Type<T>::clone() const
{
   Type<T>* t = new Type<T>(*this);
   return t;
}

template <class T> void Type<T>::print(ostream& os) const
{ os << thing; }

template <class T>
ostream& operator << (ostream& os,const Type<T>* t)
{
   t->print(os);
```

```
    return os;
}

Pair* cons(const Element* e1,const Element* e2)
{
    Pair* p = new Pair;
    p->car(e1);
    p->cdr(e2);
    return p;
}

Element* append(const Element *e1,const Element *e2)
{
    if(! e1->atom())
    {
        Pair *p = new Pair(*(Pair *)e1);
        Pair *e = p;
        while(e->cdr() != &nil) e = (Pair *)e->cdr();
        e->cdr(e2);
        return p;
    }
    cerr << "\nFirst argument of append must be a list" << endl;
    return &nil;
}

Element * car(const Element *e)
{
    if(!e->atom()) return ((Pair*)e)->car();
    cerr << "\ncar: cannot take car of an atom!" << endl;
    return &nil;
}

Element* cdr(const Element *e)
{
    if(! e->atom()) return ((Pair*)e)->cdr();
    cerr << "\ncdr: cannot take cdr of an atom!" << endl;
    return &nil;
}

Element* null(Element *e)
{
 if(e==&nil) return &t;
 return &nil;
}
```

```
int is_lisp_list(Element *l)
{
 if(l==&nil) return 1;
 if(l->atom()) return 0;
 return is_lisp_list(((Pair*)l)->cdr());
}

int is_nonempty_lisp_list(Element *l)
{
 if(l->atom()) return 0;
 return is_lisp_list(((Pair*)l)->cdr());
}

Element *cond(Element *e)
{
 if(is_nonempty_lisp_list(e))
 {
  Element *firstcase=((Pair*)e)->car();
  if(is_nonempty_lisp_list(firstcase))
  {
   Element *condition=((Pair*)firstcase)->car();
   if(condition->atom())
   {
   if(condition==&nil)
   return cond(((Pair*)e)->cdr());
   else if(condition==&t)
   return ((Pair*)firstcase)->cdr();
   else cerr<<"cond expects a case first element to be nil or t."<<endl;
   }
   else cerr<<"cond expects a case first element to be nil or t."<<endl;
  }
  else cerr<<"cond expects a list for each case."<<endl;
 }
 else cerr<<"cond expects a list for evaluation."<<endl;
 return &nil;
}
#endif
```

The following program applies the classes in Lisp.h which has just been developed.

```cpp
// lisprun.cpp

#include <iostream.h>
#include "Lisp.h"
#include "Verylong.h"

typedef Type<int> l_int;
typedef Type<char> l_char;
typedef Type<double> l_double;
typedef Type<Verylong> l_verylong;

Element *plus(Element *e)
{
 if (is_nonempty_lisp_list(e))
 {
  if (is_nonempty_lisp_list(cdr(e)))
  {
  //assume int for calculations
  Element *param1,*param2;
  param1 = car(e); param2 = car(cdr(e));
  if(param2!=&nil)
  return
  l_int(((l_int*)param1)->value()+((l_int*)param2)->value()).clone();
  }
 }
 cerr<<"Plus takes a list of two integers as arguments"<<endl;
 return &nil;
}

Element* length_list(Element *e)
{
 l_int zero(0),one(1);

 //stop C++ recursion
 if(e==&nil) return cons(&zero,&nil);

 return cond(cons(cons(null(e),cons(&zero,&nil)),
  cons(cons(&t,cons(plus(
  cons(&one,length_list(cdr(e)))),&nil)),&nil))
  );
}
```

```
void main()
{
   // Define two "atoms"
   l_int a(1);    // 1
   l_char b('a'); // a

   cout << "Examples on the cons function:" << endl;

   Pair *A = cons(&a,&nil);  // in Lisp :  ( 1 )
   Pair *B = cons(&b,&nil);  // in Lisp :  ( a )

   Pair *C = cons(&a,&b);    // in Lisp : (cons '1 'a)
   cout << "(cons '1 'a)     => " << C << endl;

   C = cons(&a,B);           // in Lisp (cons '1 '(a))
   cout << "(cons '1 '(a))   => " << C << endl;

   C = cons(A,&b);           // in Lisp (cons '(1) 'a)
   cout << "(cons '(1) 'a)   => " << C << endl;

   C = cons(A,B);            // in Lisp (cons '(1) '(a))
   cout << "(cons '(1) '(a)) => " << C << endl;
   cout << endl;

   cout << "Examples on the append function:" << endl;

   // (setq D1 '(1 2 3))
   Pair *D1 =
   cons(new l_int(1),cons(new l_int(2),cons(new l_int(3),&nil)));

   // (setq D2 '(a b c))
   Pair *D2 = cons(new l_char('a'),cons(new l_char('b'),
   cons(new l_char('c'),&nil)));

   Pair *D = (Pair*)append(D1,D2);
   cout << "(append '(1 2 3) '(a b c)) => " << D << endl;
   cout << endl;

   cout << "Examples on the car and cdr functions:" << endl;

   // setq E '((1 2 3) (a b c))
   Pair *E = cons(D1,cons(D2,&nil));
```

```
      cout << "(car '((1 2 3) (a b c)))        => " << car(E)         << endl;
      cout << "(car (car '((1 2 3) (a b c))))=> " << car(car(E)) << endl;
      cout << "(cdr '((1 2 3) (a b c)))        => " << cdr(E)         << endl;
      cout << "(car (cdr '((1 2 3) (a b c))))=> " << car(cdr(E)) << endl;
      cout << "(cdr (cdr '((1 2 3) (a b c))))=> " << cdr(cdr(E)) << endl;
      cout << "(car 'a) => " << car(&a) << endl;
      cout << endl;

      // Abstract data types
      cout << "Applications with abstract data type Verylong" << endl;

      l_verylong v("123456789012");
      l_double r(3.14159);
      Pair *Very = cons(&v, cons(&r,&nil));
      cout << "(cons 'v '(r)) => " << Very << endl;

      cout << endl << "Applications on the cond function:" << endl;

      //((nil r) (t v))
      cout << "(cond (nil 3.14159) (t 123456789012)) => "
           <<   cond(cons(cons(&nil,cons(&r,&nil)),
                 cons(cons(&t,cons(&v,&nil)),&nil)))
           << endl;
      //((t r) (nil v))
      cout << "(cond (t 3.14159) (nil 123456789012)) => "
           <<   cond(cons(cons(&t,cons(&r,&nil)),
                 cons(cons(&nil,cons(&v,&nil)),&nil)))
           << endl;

      Pair *test=cons(&a,cons(&b,cons(&v,cons(&r,&nil))));
      cout << endl << "(a b v r) => " << test << endl;
      cout << "(length_list (a b v r)) => " << length_list(test) << endl;
}

/*
Result
======

Examples on the cons function:
(cons '1 'a)     => (1 . a)
(cons '1 '(a))   => (1 . (a . nil))
(cons '(1) 'a)   => ((1 . nil) . a)
(cons '(1) '(a)) => ((1 . nil) . (a . nil))
```

Examples on the append function:
(append '(1 2 3) '(a b c)) => (1 . (2 . (3 . (a . (b . (c . nil))))))

Examples on the car and cdr functions:
```
(car '((1 2 3) (a b c)))        => (1 . (2 . (3 . nil)))
(car (car '((1 2 3) (a b c))))  => 1
(cdr '((1 2 3) (a b c)))        => ((a . (b . (c . nil))) . nil)
(car (cdr '((1 2 3) (a b c))))  => (a . (b . (c . nil)))
(cdr (cdr '((1 2 3) (a b c))))  => nil
```

car: cannot take car of an atom!
(car 'a) => nil

Applications with the abstract data type: Verylong
(cons 'v '(r)) => (123456789012 . (3.14159 . nil))

Applications on the cond function:
(cond (nil 3.14159) (t 123456789012)) => (123456789012 . nil)
(cond (t 3.14159) (nil 123456789012)) => (3.14159 . nil)

(a b v r) => (1 . (a . (123456789012 . (3.14159 . nil))))
(length_list (a b v r)) => (4 . nil)
*/

Chapter 10

Program Listing

This chapter contains the listing of the header files of the classes presented in Chapters 6 and 7. For each class, a brief description of the public member functions is given prior to the complete program listing.

10.1 Verylong Class

The public interface of the Verylong class:

- Verylong(const char* = NULL) : Constructor.
- Verylong(int) : Constructor.
- abs(const Verylong&) : Absolute value function.
- sqrt(const Verylong&) : Integer square root function.
- pow(const Verylong&,const Verylong&) : Integer power function.
- div(const Verylong&,const Verylong&) : Double division function.
- Arithmetic operators : ++, --, -(unary), +, -, *, /, %, =, +=, -=, *=, /=, %=
- Relational operators : ==, !=, <, <=, >, >=
- Type conversion operators : operator int(), operator double()
- Stream operators : >>, <<

For a detailed description of the class structure and each member function, please refer to Section 6.1.

```
// Verylong.h

#ifndef VERYLONG_H
#define VERYLONG_H

#include <assert.h>
#include <iostream.h>
#include <iomanip.h>
#include <string.h>
#include <stdlib.h>
#include <ctype.h>
#include <limits.h>
#include <math.h>

class Verylong
{
   private:
      // Data Fields
      char* vlstr;      // The string is stored in reverse order.
      int   vlen;       // Length of the Verylong string
      int   vlsign;     // Sign of Verylong: +=>0; -=>1

      // Private member functions
      char *strrev(char *s) const;       // Reverse string s
      Verylong multdigit(int) const;
      Verylong mult10(int) const;

      // Class Data
      static const Verylong zero;
      static const Verylong one;
      static const Verylong two;

   public:
      // Constructors and destructor
      Verylong(const char* = NULL);
      Verylong(int);
      Verylong(const Verylong&);
      ~Verylong();

      // Conversion operators
      operator int () const;
      operator double () const;
      operator char * () const;

      // Arithmetic operators and Relational operators
      const Verylong& operator = (const Verylong&);
      Verylong operator - () const;      // negate  operator
```

```
        Verylong operator ++ ();           // prefix  increment operator
        Verylong operator ++ (int);        // postfix increment operator
        Verylong operator -- ();           // prefix  decrement operator
        Verylong operator -- (int);        // postfix decrement operator

        Verylong operator += (const Verylong&);
        Verylong operator -= (const Verylong&);
        Verylong operator *= (const Verylong&);
        Verylong operator /= (const Verylong&);
        Verylong operator %= (const Verylong&);

        friend Verylong operator + (const Verylong&,const Verylong&);
        friend Verylong operator - (const Verylong&,const Verylong&);
        friend Verylong operator * (const Verylong&,const Verylong&);
        friend Verylong operator / (const Verylong&,const Verylong&);
        friend Verylong operator % (const Verylong&,const Verylong&);

        friend int operator == (const Verylong&,const Verylong&);
        friend int operator != (const Verylong&,const Verylong&);
        friend int operator <  (const Verylong&,const Verylong&);
        friend int operator <= (const Verylong&,const Verylong&);
        friend int operator >  (const Verylong&,const Verylong&);
        friend int operator >= (const Verylong&,const Verylong&);

        // Others functions
        friend Verylong abs(const Verylong&);
        friend Verylong sqrt(const Verylong&);
        friend Verylong pow(const Verylong&,const Verylong&);
        friend double div(const Verylong&,const Verylong&);

        // I/O stream functions
        friend ostream & operator << (ostream&,const Verylong&);
        friend istream & operator >> (istream&,Verylong &);
};

//
// Class Data
const Verylong Verylong::zero = Verylong("0");
const Verylong Verylong::one = Verylong("1");
const Verylong Verylong::two = Verylong("2");

//
// Constructors, Destructors and Conversion operators.
Verylong::Verylong(const char *value)
{
    if(value)
    {
```

```
        vlsign = (*value == '-') ? 1:0; // check for negative sign

        if(ispunct(*value))              // if the first character is a
        {                                // punctuation mark.
          vlen = strlen(value)-1;
          vlstr = new char[vlen + 1]; assert(vlstr != NULL);
          strcpy(vlstr, value+1);
        }
        else
        {
          vlen = strlen(value);
          vlstr = new char[vlen + 1]; assert(vlstr != NULL);
          strcpy(vlstr, value);
        }
        strrev(vlstr);
    }
    else   // initialize to zero
    {
        vlstr = new char[2]; assert(vlstr != NULL);
        *vlstr = '0'; *(vlstr+1) = '\0';
        vlen = 1; vlsign = 0;
    }
}

Verylong::Verylong(int n)
{
    int i;
    if(n<0)  {vlsign = 1; n = (-n); } // check for sign and convert the
    else vlsign = 0;                  // number to positive if it is negative

    if(n)
    {
        i = (int)log10(n)+2;  // check for the length of the integer
        vlstr = new char[i]; assert(vlstr != NULL);
        vlen = i-1;
        i = 0;
        while(n >= 1)       // extract the number digit by digit and store
        {                   // internally
          vlstr[i] = n%10 + '0';
          n /= 10;
          i++;
        }
        vlstr[i] = '\0';
    }
    else                       // else number is zero
    {
        vlstr = new char[2]; assert(vlstr != NULL);
```

```
      *vlstr = '0'; *(vlstr+1) = '\0';
      vlen = 1;
   }
}

Verylong::Verylong(const Verylong& x) : vlen(x.vlen), vlsign(x.vlsign)
{
   vlstr = new char[x.vlen + 1]; assert(vlstr != NULL);
   strcpy(vlstr, x.vlstr);
}

Verylong::~Verylong() { delete [] vlstr; }

Verylong::operator int() const
{
   static Verylong max0(INT_MAX);
   static Verylong min0(INT_MIN+1);
   int number, factor=1;
   if(*this > max0)
   {
   cerr << "Error: Conversion Verylong->integer is not possible" << endl;
   return INT_MAX;
   }
   else if(*this < min0)
   {
   cerr << "Error: Conversion Verylong->integer is not possible" << endl;
   return INT_MIN;
   }
   number = vlstr[0]-'0';
   for(int j=1; j<vlen; j++)
   {
      factor *= 10;
      number += (vlstr[j]-'0') * factor;
   }

   if(vlsign) return -number;
   return number;
}

Verylong::operator double() const
{
   double sum, factor = 1.0;
   sum = double(vlstr[0]-'0');
   for(int i=1; i<vlen; i++)
   {
      factor *= 10.0;
      sum += double(vlstr[i]-'0') * factor;
```

```
   }
   if(vlsign) return -sum;
   return sum;
}

Verylong::operator char * () const
{
   char *temp = new char[vlen + 1]; assert(temp != NULL);
   char *s;

   if(vlen > 0)
   {
      strcpy(temp, vlstr);
      if(vlsign)
      {
      s = new char[vlen + 2];
      strcpy(s,"-");
      }
      else
      {
      s = new char[vlen + 1];
      strcpy(s,"");
      }
      strcat(s,strrev(temp));
   }
   else
   {
      s = new char[2];
      strcpy(s,"0");
   }
   delete [] temp;
   return s;
}


//
// Various member operators
const Verylong& Verylong::operator = (const Verylong& rhs)
{
   if(this == &rhs) return *this;
   delete [] vlstr;
   vlstr = new char [rhs.vlen + 1]; assert(vlstr != NULL);
   strcpy(vlstr, rhs.vlstr);
   vlen = rhs.vlen;
   vlsign = rhs.vlsign;
   return *this;
}
```

```
// Unary - operator
Verylong Verylong::operator -() const
{
    Verylong temp(*this);
    if(temp != zero)  temp.vlsign = !vlsign;
    return temp;
}

// Prefix increment operator
Verylong Verylong::operator ++ ()
{
    return *this = *this + one;
}

// Postfix increment operator
Verylong Verylong::operator ++ (int)
{
    Verylong result(*this);

    *this = *this + one;
    return result;
}

// Prefix decrement operator
Verylong Verylong::operator -- ()
{  return *this = *this - one; }

// Postfix decrement operator
Verylong Verylong::operator -- (int)
{
    Verylong result(*this);
    *this = *this - one;
    return result;
}

Verylong Verylong::operator += (const Verylong& v)
{  return *this = *this + v; }

Verylong Verylong::operator -= (const Verylong& v)
{  return *this = *this - v; }

Verylong Verylong::operator *= (const Verylong& v)
{  return *this = *this * v; }

Verylong Verylong::operator /= (const Verylong& v)
{  return *this = *this / v; }
```

```
Verylong Verylong::operator %= (const Verylong& v)
{ return *this = *this % v; }


//
// Various friendship operators and functions.
Verylong operator + (const Verylong &u,const Verylong &v)
{
    if(u.vlsign ^ v.vlsign)
    {
        if(u.vlsign == 0) { Verylong t1 = u-abs(v); return t1;}
        else              { Verylong t2 = v-abs(u); return t2;}
    }

    int j,d1, d2, digitsum, carry = 0,
        maxlen = (u.vlen > v.vlen) ? u.vlen:v.vlen;
    char *temp = new char[maxlen+2]; assert(temp != NULL);

    for(j=0; j<maxlen; j++)     // addition starts from left
    {                           // because string is reversed
        d1 = (j > u.vlen-1) ? 0 : u.vlstr[j]-'0'; // get digit
        d2 = (j > v.vlen-1) ? 0 : v.vlstr[j]-'0'; // get digit
        digitsum = d1 + d2 + carry;          // add digits
        if(digitsum >= 10)              // if there's a carry,
        {                               // decrease sum by 10
        digitsum -= 10;
        carry = 1;
        }
        else                    // set carry to 1
        carry = 0;              // otherwise carry is 0
        temp[j] = digitsum + '0';   // insert char in string
    }

    if(carry)                       // if carry at end,
        temp[j++] = '1';            // last digit is 1

    if(u.vlsign) temp[j++] = '-';

    temp[j] = '\0';                     // terminate string
    u.strrev(temp);
    Verylong result(temp);
    delete [] temp;
    return result;
}

Verylong operator - (const Verylong& u,const Verylong& v)
{
    if(u.vlsign ^ v.vlsign)
```

```
   {
       if(u.vlsign == 0) { Verylong t1 = u+abs(v); return t1;}
       else              { Verylong t2 = -(v+abs(u)); return t2;}
   }

   int maxlen = (u.vlen>v.vlen) ? u.vlen:v.vlen,
       d, d1, d2, i, negative, borrow=0;
   char *temp = new char[maxlen+1]; assert(temp != NULL);

   Verylong w, y;

   if(u.vlsign == 0)  // both u,v are positive
     if(u < v)   { w=v; y=u; negative=1;}
       else      { w=u; y=v; negative=0;}
   else             // both u,v are negative
     if(u < v) { w=u; y=v; negative=1;}
       else    { w=v; y=u; negative=0;}

   for(i=0; i<maxlen; i++)
   {
       d1 = (i>w.vlen-1) ? 0:w.vlstr[i]-'0';
       d2 = (i>y.vlen-1) ? 0:y.vlstr[i]-'0';
       d = d1 - d2 - borrow;
       if(d < 0)       // if this is less than 0
       {               // 10 is added to d and
          d += 10;     // 1 is borrowed from next digit
          borrow = 1;
       }
       else
          borrow = 0; // else no need to borrow
       temp[i] = d + '0';
   }

   while(i-1 > 0 && temp[i-1] == '0')       // if result is zero
       --i;

   if(negative)  temp[i++] = '-';
   temp[i] = '\0';
   u.strrev(temp);

   Verylong result(temp);
   delete [] temp;
   return result;
}

Verylong operator * (const Verylong& u,const Verylong& v)
{
```

```
      Verylong pprod("1"), tempsum("0");

      for(int j=0; j<v.vlen; j++)
      {
         int digit = v.vlstr[j] - '0';    // extract a digit
         pprod = u.multdigit(digit);      // multiplied by the digit
         pprod = pprod.mult10(j);         // "adds" suitable zeros behind
         tempsum = tempsum + pprod;       // result added to tempsum
      }
      tempsum.vlsign = u.vlsign^v.vlsign; // to determine sign
      return tempsum;
}


//  This algorithm is the long division algorithm.
Verylong operator / (const Verylong& u,const Verylong& v)
{
      Verylong w,y,b,c,d,quotient=Verylong::zero;
      int len = u.vlen - v.vlen;
      if(v == Verylong::zero)
      {
      cerr << "Error : division by zero" << endl;
      return Verylong::zero;
      }
      w=abs(u); y=abs(v);
      if(w<y) return Verylong::zero;

      char *temp = new char[w.vlen+1]; assert(temp != NULL);
      strcpy(temp, w.vlstr + len);
      b.strrev(temp);                     // b is dummy for strrev()
      c = Verylong(temp);
      delete [] temp;
      for(int i=0; i<=len; i++)
      {
         quotient = quotient.mult10(1);
         b = d = Verylong::zero;          // initialize b and d to 0
         while(b < c)
         { b = b + y; d = d + Verylong::one; }
         if(c < b)            // if b>c, then
         {                    // we have added one count too many
         d = d - Verylong::one;
         }

         quotient = quotient + d;     // add to the quotient

         if(i < len)
         {
         c = (c-b).mult10(1);
```

```
    c = c + Verylong(w.vlstr[len-i-1]-'0');
    }
  }

  quotient.vlsign = u.vlsign^v.vlsign;
  return quotient;
}

Verylong operator % (const Verylong& u,const Verylong& v)
{
  return (u - v*(u/v));
}

int operator == (const Verylong& u,const Verylong& v)
{
  return (u.vlsign==v.vlsign && !strcmp(u.vlstr,v.vlstr));
}

int operator != (const Verylong& u,const Verylong& v)
{
  return !(u==v);
}

int operator < (const Verylong& u,const Verylong& v)
{
  if(u.vlsign < v.vlsign) return 0;
  else if(u.vlsign > v.vlsign) return 1;
  if(u.vlen < v.vlen) return (1^u.vlsign);      // XOR (^)
  else if(u.vlen > v.vlen) return (0^u.vlsign); // to determine sign

  int    temp;
  char *temp1 = new char[u.vlen+1], *temp2 = new char[v.vlen+1];
  assert(temp1 != NULL);
  assert(temp2 != NULL);

  strcpy(temp1, u.vlstr);
  strcpy(temp2, v.vlstr);
  u.strrev(temp1);
  u.strrev(temp2);             // u is dummy variable

  temp = strcmp(temp1,temp2);
  delete [] temp1; delete [] temp2;

  if(temp < 0)
    return (1^u.vlsign);  // use XOR (^) to determine sign
  else if (temp > 0)
    return (0^u.vlsign);
```

```
    else
        return 0;                // if u==v return 0
}

int operator <= (const Verylong& u,const Verylong& v)
{
    return (u<v || u==v);
}

int operator > (const Verylong& u,const Verylong& v)
{
    return (!(u<v) && u!=v);
}

int operator >= (const Verylong& u,const Verylong& v)
{
    return (u>v || u==v);
}

// Calculate the absolute value of a number
Verylong abs(const Verylong &v)
{
    Verylong u(v);
    if(u.vlsign) u.vlsign = 0;
    return u;
}

// Calculate the integer square root of a number
// based on the formula (a+b)^2 = a^2 + 2ab + b^2
Verylong sqrt(const Verylong& v)
{
    // if v is negative, error is reported
    if(v.vlsign) { cerr << "NaN" << endl; return Verylong::zero; }

    int j, k = v.vlen+1, num = k >> 1;
    Verylong y, z, sum, tempsum, digitsum;

    char *temp = new char[num + 1]; assert(temp != NULL);
    char *w    = new char[k];       assert(w != NULL);

    strcpy(w,v.vlstr);
    k = v.vlen-1;
    j = 1;

    // segmentate the number 2 digits by 2 digits
    if(v.vlen % 2) digitsum = Verylong(w[k--] - '0');
    else
```

```
    {
        digitsum = Verylong((w[k] - '0')*10 + w[k-1] - '0');
        k -= 2;
    }

    // find the first digit of the integer square root
    sum = z = Verylong(int(sqrt(double(digitsum))));

    // store partial result
    temp[0] = int(z) + '0';
    digitsum = digitsum - z*z;

    for(;j<num;j++)
    {
        // get next digit from the number
        digitsum = digitsum.mult10(1) + Verylong(w[k--] - '0');
        y = z + z;          // 2*a
        z = digitsum/y;
        tempsum = digitsum.mult10(1) + Verylong(w[k] - '0');
        digitsum = -y*z.mult10(1) + tempsum - z*z;

        // decrease z by 1 and re-calculate when it is over-estimated.
        while(digitsum < Verylong::zero)
        {
            --z;
            digitsum = -y*z.mult10(1) + tempsum - z*z;
        }
        --k;

        temp[j] = int(z) + '0';
        z = sum = sum.mult10(1) + z;
    }

    temp[num] = '\0';
    Verylong result(temp);
    delete [] temp; delete [] w;
    return result;
}

// Raise a number X to a power of degree
Verylong pow(const Verylong& X,const Verylong& degree)
{
    Verylong N(degree), Y("1"), x(X);
    if(N == Verylong::zero) return Verylong::one;
    if(N < Verylong::zero) return Verylong::zero;

    while(1)
```

```
   {
      if(N%Verylong::two != Verylong::zero)
      {
        Y = Y * x;
        N = N / Verylong::two;
        if(N == Verylong::zero) return Y;
      }
      else N = N / Verylong::two;
      x = x * x;
   }
}

// Double division function
double div(const Verylong& u,const Verylong& v)
{
   double qq = 0.0, qqscale = 1.0;
   Verylong w,y,b,c;
   int d, count,
       decno = 16;                  // number of significant digits

   if(v == Verylong::zero)
   {
      cerr << "ERROR : Division by zero" << endl;
      return 0.0;
   }
   if(u == Verylong::zero) return 0.0;

   w=abs(u); y=abs(v);
   while (w<y) { w = w.mult10(1); qqscale *= 0.1; }

   int len = w.vlen - y.vlen;
   char *temp = new char[w.vlen+1]; assert(temp != NULL);

   strcpy(temp,w.vlstr+len);
   w.strrev(temp);                  // w is dummy variable
   c = Verylong(temp);
   delete [] temp;

   for(int i=0; i<=len; i++)
   {
      qq *= 10.0;
      b = Verylong::zero; d = 0;       // initialize b and d to 0

      while(b < c)
      { b += y; d += 1;}

      if(c < b)                        // if b>c, then
```

```
      { b -= y; d -= 1;}                  // we have added one count too many

      qq += double(d);                    // add to the quotient

      c = (c-b).mult10(1);                // the partial remainder * 10

      if(i < len)                         // and add to next digit
         c += Verylong(w.vlstr[len-i-1]-'0');
   }

   qq *= qqscale; count = 0;

   while(c != Verylong::zero && count < decno)
   {
      qqscale *= 0.1;

      b = Verylong::zero; d = 0;          // initialize b and d to 0

      while(b < c)
      { b += y; d += 1;}

      if(c<b)                  // if b>c, then
      { b -= y; d -= 1;}       // we have added one count too many

      qq += double(d)*qqscale;

      c = (c-b).mult10(1);
      count++;
   }

   if(u.vlsign^v.vlsign) qq *= (-1.0); // check for the sign
   return qq;
}

ostream& operator << (ostream& s,const Verylong& v)
{
   char *temp = new char[v.vlen + 1]; assert(temp != NULL);
   if(v.vlen > 0)
   {
      strcpy(temp, v.vlstr);
      if(v.vlsign) s << "-";
      s << v.strrev(temp);
   }
   else  s << "0";
   delete [] temp;
   return s;
}
```

```
istream& operator >> (istream& s,Verylong& v)
{
    char temp[100000];
    s >> temp;

    delete [] v.vlstr;
    v.vlen = strlen(temp);
    v.strrev(temp);
    v.vlstr = new char[v.vlen+1]; assert(v.vlstr != NULL);
    strcpy(v.vlstr, temp);
    return s;
}

// Private member functions: strrev(), multdigit(), mult10().
//
// Reverse the string s, with the original string changed
char* Verylong::strrev(char *s) const
{
    int len = strlen(s), len1 = len - 1, index,
        limit = len >> 1;
    char t;

    for(int i=0; i<limit; i++)
    {
        index = len1-i;
        t = s[index];
        s[index] = s[i];
        s[i] = t;
    }
    return s;
}

// Multiply this Verylong number by num
Verylong Verylong::multdigit(int num) const
{
    int j, carry = 0;
    if(num)
    {
        char *temp = new char[vlen + 2]; assert(temp != NULL);
        for(j = 0; j<vlen; j++)
        {
        int d1 = vlstr[j] - '0', digitprod = d1*num + carry;

        if(digitprod >= 10)              // if there's a new carry,
        {
        carry = digitprod/10;            // carry is high digit
```

```
        digitprod -= carry*10;      // result is low digit
        }
        else carry = 0;             // otherwise carry is 0
        temp[j] = digitprod + '0';  // insert char in string
        }

        if(carry)                   // if carry at end,
            temp[j++] = carry + '0';  // it's last digit
        temp[j] = '\0';             // terminate string

        strrev(temp);
        Verylong result(temp);
        delete [] temp;
        return result;
    }
    else return zero;
}

// Multiply this Verylong number by 10*num
Verylong Verylong::mult10(int num) const
{
    if(*this != zero)
    {
        int j, dd = vlen + num, bb = vlen - 1;
        char *temp = new char [dd + 1]; assert(temp != NULL);

        for(j=0; j<vlen; j++) temp[j] = vlstr[bb-j];
        for(j=vlen; j<dd; j++) temp[j] = '0';

        temp[dd] = '\0';
        Verylong result(temp);
        delete [] temp;
        return result;
    }
    else return zero;
}
#endif
```

10.2   Rational Class

The public interface of the `Rational` class:

- `Rational()` : Default constructor.

- `Rational(T)` : Integer constructor.

- `Rational(T,T)` : Rational number constructor.

- `num()` : Numerator of the rational number.

- `den()` : Denominator of the rational number.

- `frac()` : Fractional part of the rational number.

- `normalize()` : Normalization of the rational number.

- Arithmetic operators : -(unary), +, -, *, /, =, +=, -=, *=, /=

- Relational operators : ==, !=, <, <=, >, >=

- Type conversion operator : `operator double()`

- Stream operators : >>, <<

For a detailed description of the class structure and each member function, please refer to Section 6.2.

```cpp
// Rational.h

#ifndef RATIONAL_H
#define RATIONAL_H
#include <iostream.h>
#include <stdlib.h>
#include <ctype.h>

template <class T> class Rational
{
   private:
      // Data Fields : Numerator and Denominator
      T p, q;
      // Private member function
      T gcd(T,T);
   public:
      // Constructors and Destructor
      Rational();
```

```
    Rational(T);
    Rational(T,T);
    Rational(const Rational<T>&);
    ~Rational();

    // Conversion operator
    operator double () const;

    // Member functions
    T num() const;            // numerator of r
    T den() const;            // denominator of r
    Rational<T> frac() const; // fractional part of r
    void normalize();         // normalize the rational number

    // Arithmetic operators and Relational operators
    const Rational<T>& operator = (const Rational<T>&);
    Rational<T> operator - () const;
    const Rational<T>& operator += (const Rational<T>&);
    const Rational<T>& operator -= (const Rational<T>&);
    const Rational<T>& operator *= (const Rational<T>&);
    const Rational<T>& operator /= (const Rational<T>&);

    friend Rational<T> operator + (const Rational<T>&,const Rational<T>&);
    friend Rational<T> operator - (const Rational<T>&,const Rational<T>&);
    friend Rational<T> operator * (const Rational<T>&,const Rational<T>&);
    friend Rational<T> operator / (const Rational<T>&,const Rational<T>&);
    friend int operator == (const Rational<T>&,const Rational<T>&);
    friend int operator != (const Rational<T>&,const Rational<T>&);
    friend int operator >  (const Rational<T>&,const Rational<T>&);
    friend int operator <  (const Rational<T>&,const Rational<T>&);
    friend int operator >= (const Rational<T>&,const Rational<T>&);
    friend int operator <= (const Rational<T>&,const Rational<T>&);

    // I/O stream functions
    friend ostream& operator << (ostream &,const Rational<T>&);
    friend istream& operator >> (istream &,Rational<T>&);
};

// Constructors, destructor and conversion operator.
template <class T> Rational<T>::Rational() : p(T(0)), q(T(1)) {}
template <class T> Rational<T>::Rational(T N) : p(N), q(T(1)) {}

template <class T> Rational<T>::Rational(T N,T D) : p(N), q(D)
{
    static T zero(0);
    if(D == zero)
    {
```

```
        cerr << "Zero denominator in Rational Number " << endl;
        return;
    }
    if(q < zero) { p = -p; q = -q; }
    normalize();
}


template <class T> Rational<T>::Rational(const Rational<T>& r)
    : p(r.p), q(r.q) {}


template <class T> Rational<T>::~Rational() { }


template <class T> Rational<T>::operator double() const
{ return double(p)/double(q); }


#include "Verylong.h"
Rational<Verylong>::operator double() const { return div(p,q); }


// Member functions
template <class T> T Rational<T>::num() const { return p; }
template <class T> T Rational<T>::den() const { return q; }


template <class T> Rational<T> Rational<T>::frac() const
{
    static Rational<T> zero(0), one(1);
    Rational<T> temp(*this);
    if(temp < zero)
    {
        while(temp < zero) temp = temp + one;
        return temp - one;
    }
    else
    {
        while(zero < temp) temp = temp - one;
        return temp + one;
    }
}


template <class T> void Rational<T>::normalize()
{
    static T zero(0);  static T one(1);
    T t;
    if(p < zero) t = -p;
    else         t = p;
    t = gcd(t,q);
    if(t > one)  { p /= t; q /= t; }
}
```

```
// Various operators
template <class T>
const Rational<T>& Rational<T>::operator = (const Rational<T>& r)
{ p = r.p; q = r.q; return *this; }


template <class T> Rational<T> Rational<T>::operator - () const
{ return Rational<T>(-p,q); }


template <class T>
const Rational<T>& Rational<T>::operator += (const Rational<T>& r)
{ return *this = *this + r; }


template <class T>
const Rational<T>& Rational<T>::operator -= (const Rational<T>& r)
{ return *this = *this - r; }


template <class T>
const Rational<T>& Rational<T>::operator *= (const Rational<T>& r)
{ return *this = *this * r; }


template <class T>
const Rational<T>& Rational<T>::operator /= (const Rational<T>& r)
{ return *this = *this / r; }


// Various friendship operators and functions.
template <class T>
Rational<T> operator + (const Rational<T>& r1,const Rational<T>& r2)
{ return Rational<T> (r1.p * r2.q + r2.p * r1.q, r1.q * r2.q); }


template <class T>
Rational<T> operator - (const Rational<T>& r1,const Rational<T>& r2)
{ return Rational<T> (r1.p * r2.q - r2.p * r1.q, r1.q * r2.q); }


template <class T>
Rational<T> operator * (const Rational<T>& r1,const Rational<T>& r2)
{ return Rational<T> (r1.p * r2.p, r1.q * r2.q); }


template <class T>
Rational<T> operator / (const Rational<T>& r1,const Rational<T>& r2)
{ return Rational<T> (r1.p * r2.q, r1.q * r2.p); }


template <class T>
int operator == (const Rational<T>& r1,const Rational<T>& r2)
{ return (r1.p * r2.q) == (r2.p * r1.q); }


template <class T>
```

```
int operator != (const Rational<T>& r1,const Rational<T>& r2)
{ return !(r1 == r2); }

template <class T>
int operator > (const Rational<T>& r1,const Rational<T>& r2)
{
   static T zero(0);
   return (r1.p*r2.q - r2.p*r1.q > zero);
}

template <class T>
int operator <  (const Rational<T>& r1,const Rational<T>& r2)
{
   static T zero(0);
   return (r1.p*r2.q - r2.p*r1.q < zero);
}

template <class T>
int operator >= (const Rational<T>& r1,const Rational<T>& r2)
{ return (r1>r2) || (r1==r2); }

template <class T>
int operator <= (const Rational<T>& r1,const Rational<T>& r2)
{ return (r1<r2) || (r1==r2); }

template <class T>
ostream& operator << (ostream& s,const Rational<T>& r)
{
   static T one(1);
   if(r.q == one) return s << r.p;
   return s << r.p << "/" << r.q;
}

template <class T>
istream& operator >> (istream& s,Rational<T>& r)
{
   char c;
   T n, d(1);
   s.clear();                 // set stream state to good
   s >> n;                    // read numerator
   if(! s.good()) return s;   // can't get an integer, just return
   c = s.peek();              // peek next character
   if(c == '/')
   {
      c = s.get();            // clear '/'
      s >> d;                 // read denominator
      if(!s.good())
```

```
            {
                s.clear(s.rdstate() | ios::badbit);
                return s;
            }
        }
        r = Rational<T>(n,d);
        return s;
    }

    // Private member function: gcd()
    template <class T> T Rational<T>::gcd(T a,T b)
    {
        static T zero(0);
        while(b > zero)
        {
            T m = a%b;   a = b;   b = m;
        }
        return a;
    }

    // This function is a global functions.
    // It should be in a global header file.
    template <class T> T abs(const T& x)
    {
        if(x > T(0)) return x;
        return -x;
    }
    #endif
```

10.3   Complex Class

The public interface of the Complex class:

- Complex() : Default constructor.

- Complex(T) : Real number constructor.

- Complex(T,T) : Complex number constructor.

- realPart() : Real part of the complex number.

- imagPart() : Imaginary part of the complex number.

- magnitude() : Magnitude of the complex number.

- argument() : Argument of the complex number.

- conjugate() : Conjugate of the complex number.

- negate() : Negative of the complex number.

- Arithmetic operators : -(unary), +, -, *, /, =, +=, -=, *=, /=

- Relational operators : ==, !=, <, <=, >, >=

- Stream operators : >>, <<

For a detailed description of the class structure and each member function, please refer to Section 6.3.

```
// MComplex.h

#ifndef COMPLEX_H
#define COMPLEX_H

#include <iostream.h>
#include <math.h>

template <class T> class Complex
{
   private:
      // Data Fields : real part and imaginary part
      T real;
      T imag;
   public:
      // Constructors and Destructor
      Complex();
      Complex(T);
```

```
        Complex(T,T);
        Complex(const Complex<T>&);
        ~Complex();

        // Member Functions
        T realPart() const;
        T imagPart() const;
        double magnitude() const;
        double argument() const;
        Complex<T> conjugate();
        Complex<T> negate();

        // Arithmetic operators and relational operators
        const Complex<T>& operator = (const Complex<T>&);
        Complex<T> operator - () const;
        Complex<T> operator += (const Complex<T>&);
        Complex<T> operator -= (const Complex<T>&);
        Complex<T> operator *= (const Complex<T>&);
        Complex<T> operator /= (const Complex<T>&);

        friend Complex<T> operator + (const Complex<T>&,const Complex<T>&);
        friend Complex<T> operator - (const Complex<T>&,const Complex<T>&);
        friend Complex<T> operator * (const Complex<T>&,const Complex<T>&);
        friend Complex<T> operator / (const Complex<T>&,const Complex<T>&);
        friend int operator < (const Complex<T>&,const Complex<T>&);
        friend int operator <=(const Complex<T>&,const Complex<T>&);
        friend int operator > (const Complex<T>&,const Complex<T>&);
        friend int operator >=(const Complex<T>&,const Complex<T>&);
        friend int operator ==(const Complex<T>&,const Complex<T>&);
        friend int operator !=(const Complex<T>&,const Complex<T>&);

        // I/O stream functions
        friend istream& operator >> (istream&,Complex<T>&);
        friend ostream& operator << (ostream&,const Complex<T>&);
};

// Constructors, Destructor and Copy constructor
template <class T> Complex<T>::Complex() : real(T(0)), imag(T(0)) {}
template <class T> Complex<T>::Complex(T r) : real(r), imag(T(0)) {}

template <class T> Complex<T>::Complex(T r,T i) : real(r), imag(i) {}

template <class T> Complex<T>::Complex(const Complex<T> &c)
    : real(c.real), imag(c.imag) {}

template <class T> Complex<T>::~Complex() {}
```

```
// Member Functions
template <class T> T Complex<T>::realPart() const { return real; }

template <class T> T Complex<T>::imagPart() const { return imag; }

template <class T> double Complex<T>::magnitude() const
{ return sqrt(double(real*real+imag*imag)); }

template <class T> double Complex<T>::argument() const
{ return atan2(double(imag),double(real)); }

template <class T> Complex<T> Complex<T>::conjugate()
{ return Complex<T>(real, imag = -imag); }

template <class T> Complex<T> Complex<T>::negate()
{ return Complex<T>(real = -real, imag); }

// Arithmetic Operators and Relational Operators
template <class T>
const Complex<T>& Complex<T>::operator = (const Complex<T>& c)
{
   real = c.real; imag = c.imag;
   return *this;
}

template <class T> Complex<T> Complex<T>::operator - () const
{ return Complex<T>(-real, -imag); }

template <class T>
Complex<T> Complex<T>::operator += (const Complex<T>& c)
{ return *this = *this + c; }

template <class T>
Complex<T> Complex<T>::operator -= (const Complex<T>& c)
{ return *this = *this - c; }

template <class T>
Complex<T> Complex<T>::operator *= (const Complex<T>& c)
{ return *this = *this * c; }

template <class T>
Complex<T> Complex<T>::operator /= (const Complex<T>& c)
{ return *this = *this / c; }

template <class T>
Complex<T> operator + (const Complex<T>& c1,const Complex<T>& c2)
{ return Complex<T>(c1.real+c2.real, c1.imag+c2.imag); }
```

```
template <class T>
Complex<T> operator - (const Complex<T>& c1,const Complex<T>& c2)
{ return Complex<T>(c1.real-c2.real, c1.imag-c2.imag); }

template <class T>
Complex<T> operator * (const Complex<T>& c1,const Complex<T>& c2)
{
  return Complex<T>(c1.real*c2.real - c1.imag*c2.imag,
                    c1.real*c2.imag + c1.imag*c2.real);
}

template <class T>
Complex<T> operator / (const Complex<T>& c1,const Complex<T>& c2)
{
   T modulus = c2.real * c2.real + c2.imag * c2.imag;
   static T zero(0);
   if(modulus == zero)
   {
       cerr << "Error : Zero divisor" << endl;
       return Complex<T>(zero,zero);
   }
   return Complex<T>((c1.imag * c2.imag + c1.real * c2.real)/modulus,
                     (c1.imag * c2.real - c1.real * c2.imag)/modulus);
}

template <class T>
int operator < (const Complex<T>& c1,const Complex<T>& c2)
{ return (c1.real*c1.real+c1.imag*c1.imag <
          c2.real*c2.real+c2.imag*c2.imag); }

template <class T>
int operator <= (const Complex<T>& c1,const Complex<T>& c2)
{ return (c1<c2) || (c1==c2); }

template <class T>
int operator > (const Complex<T>& c1,const Complex<T>& c2)
{ return (c1.real*c1.real+c1.imag*c1.imag >
          c2.real*c2.real+c2.imag*c2.imag); }

template <class T>
int operator >= (const Complex<T>& c1,const Complex<T>& c2)
{ return (c1>c2) || (c1==c2); }

template <class T>
int operator == (const Complex<T>& c1,const Complex<T>& c2)
{ return (c1.real==c2.real) && (c1.imag==c2.imag); }
```

```cpp
template <class T>
int operator != (const Complex<T>& c1,const Complex<T>& c2)
{ return !(c1==c2); }

// I/O Stream Functions
template <class T> istream& operator >> (istream& s,Complex<T>& z)
{
    static T zero(0);
    T r, i;
    char c;
    s >> ws;                // remove leading whitespace
    c = s.peek();           // peek next character
    if(c=='(')
    {
        c = s.get();        // Clear '('
        s >> r;             // read real part
        s >> ws;            // remove extra whitespace
        c = s.peek();       // peek next character
        if(c == ',')
        {
            c = s.get();
            s >> i;         // read imaginary part
            s >> ws;
            c = s.peek();
        }
        else i = zero;      // no imaginary part
        if(c != ')') s.clear(s.rdstate() | ios::badbit);
    }
    else                    // it is a real number
    {
        s >> r;             // read real part
        i = zero;           // initialize real part
    }
    z = Complex<T>(r,i);
    return s;
}

template <class T>
ostream& operator << (ostream& s,const Complex<T>& c)
{
    static T zero(0);
    if(c.imag==zero) return s << c.real;
    if(c.real==zero) return s << c.imag << "i";
    return s << "(" << c.real << "," << c.imag << "i)";
}
#endif
```

10.4 Quaternion Class

The public interface of the `Quaternion` class:

- `Quaternion()` : Default constructor.

- `Quaternion(T,T,T,T)` : Constructor.

- `sqr()` : Square of the quaternion.

- `conjugate()` : Conjugate of the quaternion.

- `inverse()` : Inverse of the quaternion.

- `magnitude()` : Magnitude of the quaternion.

- Arithmetic operators : -(unary), +, -, *, /, ~

- Stream operators : >>, <<

For a detailed description of the class structure and each member function, please refer to Section 6.4.

```
// Quatern.h

#ifndef QUATERNION_H
#define QUATERNION_H

#include <iostream.h>
#include <math.h>        // for sqrt()

template <class T> class Quaternion
{
   private:
      // Data Fields
      T r, i, j, k;
   public:
      // Constructors
      Quaternion();
      Quaternion(T,T,T,T);
      Quaternion(const Quaternion<T>&);
      ~Quaternion();

      // Operators
      const Quaternion<T>& operator = (const Quaternion<T>&);
      Quaternion<T> operator + (const Quaternion<T>&);
      Quaternion<T> operator - (const Quaternion<T>&);
```

```
      Quaternion<T> operator - () const;
      Quaternion<T> operator * (const Quaternion<T>&);
      Quaternion<T> operator * (T);
      Quaternion<T> operator / (const Quaternion<T>&);
      Quaternion<T> operator ~ () const;

      // Member Functions
      Quaternion<T> sqr();
      Quaternion<T> conjugate() const;
      Quaternion<T> inverse() const;
      double magnitude() const;

      // Friendship Functions
      friend Quaternion<T> operator * (T,Quaternion<T>&);

      // Streams
      friend ostream& operator << (ostream&,const Quaternion<T>&);
      friend istream& operator >> (istream&,Quaternion<T>&);
};

template <class T> Quaternion<T>::Quaternion()
   : r(T(0)), i(T(0)), j(T(0)), k(T(0)) {}

template <class T> Quaternion<T>::Quaternion(T r1,T i1,T j1,T k1)
   : r(r1), i(i1), j(j1), k(k1) {}

template <class T> Quaternion<T>::Quaternion(const Quaternion<T>& arg)
   : r(arg.r), i(arg.i), j(arg.j), k(arg.k) {}

template <class T> Quaternion<T>::~Quaternion() {}

template <class T> const Quaternion<T>&
Quaternion<T>::operator = (const Quaternion<T>& rvalue)
{
   r = rvalue.r; i = rvalue.i; j = rvalue.j; k = rvalue.k;
   return *this;
}

template <class T>
Quaternion<T> Quaternion<T>::operator + (const Quaternion<T>& arg)
{ return Quaternion<T>(r+arg.r,i+arg.i,j+arg.j,k+arg.k); }

template <class T>
Quaternion<T> Quaternion<T>::operator - (const Quaternion<T>& arg)
{ return Quaternion<T>(r-arg.r,i-arg.i,j-arg.j,k-arg.k); }

template <class T>
```

```
Quaternion<T> Quaternion<T>::operator - () const
{ return Quaternion<T>(-r,-i,-j,-k); }

template <class T>
Quaternion<T> Quaternion<T>::operator * (const Quaternion<T>& arg)
{
    return Quaternion<T>(r*arg.r - i*arg.i - j*arg.j - k*arg.k,
                         r*arg.i + i*arg.r + j*arg.k - k*arg.j,
                         r*arg.j + j*arg.r + k*arg.i - i*arg.k,
                         r*arg.k + k*arg.r + i*arg.j - j*arg.i);
}

template <class T>
Quaternion<T> Quaternion<T>::operator * (T arg)
{ return Quaternion<T>(r*arg,i*arg,j*arg,k*arg); }

template <class T>
Quaternion<T> Quaternion<T>::operator / (const Quaternion<T>& arg)
{ return *this * arg.inverse(); }

// Normalize Quaternion
template <class T>
Quaternion<T> Quaternion<T>::operator ~ () const
{
    Quaternion<T> result;
    double length = magnitude();
    result.r = r/length; result.i = i/length;
    result.j = j/length; result.k = k/length;
    return result;
}

template <class T> Quaternion<T> Quaternion<T>::sqr()
{
    Quaternion<T> result;
    T temp;
    T two = T(2);
    temp = two*r;
    result.r = r*r - i*i - j*j - k*k;
    result.i = temp*i;
    result.j = temp*j;
    result.k = temp*k;
    return result;
}

template <class T> Quaternion<T> Quaternion<T>::conjugate() const
{ return Quaternion<T>(r,-i,-j,-k); }
```

```
template <class T> Quaternion<T> Quaternion<T>::inverse() const
{
   Quaternion<T> temp1(conjugate());
   T temp2 = r*r + i*i + j*j + k*k;
   return Quaternion<T>(temp1.r/temp2,temp1.i/temp2,
                        temp1.j/temp2,temp1.k/temp2);
}

template <class T> double Quaternion<T>::magnitude() const
{ return sqrt(r*r + i*i + j*j + k*k); }

template <class T> Quaternion<T> operator * (T factor,Quaternion<T>& arg)
{ return arg * factor; }

template <class T>
ostream& operator << (ostream& s,const Quaternion<T>& arg)
{
   s << "(" << arg.r << "," << arg.i << ","
           << arg.j << "," << arg.k << ")";
   return s;
}

template <class T>
istream& operator >> (istream& s,Quaternion<T>& arg)
{
   s >> arg.r >> arg.i >> arg.j >> arg.k;
   return s;
}
#endif
```

10.5 Derive Class

The public interface of the Derive class:

- Derive() : Default constructor.
- Derive(const T) : Constructor.
- set(const T) : Specifies the point where the derivative takes place.
- exp(const Derive<T>&) : Exponential function.
- cos(const Derive<T>&) : Cosine function.
- sin(const Derive<T>&) : Sine function.
- df(const Derive<T>&) : Exact derivative of an expression.
- Arithmetic operators : -(unary), +, -, *, /, +=, -=, *=, /=
- Stream operator : <<

For a detailed description of the class structure and each member function, please refer to Section 6.5.

```
// Derive.h

#ifndef DERIVE_H
#define DERIVE_H

#include <iostream.h>
#include <math.h>

template <class T> class Derive
{
   private:
      // Data Field
      T u, du;
      // Private Constructor
      Derive(const T,const T);
   public:
      // Constructors
      Derive();
      Derive(const T);
      Derive(const Derive<T>&);

      // Member Function
```

```
      void set(const T);

      // Arithmetic Operators
      Derive<T> operator - () const;
      Derive<T> operator += (const Derive<T>&);
      Derive<T> operator -= (const Derive<T>&);
      Derive<T> operator *= (const Derive<T>&);
      Derive<T> operator /= (const Derive<T>&);

      friend Derive<T> operator + (const Derive<T>&,const Derive<T>&);
      friend Derive<T> operator - (const Derive<T>&,const Derive<T>&);
      friend Derive<T> operator * (const Derive<T>&,const Derive<T>&);
      friend Derive<T> operator / (const Derive<T>&,const Derive<T>&);
      friend Derive<T> exp(const Derive<T>&);
      friend Derive<T> sin(const Derive<T>&);
      friend Derive<T> cos(const Derive<T>&);
      friend T df(const Derive<T>&);
      friend ostream & operator << (ostream&,const Derive<T>&);
};

template <class T> Derive<T>::Derive() : u(T(0)), du(T(1)) {}

template <class T> Derive<T>::Derive(const T v) : u(v), du(T(0)) {}

template <class T>
Derive<T>::Derive(const T v,const T dv) : u(v), du(dv) {}

template <class T>
Derive<T>::Derive(const Derive<T>& r) : u(r.u), du(r.du) {}

template <class T> void Derive<T>::set(const T v) { u = v; }

template <class T> Derive<T> Derive<T>::operator - () const
{ return Derive<T>(-u,-du); }

template <class T> Derive<T> Derive<T>::operator += (const Derive<T>& r)
{ return *this = *this + r; }

template <class T> Derive<T> Derive<T>::operator -= (const Derive<T>& r)
{ return *this = *this - r; }

template <class T> Derive<T> Derive<T>::operator *= (const Derive<T>& r)
{ return *this = *this * r; }

template <class T> Derive<T> Derive<T>::operator /= (const Derive<T>& r)
{ return *this = *this / r; }
```

```
template <class T>
Derive<T> operator + (const Derive<T>& x,const Derive<T>& y)
{ return Derive<T>(x.u + y.u,x.du + y.du); }

template <class T>
Derive<T> operator - (const Derive<T>& x,const Derive<T>& y)
{ return Derive<T>(x.u - y.u,x.du - y.du); }

template <class T>
Derive<T> operator * (const Derive<T>& x,const Derive<T>& y)
{ return Derive<T>(x.u*y.u,y.u*x.du + x.u*y.du); }

template <class T>
Derive<T> operator / (const Derive<T>& x,const Derive<T>& y)
{ return Derive<T>(x.u/y.u,(y.u*x.du - x.u*y.du)/(y.u*y.u)); }

template <class T>
Derive<T> exp(const Derive<T>& x)
{ return Derive<T>(exp(x.u),x.du*exp(x.u)); }

template <class T>
Derive<T> sin(const Derive<T> &x)
{ return Derive<T>(sin(x.u),x.du*cos(x.u)); }

template <class T>
Derive<T> cos(const Derive<T>& x)
{ return Derive<T>(cos(x.u),-x.du*sin(x.u)); }

template <class T> T df(const Derive<T>& x)
{ return x.du; }

template <class T>
ostream & operator << (ostream& s,const Derive<T>& r)
{ return s << r.u; }
#endif
```

10.6   Vector Class

The public interface of the Vector class:

- Vector() : Default constructor.

- Vector(int) : Constructor.

- Vector(int,T) : Constructor.

- length() : Length of the vector.

- resize(int n) : Resizes the vector to size n.

- resize(int n,T v) : Resizes the vector to size n and fills the rest of the vector entries with value v.

- reset(int n) : Resets the vector to size n.

- reset(int n,T v) : Resets the vector to size n and initializes entries to v.

- Arithmetic operators : +(unary), -(unary), +, -, *, /, =, +=, -=, *=, /=

- Relational operators : ==, !=

- Subscript operator : []

- Dot product : |

- Cross product : %

- Stream operators : >>, <<

Auxiliary functions in VecNorm.h:

- norm1(const Vector&) : One-norm of the vector.

- norm2(const Vector&) : Two-norm of the vector.

- normI(const Vector&) : Infinite-norm of the vector.

- normalize(const Vector&) : Normalization of the vector.

For a detailed description of the class structure and each member function, please refer to Section 6.6.

```
// Vector.h

#ifndef MVECTOR_H
#define MVECTOR_H

#include <iostream.h>
#include <string.h>
#include <math.h>
#include <assert.h>

// forward declaration
template <class T> class Matrix;

// declaration class Vector
template <class T> class Vector
{
    private:
        // Data Fields
        int size;
        T *data;
    public:
        // Constructors
        Vector();
        Vector(int);
        Vector(int,T);
        Vector(const Vector<T>&);
        ~Vector();

        // Member Functions
        T & operator [] (int) const;
        int length() const;
        void resize(int);
        void resize(int,T);
        void reset(int);
        void reset(int,T);

        // Arithmetic Operators
        const Vector<T>& operator = (const Vector<T>&);
        const Vector<T>& operator = (T);
        Vector<T> operator + () const;
        Vector<T> operator - () const;
        Vector<T> operator += (const Vector<T>&);
        Vector<T> operator -= (const Vector<T>&);
        Vector<T> operator *= (const Vector<T>&);
        Vector<T> operator /= (const Vector<T>&);
        Vector<T> operator +  (const Vector<T>&) const;
        Vector<T> operator -  (const Vector<T>&) const;
```

```
            Vector<T> operator *  (const Vector<T>&) const;
            Vector<T> operator /  (const Vector<T>&) const;

            Vector<T> operator += (T);
            Vector<T> operator -= (T);
            Vector<T> operator *= (T);
            Vector<T> operator /= (T);
            Vector<T> operator +  (T) const;
            Vector<T> operator -  (T) const;
            Vector<T> operator *  (T) const;
            Vector<T> operator /  (T) const;

            T operator | (const Vector<T>&);   // Inner product
            Vector<T> operator % (const Vector<T>&); // Vector product

            // I/O stream functions
            friend class Matrix<T>;
            friend ostream& operator << (ostream&,const Vector<T>&);
            friend istream& operator >> (istream&,Vector<T>&);
};

// implementation of class Vector
template <class T> Vector<T>::Vector() : size(0), data(NULL) {}

template <class T> Vector<T>::Vector(int n) : size(n), data(new T[n])
{  assert(data != NULL); }

template <class T> Vector<T>::Vector(int n,T value)
    : size(n), data(new T[n])
{
    assert(data != NULL);
    for(int i=0; i<n; i++) data[i] = value;
}

template <class T> Vector<T>::Vector(const Vector<T> &v)
    : size(v.size), data(new T[v.size])
{
    assert(data != NULL);
    for(int i=0; i<v.size; i++) data[i] = v.data[i];
}

template <class T> Vector<T>::~Vector() {  delete [] data; }

template <class T> T & Vector<T>::operator[] (int i) const
{
    assert(i >= 0 && i < size);
    return data[i];
```

```
}

template <class T> int Vector<T>::length() const
{ return size; }

template <class T> void Vector<T>::resize(int length)
{
    int i;
    T zero(0);
    T *newData = new T[length]; assert(newData != NULL);
    if(length <= size)
    for(i=0; i<length; i++) newData[i] = data[i];
    else
    {
    for(i=0; i<size; i++)   newData[i] = data[i];
    for(i=size; i<length; i++) newData[i] = zero;
    }
    delete [] data;
    size = length;
    data = newData;
}

template <class T> void Vector<T>::resize(int length,T value)
{
    int i;
    T* newData = new T[length]; assert(newData != NULL);
    if(length <= size)
    for(i=0; i<length; i++) newData[i] = data[i];
    else
    {
    for(i=0; i<size; i++)        newData[i] = data[i];
    for(i=size; i<length; i++) newData[i] = value;
    }
    delete [] data;
    size = length;
    data = newData;
}

template <class T> void Vector<T>::reset(int length)
{
    T zero(0);
    delete [] data;
    data = new T[length]; assert(data != NULL);
    size = length;
    for(int i=0; i<size; i++) data[i] = zero;
}
```

```
template <class T> void Vector<T>::reset(int length,T value)
{
    delete [] data;
    data = new T[length]; assert(data != NULL);
    size = length;
    for(int i=0; i<size; i++) data[i] = value;
}


template <class T>
const Vector<T>& Vector<T>::operator = (const Vector<T>& v)
{
    if(this == &v) return *this;
    if(size != v.size)
    {
    delete [] data;
    data = new T[v.size]; assert(data != NULL);
    size = v.size;
    }
    for(int i=0; i<v.size; i++) data[i] = v.data[i];
    return *this;
}

template <class T> const Vector<T>& Vector<T>::operator = (T value)
{
    for(int i=0; i<size; i++) data[i] = value;
    return *this;
}

template <class T> Vector<T> Vector<T>::operator + () const
{  return *this; }

template <class T> Vector<T> Vector<T>::operator - () const
{  return *this * T(-1); }

template <class T>
Vector<T> Vector<T>::operator += (const Vector<T>& v)
{
    assert(size==v.size);
    for(int i=0; i<size; i++) data[i] += v.data[i];
    return *this;
}

template <class T>
Vector<T> Vector<T>::operator -= (const Vector<T>& v)
{
    assert(size==v.size);
    for(int i=0; i<size; i++) data[i] -= v.data[i];
```

```
    return *this;
}

template <class T>
Vector<T> Vector<T>::operator *= (const Vector<T>& v)
{
    assert(size==v.size);
    for(int i=0; i<size; i++) data[i] *= v.data[i];
    return *this;
}

template <class T>
Vector<T> Vector<T>::operator /= (const Vector<T>& v)
{
    assert(size==v.size);
    for(int i=0; i<size; i++) data[i] /= v.data[i];
    return *this;
}

template <class T>
Vector<T> Vector<T>::operator + (const Vector<T>& v) const
{
    Vector<T> result(*this);
    return  result += v;
}

template <class T>
Vector<T> Vector<T>::operator - (const Vector<T>& v) const
{
    Vector<T> result(*this);
    return  result -= v;
}

template <class T>
Vector<T> Vector<T>::operator * (const Vector<T>& v) const
{
    Vector<T> result(*this);
    return  result *= v;
}

template <class T>
Vector<T> Vector<T>::operator / (const Vector<T>& v) const
{
    Vector<T> result(*this);
    return  result /= v;
}
```

```
template <class T> Vector<T> Vector<T>::operator += (T c)
{
    for(int i=0; i<size; i++) data[i] += c;
    return *this;
}

template <class T> Vector<T> Vector<T>::operator -= (T c)
{
    for(int i=0; i<size; i++) data[i] -= c;
    return *this;
}

template <class T> Vector<T> Vector<T>::operator *= (T c)
{
    for(int i=0; i<size; i++) data[i] *= c;
    return *this;
}

template <class T> Vector<T> Vector<T>::operator /= (T c)
{
    for(int i=0; i<size; i++) data[i] /= c;
    return *this;
}

template <class T> Vector<T> Vector<T>::operator + (T c) const
{
    Vector<T> result(*this);
    return result += c;
}

template <class T> Vector<T> Vector<T>::operator - (T c) const
{
    Vector<T> result(*this);
    return result -= c;
}

template <class T> Vector<T> Vector<T>::operator * (T c) const
{
    Vector<T> result(*this);
    return result *= c;
}

template <class T> Vector<T> Vector<T>::operator / (T c) const
{
    Vector<T> result(*this);
    return result /= c;
}
```

```
template <class T> Vector<T> operator + (T c,const Vector<T>& v)
{ return v+c; }

template <class T> Vector<T> operator - (T c,const Vector<T>& v)
{ return -v+c; }

template <class T> Vector<T> operator * (T c,const Vector<T>& v)
{ return v*c; }

template <class T> Vector<T> operator / (T c,const Vector<T>& v)
{
    Vector<T> result(v.length());
    for(int i=0; i<result.length(); i++) result[i] = c/v[i];
    return result;
}

// Inner Product
template <class T> T Vector<T>::operator | (const Vector<T>& v)
{
    assert(size == v.size);
    T result(0);
    for(int i=0; i<size; i++) result = result + data[i]*v.data[i];
    return result;
}

// Vector Product
template <class T> Vector<T> Vector<T>::operator % (const Vector<T>& v)
{
    assert(size == 3 && v.size == 3);
    Vector<T> result(3);
    result.data[0] = data[1] * v.data[2] - v.data[1] * data[2];
    result.data[1] = v.data[0] * data[2] - data[0] * v.data[2];
    result.data[2] = data[0] * v.data[1] - v.data[0] * data[1];
    return result;
}

// Equality
template <class T>
int operator == (const Vector<T>& u,const Vector<T>& v)
{
    if(u.length() != v.length()) return 0;
    for(int i=0; i<u.length(); i++)
        if(u[i] != v[i]) return 0;
    return 1;
}
```

```
// Inequality
template <class T>
int operator != (const Vector<T>& u,const Vector<T>& v)
{ return !(u==v); }

template <class T> ostream & operator << (ostream &s,const Vector<T>& v)
{
    int lastnum = v.length();
    for(int i=0; i<lastnum; i++) s << "[" << v[i] << "]" << endl;
    return s;
}

template <class T> istream & operator >> (istream& s,Vector<T>& v)
{
    int i, num;
    s.clear();                  // set stream state to good
    s >> num;                   // read size of Vector
    if(!s.good()) return s;  // can't get an integer, just return
    v.resize(num);              // resize Vector v
    for(i=0; i<num; i++)
    {
        s >> v[i];              // read in entries
        if(!s.good())
        {
            s.clear(s.rdstate() | ios::badbit);
            return s;
        }
    }
    return s;
}
#endif
```

```
// VecNorm.h
// Norms of Vectors

#ifndef MVECNORM_H
#define MVECNORM_H

#include <iostream.h>
#include <math.h>

template <class T> T norm1(const Vector<T>& v)
{
    T result(0);

    for(int i=0; i<v.length(); i++)
        result = result + abs(v[i]);

    return result;
}

double norm1(const Vector<double>& v)
{
    double result(0);

    for(int i=0; i<v.length(); i++)
        result = result + fabs(v[i]);

    return result;
}

template <class T> double norm2(const Vector<T>& v)
{
    T result(0);

    for(int i=0; i<v.length(); i++)
        result = result + v[i]*v[i];

    return sqrt(double(result));
}

template <class T> T normI(const Vector<T>& v)
{
    T maxItem(abs(v[0])), temp;

    for(int i=1; i<v.length(); i++)
    {
        temp = abs(v[i]);
        if(temp > maxItem) maxItem = temp;
```

```
   }
   return maxItem;
}

double normI(const Vector<double>& v)
{
   double maxItem(fabs(v[0])), temp;

   for(int i=1; i<v.length(); i++)
   {
      temp = fabs(v[i]);
      if(temp > maxItem) maxItem = temp;
   }
   return maxItem;
}

template <class T> Vector<T> normalize(const Vector<T>& v)
{
   Vector<T> result(v.length());
   double length = norm2(v);

   for(int i=0; i<v.length(); i++)
      result[i] = v[i]/length;

   return result;
}
#endif
```

10.7 Matrix Class

The public interface of the Matrix class:

- Matrix() : Default constructor.

- Matrix(int,int) : Constructor.

- Matrix(int,int,T) : Constructor.

- Matrix(const Vector<T>&) : Constructor.

- identity() : Creates an identity matrix.

- transpose() : Transpose of the matrix.

- inverse() : Inverse of the matrix.

- trace() : Trace of the matrix.

- determinant() : Determinant of the matrix.

- rows() : Number of rows of the matrix.

- cols() : Number of columns of the matrix.

- resize(int,int) : Resizes the matrix.

- resize(int,int,T) : Resizes the matrix and initializes the rest of the entries.

- vec(const Matrix<T>&) : Vectorize operator.

- kron(const Matrix<T>&,const Matrix<T>&) : Kronecker product.

- fill(T v) : Fills the matrix with the value v.

- Arithmetic operators : +(unary), -(unary), +, -, *, /, =, +=, -=, *=, /=

- Row vector operator : []

- Column vector operator : ()

- Stream operators : >>, <<

Auxiliary functions in `MatNorm.h`:

- `norm1(const Matrix&)` : One-norm of the matrix.

- `normI(const Matrix&)` : Infinite-norm of the matrix.

- `normH(const Matrix&)` : Hilbert-Schmidt norm of the matrix.

For a detailed description of the class structure and each member function, please refer to Section 6.7.

```
// Matrix.h

#ifndef MATRIX_H
#define MATRIX_H

#include <iostream.h>
#include <math.h>
#include <assert.h>
#include "Vector.h"

// definition of class Matrix
template <class T> class Matrix
{
    protected:
        // Data Fields
        int rowNum, colNum;
        Vector<T> *mat;

    public:
        // Constructors
        Matrix();
        Matrix(int,int);
        Matrix(int,int, T);
        Matrix(const Vector<T>&);
        Matrix(const Matrix<T>&);
        ~Matrix();

        // Member Functions
        Vector<T>& operator [] (int) const;
        Vector<T>  operator () (int) const;

        Matrix<T> identity();
        Matrix<T> transpose() const;
        Matrix<T> inverse() const;
        T trace() const;
```

```
    T determinant() const;

    int rows() const;
    int cols() const;
    void resize(int,int);
    void resize(int,int,T);
    void fill(T);

    // Arithmetic Operators
    const Matrix<T>& operator = (const Matrix<T>&);
    const Matrix<T>& operator = (T);

    Matrix<T> operator + () const;
    Matrix<T> operator - () const;
    Matrix<T> operator += (const Matrix<T>&);
    Matrix<T> operator -= (const Matrix<T>&);
    Matrix<T> operator *= (const Matrix<T>&);
    Matrix<T> operator +  (const Matrix<T>&) const;
    Matrix<T> operator -  (const Matrix<T>&) const;
    Matrix<T> operator *  (const Matrix<T>&) const;
    Vector<T> operator *  (const Vector<T>&) const;

    Matrix<T> operator += (T);
    Matrix<T> operator -= (T);
    Matrix<T> operator *= (T);
    Matrix<T> operator /= (T);
    Matrix<T> operator +  (T) const;
    Matrix<T> operator -  (T) const;
    Matrix<T> operator *  (T) const;
    Matrix<T> operator /  (T) const;

    friend Vector<T> vec(const Matrix<T>&);
    friend Matrix<T> kron(const Matrix<T>&,const Matrix<T>&);
    friend ostream & operator << (ostream&,const Matrix<T>&);
    friend istream & operator >> (istream&,Matrix<T>&);
};

// implementation of class Matrix
template <class T> Matrix<T>::Matrix()
    : rowNum(0), colNum(0), mat(NULL) {}

template <class T> Matrix<T>::Matrix(int r,int c)
    : rowNum(r), colNum(c), mat(new Vector<T>[r])
{
    assert(mat != NULL);
    for(int i=0; i<r; i++) mat[i].resize(c);
}
```

```
template <class T> Matrix<T>::Matrix(int r,int c,T value)
   : rowNum(r), colNum(c), mat(new Vector<T>[r])
{
   assert(mat != NULL);
   for(int i=0; i<r; i++) mat[i].resize(c,value);
}

template <class T> Matrix<T>::Matrix(const Vector<T>& v)
   : rowNum(v.length()), colNum(1), mat(new Vector<T>[rowNum])
{
   assert(mat != NULL);
   for(int i=0; i<rowNum; i++) mat[i].resize(1,v[i]);
}

template <class T> Matrix<T>::Matrix(const Matrix<T>& m)
   : rowNum(m.rowNum), colNum(m.colNum), mat(new Vector<T>[m.rowNum])
{
   assert(mat != NULL);
   for(int i=0; i<m.rowNum; i++) mat[i] = m.mat[i];
}

template <class T> Matrix<T>::~Matrix()
{
   delete [] mat;
}

template <class T> Vector<T> & Matrix<T>::operator [] (int index) const
{
   assert(index>=0 && index<rowNum);
   return mat[index];
}

template <class T> Vector<T> Matrix<T>::operator () (int index) const
{
   assert(index>=0 && index<colNum);
   Vector<T> result(rowNum);

   for(int i=0; i<rowNum; i++) result[i] = mat[i][index];
   return result;
}

template <class T> Matrix<T> Matrix<T>::identity()
{
   for(int i=0; i<rowNum; i++)
      for(int j=0; j<colNum; j++)
         if(i==j) mat[i][j] = T(1);
```

```
              else      mat[i][j] = T(0);
    return *this;
}

template <class T> Matrix<T> Matrix<T>::transpose() const
{
    Matrix<T> result(colNum,rowNum);
    for(int i=0; i<rowNum; i++)
       for(int j=0; j<colNum; j++)
          result[j][i] = mat[i][j];
    return result;
}

// Symbolical Inverse using Leverrier's Method
template <class T> Matrix<T> Matrix<T>::inverse() const
{
    assert(rowNum == colNum);
    Matrix<T> B(*this), D, I(rowNum,colNum);
    T c0(B.trace()), c1;
    int i;
    I.identity();
    for(i=2; i<rowNum; i++)
    {
       B = *this * (B-c0*I);
       c0 = B.trace()/T(i);
    }
    D = *this * (B-c0*I);
    c1 = D.trace()/T(i);
    return (B-c0*I)/c1;
}

template <class T> T Matrix<T>::trace() const
{
    assert(rowNum == colNum);
    T result(0);
    for(int i=0; i<rowNum; i++) result += mat[i][i];
    return result;
}

// Symbolical determinant
template <class T> T Matrix<T>::determinant() const
{
    assert(rowNum==colNum);
    Matrix<T> B(*this), I(rowNum,colNum,T(0));
    T c(B.trace());
    int i;
    for(i=0; i<rowNum; i++) I[i][i] = T(1);
```

```
   // Note that determinant of int-type gives zero
   // because of division by T(i)
   for(i=2; i<=rowNum; i++)
   {
      B = *this * (B-c*I);
      c = B.trace()/T(i);
   }
   if(rowNum%2) return c;
   return -c;
}


template <class T> int Matrix<T>::rows() const
{
   return rowNum;
}


template <class T> int Matrix<T>::cols() const
{
   return colNum;
}


template <class T> void Matrix<T>::resize(int r,int c)
{
   int i;
   Vector<T> *newMat = new Vector<T>[r]; assert(newMat != NULL);
   if(r<=rowNum)
   {
      for(i=0; i<r; i++)
      {
         (mat+i) -> resize(c);
         newMat[i] = mat[i];
      }
   }
   else
   {
      for(i=0; i<rowNum; i++)
      {
         (mat+i) -> resize(c);
         newMat[i] = mat[i];
      }
      for(i=rowNum; i<r; i++) newMat[i].resize(c);
   }
   delete [] mat;
   rowNum = r; colNum = c;
   mat = newMat;
}
```

```
template <class T> void Matrix<T>::resize(int r,int c,T value)
{
    int i;
    Vector<T> *newMat = new Vector<T>[r]; assert(newMat != NULL);
    if(r<=rowNum)
    {
        for(i=0; i<r; i++)
        {
            (mat+i) -> resize(c,value);
            newMat[i] = mat[i];
        }
    }
    else
    {
        for(i=0; i<rowNum; i++)
        {
            (mat+i) -> resize(c,value);
            newMat[i] = mat[i];
        }
        for(i=rowNum; i<r; i++) newMat[i].resize(c,value);
    }
    delete [] mat;
    rowNum = r; colNum = c;
    mat = newMat;
}

template <class T> void Matrix<T>::fill(T value)
{
    for(int i=0; i<rowNum; i++)
        for(int j=0; j<colNum; j++)
            mat[i][j] = value;
}

template <class T>
const Matrix<T>& Matrix<T>::operator = (const Matrix<T>& m)
{
    if(this == &m) return *this;
    delete [] mat;
    rowNum = m.rowNum; colNum = m.colNum;
    mat = new Vector<T>[m.rowNum]; assert(mat != NULL);
    for(int i=0; i<m.rowNum; i++) mat[i] = m.mat[i];
    return *this;
}

template <class T>
const Matrix<T>& Matrix<T>::operator = (T value)
```

```
{
    for(int i=0; i<rowNum; i++) mat[i] = value;
    return *this;
}

template <class T> Matrix<T> Matrix<T>::operator + () const
{
    return *this;
}

template <class T> Matrix<T> Matrix<T>::operator - () const
{
    return *this * T(-1);
}

template <class T>
Matrix<T> Matrix<T>::operator += (const Matrix<T>& m)
{
    return *this = *this + m;
}

template <class T>
Matrix<T> Matrix<T>::operator -= (const Matrix<T>& m)
{
    return *this = *this - m;
}

template <class T>
Matrix<T> Matrix<T>::operator *= (const Matrix<T>& m)
{
    return *this = *this * m;
}

template <class T>
Matrix<T> Matrix<T>::operator + (const Matrix<T>& m) const
{
    assert(rowNum == m.rowNum && colNum == m.colNum);
    Matrix<T> result(*this);
    for(int i=0; i<rowNum; i++) result[i] += m[i];
    return result;
}

template <class T>
Matrix<T> Matrix<T>::operator - (const Matrix<T>& m) const
{
    assert(rowNum == m.rowNum && colNum == m.colNum);
    Matrix<T> result(*this);
```

```
      for(int i=0; i<rowNum; i++) result[i] -= m[i];
      return result;
}

template <class T>
Matrix<T> Matrix<T>::operator * (const Matrix<T>& m) const
{
   assert(colNum == m.rowNum);
   Matrix<T> result(rowNum, m.colNum, T(0));
   for(int i=0; i<rowNum; i++)
      for(int j=0; j<m.colNum; j++)
         for(int k=0; k<colNum; k++)
            result[i][j] += mat[i][k] * m[k][j];
   return result;
}

template <class T>
Vector<T> Matrix<T>::operator * (const Vector<T>& v) const
{
   assert(colNum == v.length());
   Vector<T> result(rowNum);

   // dot product | is used
   for (int i=0; i<rowNum; i++) result[i] = (mat[i] | v);
   return result;
}

template <class T> Matrix<T> Matrix<T>::operator += (T c)
{
   assert(rowNum == colNum);
   for(int i=0; i<rowNum; i++) mat[i][i] += c;
   return *this;
}

template <class T> Matrix<T> Matrix<T>::operator -= (T c)
{
   assert(rowNum == colNum);
   for(int i=0; i<rowNum; i++) mat[i][i] -= c;
   return *this;
}

template <class T> Matrix<T> Matrix<T>::operator *= (T c)
{
   for(int i=0; i<rowNum; i++) mat[i] *= c;
   return *this;
}
```

```
template <class T> Matrix<T> Matrix<T>::operator /= (T c)
{
   for(int i=0; i<rowNum; i++) mat[i] /= c;
   return *this;
}

template <class T>
Matrix<T> Matrix<T>::operator + (T value) const
{
   assert(rowNum == colNum);
   Matrix<T> result(*this);
   return result += value;
}

template <class T>
Matrix<T> Matrix<T>::operator - (T value) const
{
   assert(rowNum == colNum);
   Matrix<T> result(*this);
   return result -= value;
}

template <class T>
Matrix<T> Matrix<T>::operator * (T value) const
{
   Matrix<T> result(*this);
   return result *= value;
}

template <class T>
Matrix<T> Matrix<T>::operator / (T value) const
{
   Matrix<T> result(*this);
   return result /= value;
}

template <class T>
Matrix<T> operator + (T value, const Matrix<T> &m)
{
   return m + value;
}

template <class T>
Matrix<T> operator - (T value,const Matrix<T>& m)
{
   return -m + value;
}
```

```
template <class T>
Matrix<T> operator * (T value,const Matrix<T>& m)
{
    return m * value;
}


template <class T>
Matrix<T> operator / (T value,const Matrix<T>& m)
{
    Matrix<T> result(m.rows(),m.cols());
    for(int i=0; i<result.rows(); i++) result[i] = value/m[i];
    return result;
}


// Vectorize operator
template <class T> Vector<T> vec(const Matrix<T> &m)
{
    int i=0, j, k, size = m.rowNum * m.colNum;
    Vector<T> result(size);
    for(j=0; j<m.colNum; j++)
        for(k=0; k<m.rowNum; k++) result[i++] = m.mat[k][j];
    return result;
}


// Kronecker Product
template <class T>
Matrix<T> kron(const Matrix<T>& s,const Matrix<T>& m)
{
    int size1 = s.rowNum * m.rowNum,
        size2 = s.colNum * m.colNum,
        i, j, k, p;
    Matrix<T> result(size1, size2);

    for(i=0; i<s.rowNum; i++)
        for(j=0; j<s.colNum; j++)
            for(k=0; k<m.rowNum; k++)
                for(p=0; p<m.colNum; p++)
                    result[k + i*m.rowNum][p + j*m.colNum]
                        = s.mat[i][j] * m.mat[k][p];
    return result;
}


template <class T>
int operator == (const Matrix<T>& m1,const Matrix<T>& m2)
{
    if(m1.rows() != m2.rows()) return 0;
```

```
    for(int i=0; i<m1.rows(); i++)
        if(m1[i] != m2[i]) return 0;
    return 1;
}


template <class T>
int operator != (const Matrix<T>& m1,const Matrix<T>& m2)
{
    return !(m1==m2);
}


template <class T> ostream& operator << (ostream& s,const Matrix<T>& m)
{
    int t = m.cols()-1;
    for(int i=0; i<m.rows(); i++)
    {
        s << "[";
        for(int j=0; j<t; j++) s << m[i][j] << " ";
        s << m[i][t] << "]" << endl;
    }
    return s;
}


template <class T> istream& operator >> (istream& s,Matrix<T>& m)
{
    int i, j, num1, num2;
    s.clear();                       // set stream state to good
    s >> num1;                       // read in row number
    if(!s.good()) return s;          // can't get an integer, just return
    s >> num2;                       // read in column number
    if(!s.good()) return s;          // can't get an integer, just return
    m.resize(num1,num2);             // resize to Matrix into right order
    for(i=0; i<num1; i++)
        for(j=0; j<num2; j++)
        {
            s >> m[i][j];
            if(!s.good())
            {
                s.clear(s.rdstate() | ios::badbit);
                return s;
            }
        }
    return s;
}
#endif
```

```
// MatNorm.h
// Norms of Matrices

#ifndef MATNORM_H
#define MATNORM_H

#include <iostream.h>
#include <math.h>
#include "Vector.h"
#include "VecNorm.h"

template <class T> T norm1(const Matrix<T>& m)
{
    T maxItem(0), temp;
    int i,j;

    for(i=0; i<m.rows(); i++) maxItem += m[i][0];

    for(i=1; i<m.cols(); i++)
    {
        temp = T(0);
        for(j=0; j<m.rows(); j++)
            temp += abs(m[j][i]);
        if(temp > maxItem) maxItem = temp;
    }
    return maxItem;
}

template <class T> T normI(const Matrix<T>& m)
{
    T maxItem(norm1(m[0]));

    for(int i=1; i<m.rows(); i++)
        if(norm1(m[i]) > maxItem) maxItem = norm1(m[i]);
    return maxItem;
}

template <class T> T normH(const Matrix<T>& m)
{
    return sqrt((m*(m.transpose())).trace());
}
#endif
```

10.8   Array Class

The public interface of the **Array1** class:

- **Array1(int = 0)** : Constructor.

- **Array1(int,T)** : Constructor.

- **resize(int)** : Resizes the one-dimensional array.

- **resize(int,T v)** : Resizes the one-dimensional array and initializes the rest of the entries with v.

- **size(int = 0)** : Size of the array.

- Arithmetic operators : **+, -, *, =, +=, -=, *=**

- Subscript operator : **[]**

- Stream operator : **<<**

The public interface of the **Array2** class:

- **Array2(int = 0,int = 0)** : Constructor.

- **Array2(int,int,T)** : Constructor.

- **resize(int,int)** : Resizes the two-dimensional array.

- **resize(int,int,T v)** : Resizes the two-dimensional array and initializes the rest of the entries with v.

- **size(int = 0)** : Size of the array.

- Arithmetic operators : **+, -, *, =, +=, -=, *=**

- Subscript operator : **[]**

- Stream operator : **<<**

The public interface of the **Array3** class:

- **Array3(int = 0,int = 0,int = 0)** : Constructor.

- **Array3(int,int,int,T)** : Constructor.

- **resize(int,int,int)** : Resizes the three-dimensional array.

- **resize(int,int,int,T v)** : Resizes the three-dimensional array and initializes the rest of the entries with v.

- size(int = 0) : Size of the array.

- Arithmetic operators : +, -, *, =, +=, -=, *=

- Subscript operator : []

- Stream operator : <<

The public interface of the Array4 class:

- Array4(int = 0,int = 0,int = 0,int = 0) : Constructor.

- Array4(int,int,int,int,T) : Constructor.

- resize(int,int,int,int) : Resizes the four-dimensional array.

- resize(int,int,int,int,T v) : Resizes the four-dimensional array and initializes the rest of the entries with v.

- size(int = 0) : Size of the array.

- Arithmetic operators : +, -, *, =, +=, -=, *=

- Subscript operator : []

- Stream operator : <<

For a detailed description of the classes and each member function, please refer to Section 6.8.

```
// Array.h

#ifndef ARRAY_H
#define ARRAY_H

#include <iostream.h>
#include <assert.h>

template <class T>
class Array1
{
   private:
      // Data Fields
      int n_data;
      T *data;
   public:
      // Constructors
      Array1(int = 0);
      Array1(int,T);
```

```
        Array1(const Array1<T>&);
        ~Array1();

        // Member Functions
        void resize(int);
        void resize(int,T);
        T& operator [] (int) const;
        int size(int = 0) const;

        // Arithmetic Operators
        const Array1<T>& operator = (const Array1<T>&);
        const Array1<T>& operator = (T);
        Array1<T> operator *= (T);
        Array1<T> operator += (const Array1<T>&);
        Array1<T> operator -= (const Array1<T>&);
        Array1<T> operator +  (const Array1<T>&);
        Array1<T> operator -  (const Array1<T>&);

        // I/O stream functions
        friend ostream & operator << (ostream&,const Array1<T>&);
};      // end declaration class Array1

// Constructors, destructors and copy constructor.
//
template <class T> Array1<T>::Array1(int n) : n_data(n), data(new T[n])
{ assert(data != NULL); }

template <class T>
Array1<T>::Array1(int n,T num) : n_data(n), data(new T[n])
{
    assert(data != NULL);
    for(int i=0; i<n_data; i++) data[i] = num;
}

template <class T> Array1<T>::Array1(const Array1<T> &v)
    : n_data(v.n_data), data(new T[v.n_data])
{
    assert(data != NULL);
    for(int i=0; i<v.n_data; i++) data[i] = v.data[i];
}

template <class T> Array1<T>::~Array1() { delete [] data; }

// Member functions
template <class T> void Array1<T>::resize(int n)
{
    int i;
```

```
    T *newData = new T[n]; assert(newData != NULL);
    if(n <= n_data)
        for(i=0; i<n; i++)        newData[i] = data[i];
    else
        for(i=0; i<n_data; i++) newData[i] = data[i];
    delete [] data;
    n_data = n;
    data = newData;
}

template <class T> void Array1<T>::resize(int n,T value)
{
    int i;
    T *newData = new T[n]; assert(newData != NULL);
    if(n <= n_data)
        for(i=0; i<n; i++)  newData[i] = data[i];
    else
    {
        for(i=0; i<n_data; i++) newData[i] = data[i];
        for(i=n_data; i<n; i++) newData[i] = value;
    }
    delete [] data;
    n_data = n;
    data = newData;
}

// Various member operators
template <class T> T& Array1<T>::operator [](int i) const
{
    assert(i >= 0 && i < n_data);
    return data[i];
}

template <class T> int Array1<T>::size(int index) const
{
    assert(index == 0);
    return n_data;
}

template <class T>
const Array1<T>& Array1<T>::operator = (const Array1<T>& v)
{
    if(this == &v) return *this;
    if(n_data != v.n_data)
    {
        delete [] data;
        n_data = v.n_data;
```

```
      data = new T[v.n_data]; assert(data != NULL);
   }
   for(int i=0; i<v.n_data; i++) data[i] = v.data[i];
   return *this;
}


template <class T> const Array1<T>& Array1<T>::operator = (T num)
{
   for(int i=0; i<n_data; i++) data[i] = num;
   return *this;
}


template <class T> Array1<T> Array1<T>::operator *= (T num)
{
   for(int i=0; i<n_data; i++) data[i] *= num;
   return *this;
}


template <class T>
Array1<T> Array1<T>::operator += (const Array1<T>& v)
{ return *this = *this + v; }


template <class T>
Array1<T> Array1<T>::operator -= (const Array1<T>& v)
{ return *this = *this - v; }


template <class T> Array1<T>
Array1<T>::operator +  (const Array1<T>& v)
{
   assert(n_data==v.n_data);
   Array1<T> temp(n_data);
   for(int i=0; i<n_data; i++) temp[i] = data[i] + v.data[i];
   return temp;
}


template <class T>
Array1<T> Array1<T>::operator - (const Array1<T>& v)
{
   assert(n_data==v.n_data);
   Array1<T> temp(n_data);
   for(int i=0; i<n_data; i++) temp[i] = data[i] - v.data[i];
   return temp;
}


// Friendship functions
template <class T> ostream& operator << (ostream& s,const Array1<T>& v)
{
```

```
    int n_data = v.n_data-1;
    s << "[";
    for(int i=0; i<n_data; i++) s << v.data[i] << " ";
    s << v.data[n_data] << "]";
    return s;
}


template <class T>
int operator == (const Array1<T>& v1,const Array1<T>& v2)
{
    if(v1.size() != v2.size()) return 0;
    for(int i=0; i<v1.size(); i++)
    {
        if(v1[i] != v2[i]) return 0;
    }
    return 1;
}


template <class T>
int operator != (const Array1<T>& v1,const Array1<T>& v2)
{ return !(v1==v2); }


template <class T> class Array2
{
    private:
        // Data Fields
        int rows, cols;
        Array1<T> *data2D;
    public:
        // Constructors
        Array2(int = 0,int = 0);
        Array2(int,int,T);
        Array2(const Array2<T>&);
        ~Array2();

        // Member Functions
        void resize(int,int);
        void resize(int,int,T);
        Array1<T>& operator [] (int) const;
        int size(int = 0) const;

        // Arithmetic Operators
        const Array2<T>& operator = (const Array2<T>&);
        const Array2<T>& operator = (T);
        Array2<T> operator *= (T);
        Array2<T> operator += (const Array2<T>&);
        Array2<T> operator -= (const Array2<T>&);
```

```
        Array2<T> operator +  (const Array2<T>&);
        Array2<T> operator -  (const Array2<T>&);

        // I/O stream functions
        friend ostream & operator << (ostream&,const Array2<T>&);
};     // end declaration class Array2

// Constructor, destructor and copy constructor.
template <class T> Array2<T>::Array2(int r,int c)
    : rows(r), cols(c), data2D(new Array1<T>[r])
{
    assert(data2D != NULL);
    for(int i = 0; i<r; ++i) (data2D+i) -> resize(c);
}


template <class T> Array2<T>::Array2(int r,int c,T num)
    : rows(r), cols(c), data2D(new Array1<T>[r])
{
    assert(data2D != NULL);
    for(int i = 0; i<r; ++i) (data2D+i) -> resize(c,num);
}


template <class T> Array2<T>::Array2(const Array2<T>& m)
    : rows(m.rows), cols(m.cols), data2D(new Array1<T>[m.rows])
{
    assert(data2D != NULL);
    for(int i=0; i<m.rows; ++i) data2D[i] = m.data2D[i];
}


template <class T> Array2<T>::~Array2() { delete [] data2D; }


// Member functions
template <class T> void Array2<T>::resize(int r,int c)
{
    int i;
    Array1<T> *newArray = new Array1<T>[r]; assert(newArray != NULL);
    if(r <= rows)
    {
        for(i=0; i<r; i++)
        {
            (data2D+i) -> resize(c);
            newArray[i] = data2D[i];
        }
    }
    else
    {
        for(i=0; i<rows; i++)
```

```
    {
        (data2D+i) -> resize(c);
        newArray[i] = data2D[i];
    }
    for(i=rows; i<r; i++) newArray[i].resize(c);
    }
    delete [] data2D;
    rows = r; cols = c;
    data2D = newArray;
}

template <class T> void Array2<T>::resize(int r,int c,T value)
{
    int i;
    Array1<T> *newArray = new Array1<T>[r]; assert(newArray != NULL);
    if(r<=rows)
    {
        for(i=0; i<r; i++)
        {
            (data2D+i) -> resize(c,value);
            newArray[i] = data2D[i];
        }
    }
    else
    {
        for(i=0; i<rows; i++)
        {
            (data2D+i) -> resize(c,value);
            newArray[i] = data2D[i];
        }
        for(i=rows; i<r; i++) newArray[i].resize(c,value);
    }
    delete [] data2D;
    rows = r; cols = c;
    data2D = newArray;
}

// Various member operators
template <class T> Array1<T>& Array2<T>::operator [] (int i) const
{
    assert(i >= 0 && i < rows);
    return data2D[i];
}

template <class T> int Array2<T>::size(int index) const
{
    assert(index==0 || index==1);
```

```
    if(index) return cols;
    return rows;
}

template <class T>
const Array2<T>& Array2<T>::operator = (const Array2<T>& m)
{
    if(this == &m) return *this;
    delete [] data2D;
    cols = m.cols;  rows = m.rows;
    data2D = new Array1<T>[m.rows]; assert(data2D != NULL);
    for(int i=0; i<m.rows; i++) data2D[i] = m.data2D[i];
    return *this;
}

template <class T> const Array2<T>& Array2<T>::operator = (T num)
{
    for(int i=0; i<rows; i++) data2D[i] = num;
    return *this;
}

template <class T> Array2<T> Array2<T>::operator *= (T num)
{
    for(int i=0; i<rows; i++) data2D[i] *= num;
    return *this;
}

template <class T> Array2<T> Array2<T>::operator += (const Array2<T>& m)
{ return *this = *this + m; }

template <class T> Array2<T> Array2<T>::operator -= (const Array2<T>& m)
{ return *this = *this - m; }

template <class T> Array2<T> Array2<T>::operator +  (const Array2<T>& m)
{
    assert((m.cols == cols) && (m.rows == rows));
    Array2<T> temp(rows,cols);
    for(int i=0; i<rows; i++) temp[i] = data2D[i] + m.data2D[i];
    return temp;
}

template <class T> Array2<T> Array2<T>::operator -  (const Array2<T>& m)
{
    assert((m.cols == cols) && (m.rows == rows));
    Array2<T> temp(rows,cols);
    for(int i=0; i<rows; i++) temp[i] = data2D[i] - m.data2D[i];
    return temp;
```

```
}

template <class T> ostream & operator << (ostream& s,const Array2<T>& m)
{
   for(int i=0; i<m.rows; i++) s << m.data2D[i] << endl;
   return s;
}

template <class T>
int operator == (const Array2<T>& m1,const Array2<T>& m2)
{
   if(m1.size() != m2.size()) return 0;
   for(int i=0; i<m1.size(); i++)
   {
      if(m1[i] != m2[i]) return 0;
   }
   return 1;
}

template <class T>
int operator != (const Array2<T>& m1,const Array2<T>& m2)
{ return !(m1==m2); }

template <class T> class Array3
{
   private:
      // Data Fields
      int rows, cols, levs;
      Array2<T> *data3D;
   public:
      // Constructors
      Array3(int = 0,int = 0,int = 0);
      Array3(int,int,int,T);
      Array3(const Array3<T> &);
      ~Array3();

      // Member Functions
      void resize(int,int,int);
      void resize(int,int,int,T);
      Array2<T> & operator [] (int) const;
      int size(int = 0) const;

      // Arithmetic Operators
      const Array3<T>& operator = (const Array3<T>&);
      const Array3<T>& operator = (T);
      Array3<T> operator *= (T);
      Array3<T> operator += (const Array3<T>&);
```

```
        Array3<T> operator -= (const Array3<T>&);
        Array3<T> operator +  (const Array3<T>&);
        Array3<T> operator -  (const Array3<T>&);

        // I/O stream functions
        friend ostream& operator << (ostream&,const Array3<T>&);
};      // end declaration class Array3

// Constructor, destructor and copy constructor.
template <class T> Array3<T>::Array3(int r,int c,int v)
    : rows(r), cols(c), levs(v), data3D(new Array2<T>[r])
{
    assert(data3D != NULL);
    for(int i=0; i<r; ++i) (data3D+i) -> resize(c,v);
}


template <class T> Array3<T>::Array3(int r,int c,int v,T num)
    : rows(r),cols(c),levs(v),data3D(new Array2<T>[r])
{
    assert(data3D != NULL);
    int i;
    for(i=0; i<r; ++i) (data3D+i) -> resize(c,v,num);
}


template <class T> Array3<T>::Array3(const Array3<T>& m)
    : rows(m.rows), cols(m.cols), levs(m.levs), data3D(new Array2<T>[m.rows])
{
    assert(data3D != NULL);
    for(int i=0; i<m.rows; i++) data3D[i] = m.data3D[i];
}


template <class T> Array3<T>::~Array3() { delete [] data3D; }

// Member functions
template <class T> void Array3<T>::resize(int r,int c,int v)
{
    int i;
    Array2<T> *newArray = new Array2<T>[r]; assert(newArray != NULL);
    if(r <= rows)
    {
        for(i=0; i<r; i++)
        {
            (data3D+i) -> resize(c,v);
            newArray[i] = data3D[i];
        }
    }
    else
```

```
    {
        for(i=0; i<rows; i++)
        {
            (data3D+i) -> resize(c,v);
            newArray[i] = data3D[i];
        }
        for(i=rows; i<r; i++) newArray[i].resize(c,v);
    }
    delete [] data3D;
    rows = r; cols = c; levs = v;
    data3D = newArray;
}

template <class T> void Array3<T>::resize(int r,int c,int v,T value)
{
    int i;
    Array2<T> *newArray = new Array2<T>[r]; assert(newArray != NULL);
    if(r <= rows)
    {
        for(i=0; i<r; i++)
        {
            (data3D+i) -> resize(c,v,value);
            newArray[i] = data3D[i];
        }
    }
    else
    {
        for(i=0; i<rows; i++)
        {
            (data3D+i) -> resize(c,v,value);
            newArray[i] = data3D[i];
        }
        for(i=rows; i<r; i++) newArray[i].resize(c,v,value);
    }
    delete [] data3D;
    rows = r; cols = c; levs = v;
    data3D = newArray;
}

// Various member operators
template <class T> Array2<T> & Array3<T>::operator [] (int i) const
{
    assert(i >= 0 && i < rows);
    return data3D[i];
}

template <class T> int Array3<T>::size(int index) const
```

```
{
   assert(index>=0 && index<3);
   switch(index)
   {
      case 0:  return rows;
      case 1:  return cols;
      default: return levs;
   }
}


template <class T>
const Array3<T> & Array3<T>::operator = (const Array3<T>& m)
{
   if(this == &m) return *this;
   delete [] data3D;
   cols = m.cols; rows = m.rows; levs = m.levs;
   data3D = new Array2<T>[m.rows]; assert(data3D != NULL);
   for(int i=0; i<m.rows; i++) data3D[i] = m.data3D[i];
   return *this;
}


template <class T> const Array3<T>& Array3<T>::operator = (T num)
{
   for(int i=0; i<rows; i++) data3D[i] = num;
   return *this;
}


template <class T> Array3<T> Array3<T>::operator *= (T num)
{
   for(int i=0; i<rows; i++) data3D[i] *= num;
   return *this;
}


template <class T> Array3<T> Array3<T>::operator += (const Array3<T>& m)
{  return *this = *this + m; }


template <class T> Array3<T> Array3<T>::operator -= (const Array3<T>& m)
{  return *this = *this - m; }


template <class T> Array3<T> Array3<T>::operator +  (const Array3<T> &m)
{
   assert((m.cols == cols) && (m.rows == rows) && (m.levs == levs));
   Array3<T> temp(rows,cols,levs);
   for(int i=0; i<rows; i++) temp[i] = data3D[i] + m.data3D[i];
   return temp;
}
```

```
template <class T> Array3<T> Array3<T>::operator -  (const Array3<T>& m)
{
   assert((m.cols == cols) && (m.rows == rows) && (m.levs == levs));
   Array3<T> temp(rows,cols,levs);
   for(int i=0; i<rows; i++) temp[i] = data3D[i] - m.data3D[i];
   return temp;
}


template <class T> ostream& operator << (ostream& s,const Array3<T>& m)
{
   for(int i=0; i<m.rows; i++) s << m.data3D[i] << endl;
   return s;
}


template <class T>
int operator == (const Array3<T>& m1,const Array3<T>& m2)
{
   if(m1.size() != m2.size()) return 0;
   for(int i=0; i<m1.size(); i++)
   { if(m1[i] != m2[i]) return 0; }
   return 1;
}


template <class T>
int operator != (const Array3<T>& m1,const Array3<T>& m2)
{ return !(m1==m2); }


template <class T> class Array4
{
   private:
      // Data Fields
      int rows, cols, levs, blks;
      Array3<T> *data4D;
   public:
      // Constructors
      Array4(int = 0, int = 0, int = 0, int = 0);
      Array4(int,int,int,int,T);
      Array4(const Array4<T>&);
      ~Array4();

      // Member Functions
      void resize(int,int,int,int);
      void resize(int,int,int,int,T);
      Array3<T>& operator [] (int) const;
      int size(int = 0) const;

      // Arithmetic Operators
```

```
      const Array4<T>& operator = (const Array4<T>&);
      const Array4<T>& operator = (T);
      Array4<T> operator *= (T);
      Array4<T> operator += (const Array4<T>&);
      Array4<T> operator -= (const Array4<T>&);
      Array4<T> operator +  (const Array4<T>&);
      Array4<T> operator -  (const Array4<T>&);

      // I/O stream functions
      friend ostream & operator << (ostream&,const Array4<T>&);
};    // end declaration class Array4

// Constructor, destructor and copy constructor.
template <class T> Array4<T>::Array4(int r,int c,int v,int b)
   : rows(r), cols(c), levs(v), blks(b), data4D(new Array3<T>[r])
{
   assert(data4D != NULL);
   for(int i=0; i<r; ++i) (data4D+i) -> resize(c,v,b);
}

template <class T> Array4<T>::Array4(int r,int c,int v,int b,T num)
   : rows(r), cols(c), levs(v), blks(b), data4D(new Array3<T>[r])
{
   assert(data4D != NULL);
   for (int i=0; i<r; ++i) (data4D+i) -> resize(c,v,b,num);
}

template <class T> Array4<T>::Array4(const Array4<T> &m)
   : rows(m.rows), cols(m.cols), levs(m.levs), blks(m.blks),
       data4D(new Array3<T>[m.rows])
{
   assert(data4D != NULL);
   for (int i=0; i<m.rows; i++) data4D[i] = m.data4D[i];
}

template <class T> Array4<T>::~Array4() { delete [] data4D; }

// Member functions
template <class T> void Array4<T>::resize(int r,int c,int v,int b)
{
   int i;
   Array3<T> *newArray = new Array3<T>[r]; assert(newArray != NULL);
   if(r <= rows)
   {
      for(i=0; i<r; i++)
      {
         (data4D+i) -> resize(c,v,b);
```

```
                newArray[i] = data4D[i];
        }
    }
    else
    {
        for(i=0; i<rows; i++)
        {
            (data3D+i) -> resize(c,v,b);
            newArray[i] = data4D[i];
        }
        for(i=rows; i<r; i++) newArray[i].resize(c,v,b);
    }
    delete [] data4D;
    rows = r; cols = c; levs = v; blks = b;
    data4D = newArray;
}


template <class T>
void Array4<T>::resize(int r,int c,int v,int b,T value)
{
    int i;
    Array3<T> *newArray = new Array3<T>[r]; assert(newArray != NULL);
    if(r <= rows)
    {
        for(i=0; i<r; i++)
        {
            (data4D+i) -> resize(c,v,b,value);
            newArray[i] = data4D[i];
        }
    }
    else
    {
        for(i=0; i<rows; i++)
        {
            (data4D+i) -> resize(c,v,b,value);
            newArray[i] = data4D[i];
        }
        for(i=rows; i<r; i++) newArray[i].resize(c,v,b,value);
    }
    delete [] data4D;
    rows = r; cols = c; levs = v; blks = b;
    data4D = newArray;
}


// Various member operators
template <class T> Array3<T> & Array4<T>::operator [] (int i) const
{
```

```
        assert(i >= 0 && i < rows);
        return data4D[i];
}

template <class T> int Array4<T>::size(int index) const
{
        assert(index>=0 && index<4);
        switch(index)
        {
            case 0:   return rows;
            case 1:   return cols;
            case 2:   return levs;
            default: return blks;
        }
}

template <class T>
const Array4<T>& Array4<T>::operator = (const Array4<T>& m)
{
        if(this == &m) return *this;
        int i;
        delete [] data4D;
        cols = m.cols; rows = m.rows; levs = m.levs; blks = m.blks;
        data4D = new Array3<T>[m.rows];  assert(data4D != NULL);
        for(i=0; i<m.rows; i++) data4D[i] = m.data4D[i];
        return *this;
}

template <class T> const Array4<T>& Array4<T>::operator = (T num)
{
        for(i=0; i<m.rows; i++) data4D[i] = num;
        return *this;
}

template <class T> Array4<T> Array4<T>::operator *= (T num)
{
        for(int i=0; i<rows; i++) data4D[i] *= num;
        return *this;
}

template <class T>
Array4<T> Array4<T>::operator += (const Array4<T>& m)
{  return *this = *this + m; }

template <class T>
Array4<T> Array4<T>::operator -= (const Array4<T>& m)
{  return *this = *this - m; }
```

```
template <class T>
Array4<T> Array4<T>::operator +  (const Array4<T>& m)
{
   assert((m.cols == cols) && (m.rows == rows) &&
          (m.levs == levs) && (m.blks == blks));
   Array4<T> temp(rows,cols,levs,blks);
   for(int i=0; i<rows; i++) temp[i] = data4D[i] + m.data4D[i];
   return temp;
}

template <class T> Array4<T> Array4<T>::operator -  (const Array4<T>& m)
{
   assert((m.cols == cols) && (m.rows == rows) &&
          (m.levs == levs) && (m.blks == blks));
   Array4<T> temp(rows,cols,levs,blks);
   for(int i=0; i<rows; i++) temp[i] = data4D[i] - m.data4D[i];
   return temp;
}

template <class T> ostream& operator << (ostream& s,const Array4<T>& m)
{
   for(int i=0; i<m.rows; i++) s << m.data4D[i] << endl;
   return s;
}

template <class T>
int operator == (const Array4<T>& m1,const Array4<T>& m2)
{
   if(m1.size() != m2.size()) return 0;
   for(int i=0; i<m1.size(); i++)
   { if(m1[i] != m2[i]) return 0; }
   return 1;
}

template <class T>
int operator != (const Array4<T>& m1,const Array4<T>& m2)
{ return !(m1==m2); }
#endif
```

10.9   String Class

The public interface of the `String` class:

- `String()` : Default constructor.

- `String(char)` : Constructor.

- `String(int)` : Constructor.

- `String(const char*)` : Constructor.

- `String(const String&)` : Copy constructor.

- `reverse()` : Reverses the string.

- `length()` : Length of the string.

- Assignment operator : =

- Subscript operator : []

- Concatenation operator : +

- Type conversion operator : `operator char* ()`

- Relational operators : ==, !=, <, <=, >, >=

- Stream operators : >>, <<

For a detailed description of the class structure and each member function, please refer to Section 6.9.

```
// MString.h

#ifndef MSTRING_H
#define MSTRING_H

#include <iostream.h>
#include <assert.h>
#include <string.h>

class String
{
private:
    // data field
    int datalength;
    char *data;
```

```
public:
    // constructors and destructor
    String();
    String(char);
    String(int);
    String(const char*);
    String(const String&);
    ~String();

    // assignment operator
    const String& operator = (const String&);

    // member functions and index operator
    String reverse() const;
    int length() const;
    char& operator [](int) const;

    // conversion operator
    operator const char *() const;

    // concatenation
    friend String operator + (const String&,const String&);

    // friendship operators
    friend ostream& operator << (ostream&,const String&);
    friend istream& operator >> (istream&,String&);

    // relational operators
    friend int operator <  (const String&,const String&);
    friend int operator <= (const String&,const String&);
    friend int operator != (const String&,const String&);
    friend int operator == (const String&,const String&);
    friend int operator >= (const String&,const String&);
    friend int operator >  (const String&,const String&);
};

// Class implementation
// constuctor and destructor
String::String() : datalength(1), data(new char[1])
{
    assert(data != NULL);
    data[0] = '\0';
}

String::String(char c) : datalength(2), data(new char[2])
{
    assert(data != NULL);
```

```
    data[0] = c;
    data[1] = '\0';
}

String::String(int size)
{
    assert(size >= 0);

    datalength = size + 1;

    data = new char[datalength];
    assert(data != NULL);

    data[0] = '\0';          // The string is assigned as a NULL string
}

String::String(const char *s)
    : datalength(strlen(s) + 1), data(new char[datalength])
{
    assert(data != NULL);
    strcpy(data,s);
}

String::String(const String& s)
    : datalength(strlen(s.data) + 1), data(new char[datalength])
{
    assert(data != NULL);
    strcpy(data,s.data);
}

String::~String()
{ delete [] data; }

// assignment operator
const String& String::operator = (const String& s)
{
    if(&s != this)
    {
        delete [] data;

        datalength = strlen(s.data) + 1;
        data = new char[datalength];
        assert(data != NULL);
        strcpy(data,s.data);
    }
    return *this;
}
```

```
// member function and index operator
String String::reverse() const
{
    char t;
    int i, idx, len = length();
    String temp(*this);

    for(i=0; i<len/2; i++)
    {
        idx = len-1-i;
        t = temp[idx];
        temp[idx] = temp[i];
        temp[i] = t;
    }
    return temp;
}

int String::length() const { return strlen(data); }

char& String::operator [](int index) const
{ return data[index]; }

// conversion operator
String::operator const char *() const
{ return data; }

// friendship operators
ostream& operator << (ostream& out,const String& s)
{
    out << s.data;
    return out;
}

istream& operator >> (istream& in,String& str)
{
    char temp[1000];
    if(in >> temp)
        str = temp;
    else
        str = "";
    return in;
}

// concatenation operator
String operator + (const String& s1,const String& s2)
{
```

```
    String S(s1.length() + s2.length());
    strcpy(S.data,s1.data);
    strcat(S.data,s2.data);
    return S;
}

// relational operators
int operator <  (const String& left,const String& right)
{ return strcmp(left.data,right.data) <  0; }

int operator <= (const String& left,const String& right)
{ return strcmp(left.data,right.data) <= 0; }

int operator == (const String& left,const String& right)
{ return strcmp(left.data,right.data) == 0; }

int operator != (const String& left,const String& right)
{ return strcmp(left.data,right.data) != 0; }

int operator >  (const String& left,const String& right)
{ return strcmp(left.data,right.data) >  0; }

int operator >= (const String& left,const String& right)
{ return strcmp(left.data,right.data) >= 0; }
#endif
```

10.10 Bit Vector Class

The public interface of the `BitVector` class:

- `BitVector()` : Default constructor.

- `BitVector(unsigned int)` : Constructor.

- `BitVector(unsigned int,unsigned int)` : Constructor.

- `BitVector(const BitVector&)` : Copy constructor.

- `size()` : Size of the bit vector.

- `reset(unsigned int)` : Re-specifies the size of the vector.

- `reset(unsigned int, unsigned int)` : Re-specifies the size of the vector and initializies the bit field.

- `set(unsigned int)` : Turns a specific bit on.

- `clear(unsigned int)` : Turns a specific bit off.

- `test(unsigned int)` : Checks if a specific bit is set.

- `flip(unsigned int)` : Flips a specific bit value.

- `unionSet(const BitVector&)` : Takes the union set of two bit vectors.

- `intersectSet(const BitVector&)` : Takes the intersection set of two bit vectors.

- `differenceSet(const BitVector&)` : Takes the difference set of two bit vectors.

- `subset(const BitVector&)` : Checks if a bit vector is a subset of the other.

- Equality operator : `==`

- Stream operator : `<<`

For a detailed description of the class structure and each member function, please refer to Section 6.10.

```
// Bitvec.h

#ifndef BITVEC_H
#define BITVEC_H

#include <iostream.h>
#include <assert.h>
#include "Vector.h"

// A vector of binary values (0 or 1)
class BitVector
{
private:
    unsigned int bsize;
    Vector<unsigned char> data;

    // Position decoding functions
    unsigned int byteNumber(unsigned int) const;
    unsigned int mask(unsigned int) const;

public:
    // Constructors
    BitVector();
    BitVector(unsigned int);
    BitVector(unsigned int, unsigned int);
    BitVector(const BitVector&);

    unsigned int size() const;       // Size of the bit vector
    void reset(unsigned int);        // re-specify the size of the vector
    void reset(unsigned int,unsigned int);
                                     // re-specify the size of the vector
                                     // and initialize the bit field
    // Bit operations
    void set(unsigned int);
    void clear(unsigned int);
    int  test(unsigned int) const;
    void flip(unsigned int);

    // Set operations
    BitVector unionSet(const BitVector&);
    BitVector intersectSet(const BitVector&);
    BitVector differenceSet(const BitVector&);
    int  operator == (const BitVector&);
    int  subset(const BitVector&);

    // Friends
    friend ostream& operator << (ostream&,const BitVector&);
```

```
};

BitVector::BitVector() : bsize(0), data() {}

BitVector::BitVector(unsigned int num)
   : bsize(num), data((num+7)/8, 0) {}

BitVector::BitVector(unsigned int num,unsigned int value) : bsize(num)
{
   if(value) data.reset((num+7)/8, 0xFF);
   else      data.reset((num+7)/8);
}

BitVector::BitVector(const BitVector& b) : bsize(b.bsize), data(b.data) {}

// return the size of the bit vector
unsigned int BitVector::size() const { return bsize; }

// re-specify the size of the vector
void BitVector::reset(unsigned int num)
{ bsize = num; data.reset((num+7)/8, 0); }

// re-specify the size of the vector
// and initialize the rest of the bit with value (0 or 1)
void BitVector::reset(unsigned int num, unsigned int value)
{
   bsize = num;
   if(value) data.reset((num+7)/8, 0xFF);
   else      data.reset((num+7)/8, 0);
}

// set the indicated bit in the vector
void BitVector::set(unsigned int index)
{ data[byteNumber(index)] |= mask(index); }

// clear the indicated bit in the vector
void BitVector::clear(unsigned int index)
{ data[byteNumber(index)] &= ~ mask(index); }

// check the indicated bit in the vector
int BitVector::test(unsigned int index) const
{ return (data[byteNumber(index)] & mask(index)) != 0; }

// flip the indicated bit in the vector
void BitVector::flip(unsigned int index)
{ data[byteNumber(index)] ^= mask(index); }
```

```
// return the index of byte containing the specified index
// where byte number is index value divided by 8
unsigned int BitVector::byteNumber(unsigned int index) const
{ return index >> 3; }

// produce a mask for the specified index
unsigned int BitVector::mask(unsigned int index) const
{
    // compute the amount to shift by examining
    // the low order 3 bits of the index
    const int shiftAmount = index & 07;

    // make a mask by shifting the value '1' left by the given amount
    return 1 << shiftAmount;
}

// form the union of set with argument set
BitVector BitVector::unionSet(const BitVector& b)
{
    assert(bsize == b.bsize);
    BitVector result(bsize);
    int i, total = (bsize+7)/8;
    for(i=0; i<total; i++)
        result.data[i] = data[i] | b.data[i];
    return result;
}

// form the intersection of set with argument set
BitVector BitVector::intersectSet(const BitVector &b)
{
    assert(bsize == b.bsize);
    BitVector result(bsize);
    int i, total = (bsize+7)/8;
    for(i=0; i<total; i++)
        result.data[i] = data[i] & b.data[i];
    return result;
}

// form the difference of set from argument set
BitVector BitVector::differenceSet(const BitVector& b)
{
    assert(bsize == b.bsize);
    BitVector result(bsize);
    int i, total = (bsize+7)/8;
    for(i=0; i<total; i++)
        result.data[i] = data[i] & (~b.data[i]);
    return result;
```

```
}

// check if two sets are the same
int BitVector::operator == (const BitVector& b)
{
   assert(bsize == b.bsize);
   // check if every position is equal to the argument
   for(int i=0; i<bsize; i++)
      if(data[i] != b.data[i])
         return 0;
   return 1;
}

// return true if set is subset of argument
int BitVector::subset(const BitVector &b)
{
   assert(bsize == b.bsize);
   // check if every position of the argument
   // is a subset of the corresponding receiver position
   for(int i=0; i<bsize; i++)
      if(b.data[i] != (data[i] & b.data[i])) return 0;
   return 1;
}

// output the bit vector to the stream
ostream& operator << (ostream& s,const BitVector& b)
{
   for(int i=0; i<b.bsize; i++) s << b.test(i);
   return s;
}
#endif
```

10.11   Linked List Class

The public interface of the List class:

- List() : Default constructor.

- List(const List<T>&) : Copy constructor.

- add(T) : Adds an element in front of the list.

- duplicate() : Makes a duplication of the list.

- first_Node() : First element of the list.

- is_Empty() : Checks if the list is empty.

- deleteAllNodes() : Removes all the nodes in the list.

- is_Include(T v) : Checks if v exists in the list.

- delete_First() : Removes the first element of the list.

The public interface of the ListIterator class:

- ListIterator(List<T>&) : Constructor.

- init() : Initializes the iterator.

- delete_Current() : Removes the element pointed to by the iterator.

- add_Before(const T v) : Adds the element v before the node pointed to by the iterator.

- add_After(const T v) : Adds the element v after the node pointed to by the iterator.

- Current value operator : ()

- End-of-list operator : !

- Increment operator : ++

- Assignment operator : =

For a detailed description of the class structure and each member function, please refer to Section 6.11.

```
// MList.h

#ifndef MITERATOR_H
#define MITERATOR_H

template <class T> class Iterator
{
public:
    virtual int  init() = 0;          // Initialization
    virtual int  operator !() = 0;    // Check if a current element exists
    virtual T    operator ()() = 0;   // return current element
    virtual int  operator ++() = 0;   // Increment operator
    virtual void operator = (const T) = 0; // Assignment operator
};
#endif

#ifndef MLIST_H
#define MLIST_H

#include <assert.h>

// Forward declarations
template <class T> class Link;
template <class T> class ListIterator;

template <class T> class List
{
protected:
    // Data field
    Link<T> *head;  // The head pointer to the first link node

    friend class ListIterator<T>;
public:
    // Constructors
    List();
    List(const List<T>&);
    virtual ~List();

    // Member functions
    void      add(const T);
    List<T>* duplicate() const;
    T         first_Node() const;
    int       is_Empty() const;

    // Virtual functions
    virtual void deleteAllNodes();
    virtual int  is_Include(T) const;
```

```
   virtual void  delete_First();
   virtual const List<T>& operator = (const List<T>&);
};


template <class T> class Link
{
private:
   // Constructors
   Link(const T,Link<T>*);
   Link(const Link<T>&);

   Link<T>* duplicate() const;

   // Data Fields
   T        data;
   Link<T> *next;

   friend class List<T>;
   friend class ListIterator<T>;
public:
   Link<T>* insert(const T);      // Insert a new element after
                                  // the current value

};


template <class T> class ListIterator : public Iterator<T>
{
protected:
   // Data Fields
   Link<T> *current;
   Link<T> *previous;
   List<T>& list;
public:
   // Constructors
   ListIterator(List<T>&);
   ListIterator(const ListIterator<T>&);

   // Iterator protocol
   virtual int  init();
   virtual T    operator ()();
   virtual int  operator !();
   virtual int  operator ++();
   virtual void operator = (const T);

   // New functions specific to list iterators
   void delete_Current();
   void add_Before(const T);
   void add_After(const T);
```

```
};

// class List implementation
template <class T> List<T>::List() : head(NULL) {}

// Empty all elements from the list
template <class T> List<T>::~List()
{ deleteAllNodes(); }

// Add a new value to the front of a linked list
template <class T> void List<T>::add(const T val)
{
    head = new Link<T>(val,head);
    assert(head != NULL);
}

// Clear all items from the list
template <class T> void List<T>::deleteAllNodes()
{
    Link<T> *nxt;
    nxt = NULL;
    // delete the element pointed to by p
    for(Link<T> *p = head; p; p = nxt)
    {
        nxt = p->next;
        p->next = NULL;
        delete p;
    }
    head = NULL;
}

// Duplicate a linked list
template <class T> List<T> * List<T>::duplicate() const
{
    List<T> *newlist = new List<T>; assert(newlist != NULL);

    // copy list
    if(head) newlist->head = head->duplicate();
    return newlist;
}

// Copy constructor
template <class T> List<T>::List(const List<T>& lst)
{
    // duplicate elements from lst list
    if(lst.is_Empty()) head = NULL;
    else
```

```
    {
        Link<T> *firstLink = lst.head;
        head = firstLink->duplicate();
    }
}

template <class T>
const List<T> & List<T>::operator = (const List<T>& lst)
{
    if(this != &lst)
    {
        // duplicate elements from lst list
        if(lst.is_Empty()) head = NULL;
        else
        {
            Link<T> *firstLink = lst.head;
            head = firstLink->duplicate();
        }
    }
    return *this;
}

// Return first value in list
template <class T> T List<T>::first_Node() const
{
    assert(head != NULL); return head->data;
}

// Check if v exists in the list
template <class T> int List<T>::is_Include(T v) const
{
    for(Link<T> *p = head; p; p = p->next)
        if(v == p->data) return 1;
    return 0;
}

// Check if the list is empty
template <class T> int List<T>::is_Empty() const
{ return head == NULL; }

// Remove the first element from the list
template <class T> void List<T>::delete_First()
{
    assert(head != NULL);
    Link<T> *p = head;
    head = p->next;
    delete p;
```

```
}

// class Link implementation
// Insert a new link behind current node
template <class T> Link<T>* Link<T>::insert(const T val)
{
   next = new Link<T>(val, next);
   assert(next != NULL);
   return next;
}

// Create and initialize a new link field
template <class T> Link<T>::Link(const T val,Link<T> *nxt)
   : data(val), next(nxt) {}

// Copy constructor
template <class T> Link<T>::Link(const Link<T> &lst)
   : data(lst.data), next(lst.next) {}

// duplicate the link
template <class T> Link<T>* Link<T>::duplicate() const
{
   Link<T> *newlink;

   // if there is a next field copy remainder of list
   if(next != NULL)
      newlink = new Link<T>(data,next->duplicate());
   else
      newlink = new Link<T>(data, NULL);

   // check that allocation was successful
   assert(newlink != NULL);
   return newlink;
}

// class ListIterator implementation
// Create and initialize a new list
template <class T> ListIterator<T>::ListIterator(List<T>& aList)
   : list(aList)
{ init(); }

// Copy constructor
template <class T>
ListIterator<T>::ListIterator(const ListIterator<T>& x) : list(x.list)
{ init(); }

// Set the iterator to the first element in the list
```

```cpp
template <class T> int ListIterator<T>::init()
{
   previous = NULL; current = list.head;
   return current != NULL;
}


// Return value of the current element
template <class T> T ListIterator<T>::operator ()()
{
   assert(current != NULL);
   return current->data;
}


// Check for the termination of the iterator
template <class T> int ListIterator<T>::operator !()
{
   // if current link references a removed value,
   // update current to point to next position
   if(current == NULL)
      if(previous != NULL) current = previous->next;

   // check if current is valid
   return current != NULL;
}


// The increment operator that move current pointer to next element
template <class T> int ListIterator<T>::operator ++()
{
   // if current link is deleted
   if(current == NULL)
   {
      if(previous == NULL) current = list.head;
      else                    current = previous->next;
   }
   else // advance pointer
   {
      previous = current;
      current = current->next;
   }

   // return true if current element is valid
   return current != NULL;
}


// Assignment operator : modify value of the current element
template <class T> void ListIterator<T>::operator = (const T val)
{
```

```
      assert(current != NULL);
      current->data = val;
}

// Remove the current element from a list
template <class T> void ListIterator<T>::delete_Current()
{
      assert(current != NULL);

      // remove the first element
      if(previous == NULL) list.head = current->next;
      else previous->next = current->next;

      // delete current node and set current pointer to null
      delete current;
      current = NULL;
}

// Add a new element to the list before current value
template <class T> void ListIterator<T>::add_Before(const T val)
{
      if(previous)                         // not at the beginning
      previous = previous->insert(val);
      else                                 // at the beginning of the list
      {
         list.add(val);
         previous = list.head;
         current = previous->next;
      }
}

// Add a new element to the list after current value
template <class T> void ListIterator<T>::add_After(const T val)
{
      if(current != NULL)                  // not at the beginning
         current->insert(val);
      else if(previous != NULL)            // at the end of list
         current = previous->insert(val);
      else                                 // at the beginning of the list
         list.add(val);
}
#endif
```

10.12   Polynomial Class

The public interface of the `Polyterm` class:

- `Polyterm(char*)` : Constructor giving the variable name (default is `"x"`).

- `Polyterm(const Polyterm<T>&)` : Copy constructor.

- `Polyterm<T>& operator+=(Polyterm<T>)`

- `Polyterm<T>& operator-=(Polyterm<T>)`

- `Polyterm<T>& operator*=(Polyterm<T>)`

- `Polyterm<T>& operator*=(T)`

- `Polyterm<T>& operator/=(Polyterm<T>)`

- `Polyterm<T>& operator/=(T)`

- `Polyterm<T> operator/(Polyterm<T>)` : Division neglecting remainder.

- `Polyterm<T> operator/(T)`

- `int operator>(Polyterm<T>)` : Comparison for ordering in `Polynomial`.

- `int operator==(Polyterm<T>)`

- `int operator!=(Polyterm<T>)`

- `Polyterm<T>& operator=(Polyterm<T>)`

- `Polyterm<T>& operator=(T)`

- `int equal(Polyterm<T>)` : Check if the `Polyterms` represent the same variable.

- `Polyterm<T> operator*(Polyterm<T>)`

- `Polyterm<T> operator^(unsigned int)` : Set the term's exponent.

- `Polyterm<T> variable()` : Return a simple `Polyterm` for the variable with coefficient and exponent 1.

- `T value(T t)` : Calculate the value of the term for t.

- `~Polyterm()` : Destructor.

The friend functions of the `Polyterm` class:

- `Polyterm<T> operator*(T,Polyterm<T>)`

- `Polyterm<T> operator*(Polyterm<T>,T)`

- `Polyterm<T> operator-(Polyterm<T>)`

- `Polyterm<T> operator+(Polyterm<T>)`

- `Polyterm<T> Diff(Polyterm<T>)` : Differentiate the term.

- `Polyterm<T> Int(Polyterm<T>)` : Integrate the term neglecting constant of integration.

- `ostream& operator<<(ostream&,Polyterm<T>&)`

The public interface of the `Polynomial` class:

- `Polynomial(Polyterm<T> x)` : Constructor for polynomial with variable x.

- `Polynomial(char*)` : Constructor giving the variable name.

- `Polynomial(Polynomial<T>)` : Copy constructor.

- `Polynomial(int)` : Constant integer polynomial constructor.

- `Polynomial<T>& operator=(Polynomial<T>)`

- `Polynomial<T>& operator+=(Polyterm<T>)`

- `Polynomial<T>& operator-=(Polyterm<T>)`

- `Polynomial<T> operator+(Polyterm<T>)`

- `Polynomial<T> operator-(Polyterm<T>)`

- `Polynomial<T> operator+=(T)`

- `Polynomial<T> operator-=(T)`

- `Polynomial<T> operator+(Polynomial<T>)`

- `Polynomial<T> operator-(Polynomial<T>)`

- `Polynomial<T>& operator+=(Polynomial<T>)`

- `Polynomial<T>& operator-=(Polynomial<T>)`

- `Polynomial<T>& operator=(Polyterm<T>)`

- `Polynomial<T> operator+(T)`

- `Polynomial<T> operator-(T)`

- `Polynomial<T>& operator=(T)`

- `Polynomial<T>& operator*=(T)`

- `Polynomial<T>& operator*=(Polyterm<T>)`

- `Polynomial<T>& operator*=(Polynomial<T>)`

- `Polynomial<T> operator*(T)`

- `Polynomial<T> operator*(Polyterm<T>)`

- `Polynomial<T> operator*(Polynomial<T>)`

- `Polynomial<T> operator^(unsigned int)`

- `Polynomial<T>& operator/=(T)`

- `Polynomial<T>& operator/=(Polyterm<T>)` : Division neglecting remainder.

- `Polynomial<T>& operator/=(Polynomial<T>)` : Division neglecting remainder.

- `Polynomial<T> operator/(T)`

- `Polynomial<T> operator/(Polyterm<T>)` : Division neglecting remainder.

- `Polynomial<T> operator/(Polynomial<T>)` : Division neglecting remainder.

- `Polynomial<T>& operator%=(T)`

- `Polynomial<T>& operator%=(Polyterm<T>)` : Remainder after division.

- `Polynomial<T>& operator%=(Polynomial<T>)` : Remainder after division.

- `Polynomial<T> operator%(T)`

- `Polynomial<T> operator%(Polyterm<T>)` : Remainder after division.

- `Polynomial<T> operator%(Polynomial<T>)` : Remainder after division.

- `int operator==(Polynomial<T>)`

- `int operator==(Polyterm<T>)`

- `int operator==(T)`

- `int operator!=(Polynomial<T>)`

- `int operator!=(Polyterm<T>)`

- `int operator!=(T)`

- `T value(T t)` : Calculate the polynomial's value at `t`.

- `T operator()(T t)` : Calculate the polynomial's value at `t`.

- `~Polynomial()` : Destructor.

The friend functions of the `Polynomial` class:

- `Polynomial<T> operator-(Polynomial<T>)`

- `Polynomial<T> operator+(Polynomial<T>)`

- `Polynomial<T> operator+(Polyterm<T>,Polynomial<T>)`

- `Polynomial<T> operator-(Polyterm<T>,Polynomial<T>)`

- `Polynomial<T> Diff(Polynomial<T>)` : Differentiate the polynomial.

- `Polynomial<T> Int(Polynomial<T>)` : Integrate the polynomial neglecting constant of integration.

- `Polynomial<T> operator+(T,Polynomial<T>)`

- `Polynomial<T> operator-(T,Polynomial<T>)`

- `Polynomial<T> operator+(Polyterm<T>,Polyterm<T>)` : Sum of `Polyterm`s is a `Polynomial`.

- `Polynomial<T> operator-(Polyterm<T>,Polyterm<T>)` : Difference of `Polyterm`s is a `Polynomial`.

- `Polynomial<T> operator+(T,Polyterm<T>)`

- `Polynomial<T> operator-(T,Polyterm<T>)`

- `Polynomial<T> operator+(Polyterm<T>,T)` : Sum of a constant and `Polyterm` is a `Polynomial`.

- `Polynomial<T> operator-(Polyterm<T>,T)` : Difference of a constant and `Polyterm` is a `Polynomial`.

- `Polynomial<T> operator*(T,Polynomial<T>)`

- `Polynomial<T> operator*(Polyterm<T>,Polynomial<T>)`

- `int operator==(Polyterm<T>,Polynomial<T>)`

- `int operator==(T,Polynomial<T>)`

- `int operator!=(Polyterm<T>,Polynomial<T>)`

- `int operator!=(T,Polynomial<T>)`

- `ostream& operator<<(ostream&,Polynomial<T>&)`

For a detailed description of the class structure and each member function, please refer to Section 6.12.

```
// poly.h

#ifndef _POLYNOMIAL
#define _POLYNOMIAL

#include <string.h>
#include <iostream.h>

// Assumption : T has typecasts defined
// and can act as a numeric data type

template <class P>
P __poly__power(P x,unsigned int y)
{
 P result(1);
 for(int i=1;i<=y;i++) result*=x;
 return result;
}

//Polyterm class for monomials

template <class T> class Polyterm
{
 public:
   Polyterm(char *sym="x");
   Polyterm(const Polyterm<T> &);
   Polyterm<T>& operator+=(const Polyterm<T>&);
   Polyterm<T>& operator-=(const Polyterm<T>&);
   Polyterm<T>& operator*=(const Polyterm<T>&);
   Polyterm<T>& operator*=(T);
   Polyterm<T>& operator/=(const Polyterm<T>&);
   Polyterm<T>& operator/=(T);
   Polyterm<T> operator/(const Polyterm<T>&) const;
   Polyterm<T> operator/(T) const;
   int operator>(const Polyterm<T>&) const;
   int operator==(const Polyterm<T>&) const;
   int operator!=(const Polyterm<T>&) const;
   Polyterm<T> &operator=(const Polyterm<T>&);
   Polyterm<T> &operator=(T);
   int equal(const Polyterm<T>&) const;
   Polyterm<T> operator*(const Polyterm<T>&) const;
   friend Polyterm<T> operator*(T,const Polyterm<T>&);
```

```
    friend Polyterm<T> operator*(const Polyterm<T>&,T);
    Polyterm<T> operator^(unsigned int) const;
    friend Polyterm<T> operator-(const Polyterm<T>&);
    friend Polyterm<T> operator+(const Polyterm<T>&);
    Polyterm<T> variable() const;
    friend Polyterm<T> Diff(Polyterm<T>);
    friend Polyterm<T> Int(Polyterm<T>);
    T value(T);
    friend ostream &operator<<(ostream&,const Polyterm<T>&);
    ~Polyterm();
  protected:
    Polyterm(T coeff,char *sym,unsigned int exp);
    char *symbol;
    T coefficient;
    unsigned int exponent;
};

template <class T>
Polyterm<T>::Polyterm(char *sym) : coefficient(T(1))
{
  exponent=1;
  symbol=new char[strlen(sym)+1];
  strcpy(symbol,sym);
}

template <class T>
Polyterm<T>::Polyterm(const Polyterm<T> &p) : coefficient(p.coefficient)
{
  symbol=new char[strlen(p.symbol)+1];
  strcpy(symbol,p.symbol);
  exponent=p.exponent;
}

template <class T>
Polyterm<T>& Polyterm<T>::operator+=(const Polyterm<T> &p)
{
  if((p.exponent==exponent) && (!strcmp(symbol,p.symbol)||(exponent==0)))
    coefficient+=p.coefficient;
  return *this;
}

template <class T>
Polyterm<T>& Polyterm<T>::operator-=(const Polyterm<T> &p)
{
  if((p.exponent==exponent) && (!strcmp(symbol,p.symbol)||(exponent==0)))
    coefficient-=p.coefficient;
  return *this;
```

```
}

template <class T>
Polyterm<T>& Polyterm<T>::operator*=(const Polyterm<T> &p)
{
 char *sym;
 if(!strcmp(symbol,p.symbol)||(exponent==0)||(p.exponent==0))
 {
  if(exponent!=0||p.exponent==0) sym=symbol;
  else sym=p.symbol;
  return *this=Polyterm<T>(coefficient*p.coefficient,
                          sym,exponent+p.exponent);
 }
 return *this;
}

template <class T>
Polyterm<T>& Polyterm<T>::operator*=(T c)
{
 coefficient*=c;
 return *this;
}

template <class T>
Polyterm<T>& Polyterm<T>::operator/=(const Polyterm<T> &x)
{
 if(x.coefficient==T(0))
 {
  *this=T(0);
  return *this;
 }
 coefficient/=x.coefficient;
 if(exponent>=x.exponent) exponent-=x.exponent;
 else coefficient=T(0);
 return *this;
}

template <class T> Polyterm<T>& Polyterm<T>::operator/=(T c)
{
 if(c==T(0))
 {
  *this=T(0);
  return *this;
 }
 coefficient/=c;
 return *this;
}
```

```
template <class T>
Polyterm<T> Polyterm<T>::operator/(const Polyterm<T> &x) const
{
 Polyterm<T> x1(*this);
 x1/=x;
 return x1;
}


template <class T> Polyterm<T> Polyterm<T>::operator/(T c) const
{
 Polyterm<T> x1(*this);
 x1/=c;
 return x1;
}


template <class T>
int Polyterm<T>::operator>(const Polyterm<T> &p) const
{
 return (exponent>p.exponent);
}


template <class T>
int Polyterm<T>::operator==(const Polyterm<T> &p) const
{
 if((coefficient==T(0))&&(coefficient==p.coefficient)) return 1;
 return ((exponent==p.exponent)&&
         (!strcmp(symbol,p.symbol)||(exponent==0))&&
         (coefficient==p.coefficient));
}


template <class T>
int Polyterm<T>::operator!=(const Polyterm<T> &p) const
{
 return !(*this==p);
}


template <class T>
Polyterm<T> &Polyterm<T>::operator=(const Polyterm<T> &p)
{
 delete[] symbol;
 coefficient=p.coefficient;
 symbol=new char[strlen(p.symbol)+1];
 strcpy(symbol,p.symbol);
 exponent=p.exponent;
 return *this;
}
```

```
template <class T> Polyterm<T> &Polyterm<T>::operator=(T c)
{
 exponent=0;
 coefficient=c;
 return *this;
}

template <class T>
int Polyterm<T>::equal(const Polyterm<T> &p) const
{
 return ((!strcmp(symbol,p.symbol))||
         (exponent==0)||(p.exponent==0));
}

template <class T>
Polyterm<T> Polyterm<T>::operator*(const Polyterm<T> &p) const
{
 if(strcmp(symbol,p.symbol)&&(exponent!=0)&&(p.exponent!=0))
   return Polyterm<T>(T(0),symbol,0);
 return Polyterm<T>(coefficient*p.coefficient,symbol,
                    exponent+p.exponent);
}

template <class T> Polyterm<T> operator*(T c,const Polyterm<T> &p)
{
 return Polyterm<T>(c*p.coefficient,p.symbol,p.exponent);
}

template <class T>
Polyterm<T> operator*(const Polyterm<T> &p,T c)
{
 return Polyterm<T>(c*p.coefficient,p.symbol,p.exponent);
}

template <class T>
Polyterm<T> Polyterm<T>::operator^(unsigned int y) const
{
 return Polyterm<T>(coefficient,symbol,y);
}

template <class T> Polyterm<T> operator-(const Polyterm<T> &x)
{
 return (Polyterm<T>(T(0)*x)-=x);
}

template <class T> Polyterm<T> operator+(const Polyterm<T> &x)
```

```
{ return x; }

template <class T> Polyterm<T> Polyterm<T>::variable() const
{ return Polyterm<T>(T(1),this->symbol,1); }

template <class T> Polyterm<T> Diff(Polyterm<T> p)
{
 p.coefficient*=T(p.exponent);
 p.exponent--;
 return p;
}

template <class T> Polyterm<T> Int(Polyterm<T> p)
{
 p.coefficient/=T(p.exponent+1);
 p.exponent++;
 return p;
}

template <class T> T Polyterm<T>::value(T x)
{
 return coefficient*__poly__power(x,exponent);
}

template <class T>
ostream &operator<<(ostream &o,const Polyterm<T> &p)
{
 if(((p.coefficient!=T(1))&&(p.coefficient!=T(-1)))||
    (p.exponent==0))
  o<<"("<<p.coefficient<<")";
 if(p.exponent) o<<p.symbol;
 if(p.exponent>1) o<<"^"<<p.exponent;
 return o;
}

template <class T> Polyterm<T>::~Polyterm()
{ delete[] symbol; }

template <class T>
Polyterm<T>::Polyterm(T coeff,char *sym,unsigned int exp)
              : coefficient(coeff)
{
 exponent=exp;
 symbol=new char[strlen(sym)+1];
 strcpy(symbol,sym);
}
```

```cpp
// structure to support list for polynomial
template <class T> struct polyListItem
{
 Polyterm<T> *item;
 polyListItem<T> *next;
 polyListItem<T> *previous;
};


// Polynomial class
template <class T> class Polynomial
{
 public:
   Polynomial(Polyterm<T> x): variable(x.variable()), head(0) {};
   Polynomial(char *sym): variable(Polyterm<T>(sym)), head(0) {};
   Polynomial(const Polynomial<T>&);
   Polynomial(int);
   Polynomial<T>& operator=(const Polynomial<T>&);
   friend Polynomial<T> operator+(const Polynomial<T>&);
   friend Polynomial<T> operator-(const Polynomial<T>&);
   Polynomial<T>& operator+=(const Polyterm<T>&);
   Polynomial<T>& operator-=(const Polyterm<T>&);
   Polynomial<T> operator+(const Polyterm<T>&) const;
   Polynomial<T> operator-(const Polyterm<T>&) const;
   Polynomial<T> operator+=(T);
   Polynomial<T> operator-=(T);
   friend Polynomial<T> operator+(const Polyterm<T>&,
                                  const Polynomial<T>&);
   friend Polynomial<T> operator-(const Polyterm<T>&,
                                  const Polynomial<T>&);
   Polynomial<T> operator+(const Polynomial<T>&) const;
   Polynomial<T> operator-(const Polynomial<T>&) const;
   Polynomial<T>& operator+=(const Polynomial<T>&);
   Polynomial<T>& operator-=(const Polynomial<T>&);
   Polynomial<T>& operator=(const Polyterm<T>&);
   friend Polynomial<T> Diff(Polynomial<T>);
   friend Polynomial<T> Int(Polynomial<T>);
   Polynomial<T> operator+(T) const;
   Polynomial<T> operator-(T) const;
   friend Polynomial<T> operator+(T,const Polynomial<T>&);
   friend Polynomial<T> operator-(T,const Polynomial<T>&);
   Polynomial<T>& operator=(T);
   friend Polynomial<T> operator+(const Polyterm<T>&,
                                  const Polyterm<T>&);
   friend Polynomial<T> operator-(const Polyterm<T>&,
                                  const Polyterm<T>&);
   friend Polynomial<T> operator+(T,const Polyterm<T>&);
   friend Polynomial<T> operator-(T,const Polyterm<T>&);
```

```
    friend Polynomial<T> operator+(const Polyterm<T>&,T);
    friend Polynomial<T> operator-(const Polyterm<T>&,T);
    Polynomial<T>& operator*=(T);
    Polynomial<T>& operator*=(const Polyterm<T>&);
    Polynomial<T>& operator*=(const Polynomial<T>&);
    Polynomial<T> operator*(T) const;
    Polynomial<T> operator*(const Polyterm<T>&) const;
    Polynomial<T> operator*(const Polynomial<T>&) const;
    Polynomial<T> operator^(unsigned int);
    friend Polynomial<T> operator*(T,const Polynomial<T>&);
    friend Polynomial<T> operator*(const Polyterm<T>&,
                                   const Polynomial<T>&);
    Polynomial<T>& operator/=(T);
    Polynomial<T>& operator/=(const Polyterm<T>&);
    Polynomial<T>& operator/=(const Polynomial<T>&);
    Polynomial<T> operator/(T) const;
    Polynomial<T> operator/(const Polyterm<T>&) const;
    Polynomial<T> operator/(const Polynomial<T>&) const;
    Polynomial<T>& operator%=(T);
    Polynomial<T>& operator%=(const Polyterm<T>&);
    Polynomial<T>& operator%=(const Polynomial<T>&);
    Polynomial<T> operator%(T) const;
    Polynomial<T> operator%(const Polyterm<T>&) const;
    Polynomial<T> operator%(const Polynomial<T>&) const;
    int operator==(T) const;
    int operator==(const Polyterm<T>&) const;
    int operator==(const Polynomial<T>&) const;
    friend int operator==(T,const Polynomial<T>&);
    friend int operator==(const Polyterm<T>&,const Polynomial<T>&);
    int operator!=(T) const;
    int operator!=(const Polyterm<T>&) const;
    int operator!=(const Polynomial<T>&) const;
    friend int operator!=(T,const Polynomial<T>&);
    friend int operator!=(const Polyterm<T>&,const Polynomial<T>&);
    T value(T);
    T operator()(T);
    friend ostream &operator<<(ostream&,const Polynomial<T> &);
    ~Polynomial();
protected:
    void tidy(void);
    Polyterm<T> variable;
    polyListItem<T> *head;
};

template <class T> Polynomial<T>::Polynomial(const Polynomial<T> &p)
{
  head=NULL;
```

```
 polyListItem<T> *temp1=p.head,*temp2;
 variable=p.variable;
 if(temp1)
 {
  head=new polyListItem<T>;
  head->item=new Polyterm<T>(*(temp1->item));
  head->previous=NULL;
  head->next=NULL;
  temp1=temp1->next;
 }
 temp2=head;
 while(temp1)
 {
  temp2->next=new polyListItem<T>;
  (temp2->next)->previous=temp2;
  temp2=temp2->next;
  temp2->item=new Polyterm<T>(*(temp1->item));
  temp2->next=NULL;
  temp1=temp1->next;
 }
}

template <class T> Polynomial<T>::Polynomial(int c)
{
 if(c!=0)
 {
  variable=Polyterm<T>("x");
  head=new polyListItem<T>;
  head->item=new Polyterm<T>(T(c)*variable^0);
  head->previous=NULL;
  head->next=NULL;
 }
 else
 {
  variable=Polyterm<T>("x");
  head=NULL;
 }
}

template <class T>
Polynomial<T>& Polynomial<T>::operator=(const Polynomial<T> &p)
{
 if(head==p.head) return *this;
 //destroy Polynomial
 polyListItem<T> *temp=head;

 variable=p.variable;
```

```
 while (temp)
 {
  head=temp->next;
  delete temp->item;
  delete temp;
  temp=head;
 }
//construct new Polynomial
 polyListItem<T> *temp1=p.head,*temp2;
 variable=p.variable;
 if(temp1)
 {
  head=new polyListItem<T>;
  head->item=new Polyterm<T>(*(temp1->item));
  head->previous=NULL;
  head->next=NULL;
  temp1=temp1->next;
 }
 temp2=head;
 while(temp1)
 {
  temp2->next=new polyListItem<T>;
  (temp2->next)->previous=temp2;
  temp2=temp2->next;
  temp2->item=new Polyterm<T>(*(temp1->item));
  temp2->next=NULL;
  temp1=temp1->next;
 }
 return *this;
}

template <class T> Polynomial<T> operator+(const Polynomial<T> &p)
{ return p; }

template <class T> Polynomial<T> operator-(const Polynomial<T> &p)
{
 return Polynomial<T>(p.variable)-p;
}

template <class T>
Polynomial<T>& Polynomial<T>::operator+=(const Polyterm<T> &x)
{
 polyListItem<T> *temp=head,*temp2;

 if(!variable.equal(x)) return *this;
 if(head==NULL)
 {
```

```
   head=new polyListItem<T>;
   head->item=new Polyterm<T>(x);
   head->previous=NULL;
   head->next=NULL;
 }
 else
 {
  while((temp->next!=NULL)&&(*(temp->item)>x))
  {
   temp=temp->next;
  }
  if((*(temp->item)+=x)==(*(temp->item))&&(x>*(temp->item)))
  {
   temp2=new polyListItem<T>;
   temp2->next=temp;
   temp2->previous=temp->previous;
   if(temp->previous==NULL) head=temp2;
   else (temp->previous)->next=temp2;
   temp->previous=temp2;
   temp2->item=new Polyterm<T>(x);
  }
  else if(*(temp->item)>x)
  {
   temp2=new polyListItem<T>;
   temp->next=temp2;
   temp2->next=NULL;
   temp2->previous=temp;
   temp2->item=new Polyterm<T>(x);
  }
 }
 tidy();
 return *this;
}


template <class T>
Polynomial<T>& Polynomial<T>::operator-=(const Polyterm<T> &x)
{
 *this+=-x;
 return *this;
}


template <class T> Polynomial<T>
Polynomial<T>::operator+(const Polyterm<T> &x) const
{
 if(head==NULL) return (Polynomial<T>(x)+=x);
 return (Polynomial<T>(*this)+=x);
}
```

```
template <class T> Polynomial<T>
Polynomial<T>::operator-(const Polyterm<T> &x) const
{
 if(head==NULL) return (Polynomial<T>(x)-=x);
 return (Polynomial<T>(*this)-=x);
}

template <class T> Polynomial<T> Polynomial<T>::operator+=(T x)
{
 Polyterm<T> y;
 y=x;
 *this+=y;
 return *this;
}

template <class T> Polynomial<T> Polynomial<T>::operator-=(T x)
{
 Polyterm<T> y;
 y=x;
 *this-=y;
 return *this;
}

template <class T> Polynomial<T>
operator+(const Polyterm<T> &x,const Polynomial<T> &p)
{ return p+x; }

template <class T> Polynomial<T>
operator-(const Polyterm<T> &x,const Polynomial<T> &p)
{ return x+(-p); }

template <class T> Polynomial<T>
Polynomial<T>::operator+(const Polynomial<T> &p) const
{
 if(head==NULL) return p;
 Polynomial<T> p1(*this);
 polyListItem<T> *temp=p.head;
 while(temp)
 {
  p1+=*(temp->item);
  temp=temp->next;
 }
 return p1;
}

template <class T> Polynomial<T>
```

```
Polynomial<T>::operator-(const Polynomial<T> &p) const
{
 if(head==NULL) return T(-1)*p;
 Polynomial<T> p1(*this);
 polyListItem<T> *temp=p.head;
 while(temp)
 {
  p1-=*(temp->item);
  temp=temp->next;
 }
 return p1;
}

template <class T> Polynomial<T>&
Polynomial<T>::operator+=(const Polynomial<T> &p)
{ return (*this=*this+p); }

template <class T> Polynomial<T>&
Polynomial<T>::operator-=(const Polynomial<T> &p)
{ return (*this=*this-p); }

template <class T> Polynomial<T>&
Polynomial<T>::operator=(const Polyterm<T> &x)
{ return *this=Polynomial<T>(variable)+x; }

template <class T>
Polynomial<T> Diff(Polynomial<T> p)
{
 Polynomial<T> p1(p);
 polyListItem<T> *temp=p1.head;
 while(temp)
 {
  *(temp->item)=Diff(*(temp->item));
  temp=temp->next;
 }
 p1.tidy();
 return p1;
}

template <class T> Polynomial<T> Int(Polynomial<T> p)
{
 Polynomial<T> p1(p);
 polyListItem<T> *temp=p1.head;
 while(temp)
 {
  *(temp->item)=Int(*(temp->item));
  temp=temp->next;
```

```
 }
 p1.tidy();
 return p1;
}

template <class T> Polynomial<T> Polynomial<T>::operator+(T c) const
{ return (*this)+(c*(variable^0)); }

template <class T> Polynomial<T> Polynomial<T>::operator-(T c) const
{ return (*this)-(c*(variable^0)); }

template <class T> Polynomial<T> operator+(T c,const Polynomial<T> &p)
{ return p+(c*(p.variable^0)); }

template <class T> Polynomial<T> operator-(T c,const Polynomial<T> &p)
{ return (c*(p.variable^0))-p; }

template <class T> Polynomial<T>& Polynomial<T>::operator=(T c)
{ return *this=Polynomial<T>(variable).operator+(c); }

template <class T> Polynomial<T>
operator+(const Polyterm<T> &x1,const Polyterm<T> &x2)
{
 Polynomial<T> p1(x1.variable());
 return p1+x1+x2;
}

template <class T> Polynomial<T>
operator-(const Polyterm<T> &x1,const Polyterm<T> &x2)
{
 Polynomial<T> p1(x1.variable());
 return p1+x1-x2;
}

template <class T> Polynomial<T>
operator+(T x1,const Polyterm<T> &x2)
{
 Polynomial<T> p1(x2.variable());
 return p1+x1+x2;
}

template <class T> Polynomial<T>
operator-(T x1,const Polyterm<T> &x2)
{
 Polynomial<T> p1(x2.variable());
 return p1+x1-x2;
}
```

```
template <class T> Polynomial<T>
operator+(const Polyterm<T> &x1,T x2)
{
 Polynomial<T> p1(x1.variable());
 return (p1.operator+(x1)).operator+(x2);
}


template <class T> Polynomial<T>
operator-(const Polyterm<T> &x1,T x2)
{
 Polynomial<T> p1(x1.variable());
 return (p1.operator+(x1)).operator-(x2);
}


template <class T> Polynomial<T>&
Polynomial<T>::operator*=(T c)
{
 Polyterm<T> x;
 x=c;
 return *this*=x;
}


template <class T> Polynomial<T>&
Polynomial<T>::operator*=(const Polyterm<T> &x)
{
 polyListItem<T> *temp=head;
 while(temp)
 {
  *(temp->item)*=x;
  temp=temp->next;
 }
 tidy();
 return *this;
}


template <class T> Polynomial<T>&
Polynomial<T>::operator*=(const Polynomial<T> &p)
{
 Polynomial<T> p1(this->variable),p2(this->variable);
 polyListItem<T> *temp=p.head;
 while(temp)
 {
  p1=*this;
  p1*=*(temp->item);
  p2+=p1;
  temp=temp->next;
```

```
  }
 *this=p2;
 return *this;
}

template <class T> Polynomial<T>
Polynomial<T>::operator*(T c) const
{
 Polynomial<T> p1(*this);
 p1*=c;
 return p1;
}

template <class T> Polynomial<T>
Polynomial<T>::operator*(const Polyterm<T> &x) const
{
 Polynomial<T> p1(*this);
 p1*=x;
 return p1;
}

template <class T> Polynomial<T>
Polynomial<T>::operator*(const Polynomial<T> &p) const
{
 Polynomial<T> p1(*this);
 p1*=p;
 return p1;
}

template <class T> Polynomial<T>
Polynomial<T>::operator^(unsigned int n)
{ return __poly__power(*this,n); }

template <class T> Polynomial<T>
operator*(T c,const Polynomial<T> &p)
{ return p*c; }

template <class T> Polynomial<T>
operator*(const Polyterm<T> &x,const Polynomial<T> &p)
{ return p*x; }

template <class T> Polynomial<T>& Polynomial<T>::operator/=(T c)
{
 polyListItem<T> *temp=head;
 while(temp)
 {
  *(temp->item)/=c;
```

```
  temp=temp->next;
 }
 tidy();
 return *this;
}

template <class T> Polynomial<T>&
Polynomial<T>::operator/=(const Polyterm<T> &x)
{
 polyListItem<T> *temp=head;
 while(temp)
 {
  *(temp->item)/=x;
  temp=temp->next;
 }
 tidy();
 return *this;
}

template <class T> Polynomial<T>&
Polynomial<T>::operator/=(const Polynomial<T> &p)
{
 polyListItem<T> *temp1=head,*temp2=p.head;
 Polynomial<T> result(this->variable);
 Polyterm<T> temp(this->variable),zero;
 zero=T(0);
 if(temp2==NULL) return *this;
 while((temp!=zero)&&(temp1!=NULL))
 {
  temp=(*(temp1->item))/(*(temp2->item));
  *this-=temp*p;
  result+=temp;
  temp1=head;
 }
 *this=result;
 return *this;
}

template <class T> Polynomial<T>
Polynomial<T>::operator/(T c) const
{
 Polynomial<T> p1(*this);
 p1/=c;
 return p1;
}

template <class T> Polynomial<T>
```

```
Polynomial<T>::operator/(const Polyterm<T> &x) const
{
 Polynomial<T> p1(*this);
 p1/=x;
 return p1;
}

template <class T> Polynomial<T>
Polynomial<T>::operator/(const Polynomial<T> &p) const
{
 Polynomial<T> p1(*this);
 p1/=p;
 return p1;
}

template <class T> Polynomial<T>& Polynomial<T>::operator%=(T c)
{ return *this-=c*(*this/c); }

template <class T> Polynomial<T>&
Polynomial<T>::operator%=(const Polyterm<T> &x)
{ return *this-=x*(*this/x); }

template <class T> Polynomial<T>&
Polynomial<T>::operator%=(const Polynomial<T> &p)
{ return *this-=p*(*this/p); }

template <class T> Polynomial<T>
Polynomial<T>::operator%(T c) const
{ return *this-c*(*this/c); }

template <class T> Polynomial<T>
Polynomial<T>::operator%(const Polyterm<T> &x) const
{
 return *this-x*(*this/x);
}

template <class T> Polynomial<T>
Polynomial<T>::operator%(const Polynomial<T> &p) const
{ return *this-p*(*this/p); }

template <class T> int Polynomial<T>::operator==(T c) const
{
 Polyterm<T> x;
 x=c;
 return *this==x;
}
```

```
template <class T>
int Polynomial<T>::operator==(const Polyterm<T> &x) const
{
 Polynomial<T> p(x);
 p=x;
 return *this==p;
}

template <class T>
int Polynomial<T>::operator==(const Polynomial<T> &p) const
{ return ((*this-p).head==NULL); }

template <class T> int operator==(T c,const Polynomial<T> &p)
{ return p==c; }

template <class T>
int operator==(const Polyterm<T> &x,const Polynomial<T> &p)
{ return p==x; }

template <class T> int Polynomial<T>::operator!=(T c) const
{ return !(*this==c); }

template <class T>
int Polynomial<T>::operator!=(const Polyterm<T> &x) const
{ return !(*this==x); }

template <class T>
int Polynomial<T>::operator!=(const Polynomial<T> &p) const
{ return !(*this==p); }

template <class T> int operator!=(T c,const Polynomial<T> &p)
{ return p!=c; }

template <class T>
int operator!=(const Polyterm<T> &x,const Polynomial<T> &p)
{ return p!=x; }

template <class T> T Polynomial<T>::value(T c)
{
 polyListItem<T> *temp=head;
 T sum=T(0);
 while(temp)
 {
  sum+=(temp->item)->value(c);
  temp=temp->next;
 }
 return sum;
```

```
}

template <class T> T Polynomial<T>::operator()(T c)
{ return value(c); }

template <class T>
ostream &operator<<(ostream &o,const Polynomial<T> &p)
{
 polyListItem<T> *temp=p.head;
 T zero(0);
 if(p.head==0)
 {
  o<<zero;
  return o;
 }
 while(temp)
 {
  o<<*((temp)->item);
  temp=((temp)->next);
  if(temp) o<<"+";
 }
 return o;
}

template <class T> Polynomial<T>::~Polynomial()
{
 polyListItem<T> *temp=head;
 while(temp)
 {
  head=temp->next;
  delete temp->item;
  delete temp;
  temp=head;
 }
}

template <class T> void Polynomial<T>::tidy(void)
{
 polyListItem<T> *temp,*temp2;
 while ((head!=NULL)&&((T(0)*(*(head->item)))==(*(head->item))))
 {
  temp=head;
  head=head->next;
  if(head) head->previous=NULL;
  delete temp->item;
  delete temp;
 }
```

```
temp=head;
while(temp)
{
 temp2=temp->next;
 if(((T(0)*(*(temp->item)))==(*(temp->item))))
 {
 if(temp->previous) (temp->previous)->next=temp->next;
 if(temp->next) (temp->next)->previous=temp->previous;
 delete temp->item;
 delete temp;
 }
 temp=temp2;
 }
}
#endif
```

10.13 Set Class

The public interface of the Set class:

- Set() : Default Constructor.
- Set(const T) : Constructor for a single element.
- ~Set() : Destructor.
- Set(const Set<T>&) : Copy constructor.
- Set<T>& operator = (const Set<T>&) : Assignment operator.
- int cardinality() const : Number of elements in set.

The friend functions of the Set class:

- friend int operator == (Set<T>&,Set<T>&) : Equals operator.
- friend Set<T> operator + (Set<T>&,Set<T>&) : Union of two sets.
- friend Set<T> operator * (Set<T>&,Set<T>&) : intersection of two sets.
- ostream& operator << (ostream&,Set<T>&) : Output stream operator.

For a detailed description of the class structure and each member function, please refer to Section 6.13.

```
// set.h

#ifndef _SET
#define _SET

#include <iostream>
#include <stdlib.h>      // for exit(0)

using namespace std;

    // Declaration class Set
template <class T> class Set
{
   public:
     Set();                   // default constructor
     Set(const T);            // constructor single element
     ~Set();                  // destructor

     Set(const Set<T>&);      // copy constructor
     Set<T>& operator = (const Set<T>&); // assignment operator
```

```
      int cardinality() const;    // number of elements

      friend Set<T> operator + (Set<T>&,Set<T>&); // union
      friend Set<T> operator * (Set<T>&,Set<T>&); // intersection
      friend int operator == (Set<T>&,Set<T>&);

      friend ostream& operator << (ostream&,Set<T>&);

       struct Box {        // element of linked list
       T value;            // the value of the item in the set
       Box* next;          // pointer to next Box
       };

       struct Set_DESC {   // the descriptor
       Box* root;          // root of list of values
       Box* last;          // last element
       int no_of_elements; // number of elements in list
       int references;     // references to this item
       };

   private:
      void first_element();      // iterate set first element
      void next_element();       // iterate move to next
      T current_element() const; // iterate return current item
      int is_at_end() const;     // iterate at end
      void add_at_end(const T);  // add to list at end
      void release();            // release sets storage
      void fail(const char[]) const; // give up
      Set_DESC* the_desc;        // description of set
      Box* the_current;          // current element looked at
};


      // defintion class Set
template <class T> Set<T>::Set()
{
   the_desc = new Set_DESC[1];
   if(the_desc == (Set_DESC*) NULL) fail("Cannot create Set - descriptor");
   the_desc -> no_of_elements = 0;
   the_desc -> references = 1;
   the_desc -> root = (Box*) NULL;
   the_desc -> last = (Box*) NULL;
}


template <class T> Set<T>::Set(const T item)
{
   the_desc = new Set_DESC[1];
```

```
    if(the_desc == (Set_DESC*) NULL) fail("Cannot create Set-descriptor");

    the_desc -> no_of_elements = 1;
    the_desc -> references = 1;
    the_desc -> root = new Box[1];
    if(the_desc -> root == (Box*) NULL) fail("Cannot create Box-descriptor");

    the_desc -> last = the_desc -> root;
    the_desc -> root->value = item;
    the_desc -> root->next = (Box*) NULL;
}

template <class T> void Set<T>::release()
{
    if(--(the_desc->references)==0)
    {
        Box* p_active = the_desc->root;
        while(p_active != (Box*) NULL)
        {
        Box* p_remove = p_active;
        p_active = p_active->next;
        delete [] p_remove;
        }
        delete [] the_desc;
    }
}

template <class T> Set<T>::~Set()
{ release(); }

template <class T> void Set<T>::fail(const char mes[]) const
{
    cout << "\n" << "Error in class : Set" << "\n";
    cout << mes << "\n";
    exit(-1);
}

template <class T> Set<T>::Set(const Set <T>& copy)
{
    copy.the_desc -> references++;  // copied item reference count
    the_desc = copy.the_desc;       // copy pointer to desc.
    the_current = copy.the_desc -> root;  // current element
}

template <class T> int Set<T>::cardinality() const
{ return the_desc -> no_of_elements; }
```

```
template <class T> void Set<T>::first_element()
{ the_current = the_desc -> root; }

template <class T> void Set<T>::next_element()
{
    if(the_current != (Box*) NULL)
        the_current = the_current->next;
}

template <class T> T Set<T>::current_element(void) const
{
    if(the_current == (Box*) NULL)
    {
        fail("Selecting item beyond end");
    }
    return the_current -> value;
}

template <class T> int Set<T>::is_at_end() const
{ return the_current == (Box*) NULL; }

template <class T> void Set<T>::add_at_end(const T item)
{
    Box* p_add = new Box[1];
    if(p_add == (Box*) NULL) fail("Cannot create new element");
    p_add -> value = item; p_add->next = (Box*) NULL;
    if(the_desc->root == (Box*) NULL)
    {
        the_desc->root = p_add;
        the_desc->last = p_add;
    } else {
        the_desc -> last -> next = p_add;
        the_desc -> last = p_add;
    }
    the_desc -> no_of_elements++;
}


template <class T>
Set<T>& Set<T>::operator = (const Set<T>& s1)
{
    s1.the_desc->references++; // copied items ref count
    release();                 // storage for overwritten element
    the_desc = s1.the_desc;
    return *this;
}

template <class T> Set<T> operator + (Set<T>& s1,Set<T>& s2)
```

```
{
   Set<T> result;
   s1.first_element(); s2.first_element();
   while(!s1.is_at_end() || !s2.is_at_end())
   {
      if(s1.is_at_end())
      {
      result.add_at_end(s2.current_element());
      s2.next_element();
      } else
      if(s2.is_at_end())
      {
      result.add_at_end(s1.current_element());
      s1.next_element();
      } else
      if(s1.current_element() < s2.current_element())
      {
      result.add_at_end(s1.current_element());
      s1.next_element();
      } else
      if(s1.current_element() > s2.current_element())
      {
      result.add_at_end(s2.current_element());
      s2.next_element();
      } else
      if(s1.current_element() == s2.current_element())
      {
      result.add_at_end(s1.current_element());
      s1.next_element(); s2.next_element();
      }
   }
   return result;
}

template <class T> Set<T> operator * (Set<T>& s1,Set<T>& s2)
{
   int flag = 0;
   Set<T> intersection;
   s1.first_element();  s2.first_element();
   while(!s1.is_at_end())
   {
   while((!s2.is_at_end()) && (!flag))
   {
   if(s1.current_element() == s2.current_element())
   {
   intersection.add_at_end(s1.current_element());
   flag = 1;
```

```
   }
   s2.next_element();
   }
   flag = 0;
   s1.next_element();
   }
   return intersection;
}

template <class T> int operator == (Set<T>& s1,Set<T>& s2)
{
   s1.first_element(); s2.first_element();
   if(s1.cardinality() != s2.cardinality())
   return 0;
   if(s1.cardinality() == 0) return 1;
   while(!s1.is_at_end())
   {
   if(s1.current_element() != s2.current_element())
   return 0;
   s1.next_element(); s2.next_element();
   }
   return 1;
}

template <class T> ostream& operator << (ostream& s,Set<T>& s1)
{
   int i;
   s1.first_element(); s << "(";
   for(i=0; i < s1.cardinality(); i++ )
   {
   s << s1.current_element();
   s1.next_element();
   if(i != s1.cardinality()-1) s << ",";
   }
   s << ")";
   return s;
}
#endif
```

10.14 Symbolic Class

The header file `Mall.h` includes the classes `Terms`, `Magnitude`, `Variable`, `Number`, `Function`, `Fsqrt`, `Fdf`, `FInt`, `Fexp`, `Fsinh`, `Fcosh`, `Fsin`, `Fcos`, `Fcos`, `Fln`. The header file `MSymbol.h` includes the classes `Symbol`, `Sum`, `Product`.

The public interface of the `Variable` class:

- `Variable()` : Default constructor.

- `Variable(String)` : Constructor.

- `Variable(String,T)` : Constructor.

- `type()` : Returns the character `'V'` to indicate the object type.

- `varName()` : Returns the variable name.

- `oprint(ostream&)` : Puts the `Variable` on the output stream.

- `val()` : Numerical value of the `Variable`.

- `set(const T)` : Sets the numerical value of the `Variable`.

The public interface of the `Number` class:

- `Number()` : Default constructor.

- `Number(T)` : Constructor.

- Type conversion operator : `operator T()`

- Assignment operator : `=`

- `type()` : Returns the character `'N'` to indicate the object type.

- `varName()` : Returns a NULL string.

- `oprint(ostream&)` : Puts the `Number` on the output stream.

- `val()` : Numerical value of the `Number`.

- `set(const T)` : Sets the numerical value of the `Number`.

The public interface of the classes `Fsqrt`, `Fdf`, `FInt`, `Fexp`, `Fsinh`, `Fcosh`, `Fsin`, `Fcos`, `Fln`:

- `varName()` : Returns the function name `Fname`.

- `oprint()` : Puts the function name on the output stream.

- `f(const T&)` : Numerical value of the function.

- `type()` : Returns a character to indicate different functions.

The public interface of the Sum class:

- `Sum()` : Default constructor.

- `Sum(const T)` : Constructor for numeric numbers.

- `Sum(String, int)` : Constructor for symbolic variables.

- `clear()` : Removes the value and dependency of a variable.

- `expand()` : Distributive law and binomial/multinomial expansion.

- `set(const T)` : Assigns a numerical value to a variable.

- `put(const Sum<T>&, const Sum<T> &)` : Replaces an expression by another.

- `nvalue()` : Numeric value of an expression.

- `value()` : Symbolic value of an expression.

- `coeff(int)` : Coefficient of constant term.

- `coeff(const Sum<T>&)` : Coefficient of a term.

- `coeff(const Sum<T>&, int n)` : Coefficient of a term with degree n.

- `depend(const Sum<T>&)` : Declares dependency of variables.

- `nodepend(const Sum<T>&)` : Declares no-dependency of variables.

- `isdepend(const Sum<T>&)` : Checks dependency of variables.

- `is_Number()` : Checks if a variable is a numeric number.

- `df(const Sum<T>&, const Sum<T>&)` : The first derivative of an expression.

- `df(const Sum<T>&, const Sum<T>&, int n)` : The n-th derivative of an expression.

- `Int(const Sum<T>&, const Sum<T>&)` : The integral of an expression.

- `Int(const Sum<T>&, const Sum<T>&, iny n)` : The n-th integral of an expression.

- `power(const Sum<T>&, T)` : Defines an integer power function.

- `exp(const Sum<T>&)` : Defines an exponential function.

- `cosh(const Sum<T>&)` : Defines a hyperbolic cosine function.

- `sinh(const Sum<T>&)` : Defines a hyperbolic sine function.

- `cos(const Sum<T>&)` : Defines a cosine function.

- `sin(const Sum<T>&)` : Defines a sine function.

- `ln(const Sum<T>&)` : Defines a logarithm function.

- `sqrt(const Sum<T>&)` : Defines a square root function.

- Arithmetic operators : +(unary), -(unary), +, -, *, /, =, +=, -=, *=, /=

- Relational operators : ==, !=

- Stream operators : >>, <<

For a detailed description of the class structure and each member function, please refer to Chapter 7.

```
// Mall.h

#ifndef ALL_H
#define ALL_H

#include <iostream.h>
#include <math.h>
#include "MString.h"

template <class T> class Terms
{
public:
    virtual ~Terms();
    virtual char type() const = 0;
    virtual String varName() const = 0;
    virtual void oprint(ostream&) const = 0;
};

template <class T> class Magnitude : public Terms<T>
{
public:
    virtual ~Magnitude();
    virtual T val() const = 0;
    virtual void set(const T) = 0;
```

```
};

template <class T> class Variable : public Magnitude<T>
{
private:
   String name;
   T value;
public:
   Variable();
   Variable(String);
   Variable(String, T);
   Variable(const Variable<T>&);
   virtual ~Variable();
   virtual char type() const;
   virtual String varName() const;
   virtual void oprint(ostream&) const;
   virtual T val() const;
   virtual void set(const T);
};

template <class T> class Number : public Magnitude<T>
{
private:
   T data;
public:
   // Constructor
   Number();
   Number(T);
   Number(const Number<T>&);
   virtual ~Number();

   // Conversion operator
   operator T () const;

   // Member functions
   Number<T> & operator = (const Number<T>&);
   virtual char type() const;
   virtual String varName() const;
   virtual void oprint(ostream&) const;
   virtual T val() const;
   virtual void set(const T);
};

template <class T> class Function : public Terms<T>
{
public:
   virtual ~Function();
```

```
   virtual double f(const T&) const = 0;
};

template <class T> class Fsqrt : public Function<T>
{
private:
   static const char* const Fname;
public:
   Fsqrt();
   virtual ~Fsqrt();
   virtual String varName() const;
   virtual void oprint(ostream&) const;
   virtual char type() const;
   virtual double f(const T&) const;
};

template <class T> class Fdf : public Function<T>
{
private:
   static const char* const Fname;
public:
   Fdf();
   virtual ~Fdf();
   virtual String varName() const;
   virtual void oprint(ostream&) const;
   virtual char type() const;
   virtual double f(const T&) const;
};

template <class T> class FInt : public Function<T>
{
private:
   static const char* const Fname;
public:
   FInt();
   virtual ~FInt();
   virtual String varName() const;
   virtual void oprint(ostream&) const;
   virtual char type() const;
   virtual double f(const T&) const;
};

template <class T> class Fexp : public Function<T>
{
private:
   static const char* const Fname;
public:
```

```
   Fexp();
   virtual ~Fexp();
   virtual String varName() const;
   virtual void oprint(ostream&) const;
   virtual char type() const;
   virtual double f(const T&) const;
};


template <class T> class Fsinh : public Function<T>
{
private:
   static const char* const Fname;
public:
   Fsinh();
   virtual ~Fsinh();
   virtual String varName() const;
   virtual void oprint(ostream&) const;
   virtual char type() const;
   virtual double f(const T&) const;
};


template <class T> class Fcosh : public Function<T>
{
private:
   static const char* const Fname;
public:
   Fcosh();
   virtual ~Fcosh();
   virtual String varName() const;
   virtual void oprint(ostream&) const;
   virtual char type() const;
   virtual double f(const T&) const;
};


template <class T> class Fsin : public Function<T>
{
private:
   static const char* const Fname;
public:
   Fsin();
   virtual ~Fsin();
   virtual String varName() const;
   virtual void oprint(ostream&) const;
   virtual char type() const;
   virtual double f(const T&) const;
};
```

```
template <class T> class Fcos : public Function<T>
{
private:
    static const char* const Fname;
public:
    Fcos();
    virtual ~Fcos();
    virtual String varName() const;
    virtual void oprint(ostream&) const;
    virtual char type() const;
    virtual double f(const T&) const;
};

template <class T> class Fln : public Function<T>
{
private:
    static const char* const Fname;
public:
    Fln();
    virtual ~Fln();
    virtual String varName() const;
    virtual void oprint(ostream&) const;
    virtual char type() const;
    virtual double f(const T&) const;
};

// Terms class
template <class T> Terms<T>::~Terms() {}

// Magnitude class
template <class T> Magnitude<T>::~Magnitude() {}

// Variable class
template <class T> Variable<T>::Variable() : name(""), value(T(0)) {}
template <class T> Variable<T>::Variable(String nm)
        : name(nm), value(T(0)) {}
template <class T> Variable<T>::Variable(String nm,T v)
        : name(nm), value(v) {}
template <class T> Variable<T>::Variable(const Variable<T> & v)
        : name(v.name), value(v.value) {}
template <class T> Variable<T>::~Variable() {}
template <class T> void   Variable<T>::set(const T num) { value = num; }
template <class T> char   Variable<T>::type()   const { return 'V'; }
template <class T> String Variable<T>::varName() const { return name; }
template <class T> void   Variable<T>::oprint(ostream& os) const
        { os << name;}
template <class T> T      Variable<T>::val()      const { return value; }
```

```
// Number class
template <class T> Number<T>::Number() : data(T(0)) {}
template <class T> Number<T>::Number(T num) : data(num) {}
template <class T> Number<T>::Number(const Number<T>& num)
         : data(num.data) {}
template <class T> Number<T>::~Number() {}

template <class T> Number<T> & Number<T>::operator = (const Number<T>& num)
{
   if(this != &num) data = num.data;
   return *this;
}

template <class T> Number<T>::operator T () const    { return data; }
template <class T> char    Number<T>::type()    const { return 'N'; }
template <class T> String Number<T>::varName() const { return ""; }
template <class T> void    Number<T>::oprint(ostream& os) const
         { os << data; }
template <class T> T       Number<T>::val()     const { return data; }
template <class T> void    Number<T>::set(const T num) { data = num; }

// Function class
template <class T> Function<T>::~Function() {}

// Fsqrt class
template <class T> const char* const Fsqrt<T>::Fname = "sqrt";
template <class T> Fsqrt<T>::Fsqrt() {}
template <class T> Fsqrt<T>::~Fsqrt() {}
template <class T> char    Fsqrt<T>::type()    const { return 'q'; }
template <class T> String Fsqrt<T>::varName() const { return Fname; }
template <class T> void    Fsqrt<T>::oprint(ostream& os) const
         { os << Fname; }
template <class T> double Fsqrt<T>::f(const T& x) const
        { return sqrt(double(x)); }

// Fdf class
template <class T> const char* const Fdf<T>::Fname = "df";
template <class T> Fdf<T>::Fdf() {}
template <class T> Fdf<T>::~Fdf() {}
template <class T> char    Fdf<T>::type()    const { return 'd'; }
template <class T> String Fdf<T>::varName() const { return Fname; }
template <class T> void    Fdf<T>::oprint(ostream& os) const
         { os << Fname; }
template <class T> double Fdf<T>::f(const T&) const
        { return 0.0; }
```

```
// FInt class
template <class T> const char* const FInt<T>::Fname = "Int";
template <class T> FInt<T>::FInt() {}
template <class T> FInt<T>::~FInt() {}
template <class T> char    FInt<T>::type()     const { return 'i'; }
template <class T> String FInt<T>::varName() const { return Fname; }
template <class T> void    FInt<T>::oprint(ostream& os) const
        { os << Fname; }
template <class T> double FInt<T>::f(const T&) const
        { return 0.0; }

// Fexp class
template <class T> const char* const Fexp<T>::Fname = "exp";
template <class T> Fexp<T>::Fexp() {}
template <class T> Fexp<T>::~Fexp() {}
template <class T> char    Fexp<T>::type()     const { return 'e'; }
template <class T> String Fexp<T>::varName() const { return Fname; }
template <class T> void    Fexp<T>::oprint(ostream& os) const
        { os << Fname; }
template <class T> double Fexp<T>::f(const T& x) const
   { return exp(double(x)); }

// Fsinh class
template <class T> const char* const Fsinh<T>::Fname = "sinh";
template <class T> Fsinh<T>::Fsinh() {}
template <class T> Fsinh<T>::~Fsinh() {}
template <class T> char    Fsinh<T>::type()     const { return 'j'; }
template <class T> String Fsinh<T>::varName() const { return Fname; }
template <class T> void    Fsinh<T>::oprint(ostream& os) const
        { os << Fname; }
template <class T> double Fsinh<T>::f(const T& x) const
   { return sinh(double(x)); }

// Fcosh class
template <class T> const char* const Fcosh<T>::Fname = "cosh";
template <class T> Fcosh<T>::Fcosh() {}
template <class T> Fcosh<T>::~Fcosh() {}
template <class T> char    Fcosh<T>::type()     const { return 'w'; }
template <class T> String Fcosh<T>::varName() const { return Fname; }
template <class T> void    Fcosh<T>::oprint(ostream& os) const
        { os << Fname; }
template <class T> double Fcosh<T>::f(const T& x) const
   { return cosh(double(x)); }

// Fsin class
template <class T> const char* const Fsin<T>::Fname = "sin";
template <class T> Fsin<T>::Fsin() {}
```

```
template <class T> Fsin<T>::~Fsin() {}
template <class T> char   Fsin<T>::type()     const { return 'z'; }
template <class T> String Fsin<T>::varName() const { return Fname; }
template <class T> void   Fsin<T>::oprint(ostream& os) const
         { os << Fname; }
template <class T> double Fsin<T>::f(const T & x) const
   { return sin(double(x)); }


// Fcos class
template <class T> const char* const Fcos<T>::Fname = "cos";
template <class T> Fcos<T>::Fcos() {}
template <class T> Fcos<T>::~Fcos() {}
template <class T> char   Fcos<T>::type()     const { return 'c'; }
template <class T> String Fcos<T>::varName() const { return Fname; }
template <class T> void   Fcos<T>::oprint(ostream& os) const
         { os << Fname; }
template <class T> double Fcos<T>::f(const T& x) const
   { return cos(double(x)); }


// Fln class
template <class T> const char* const Fln<T>::Fname = "ln";
template <class T> Fln<T>::Fln() {}
template <class T> Fln<T>::~Fln() {}
template <class T> char   Fln<T>::type()     const { return 'l'; }
template <class T> String Fln<T>::varName() const { return Fname; }
template <class T> void   Fln<T>::oprint(ostream& os) const
         { os << Fname; }
template <class T> double Fln<T>::f(const T& x) const
   { return log(double(x)); }
#endif
```

```
// MSymbol.h

#ifndef SYMBOLIC_H
#define SYMBOLIC_H

#include <assert.h>
#include <iostream.h>
#include <stdlib.h>
#include <ctype.h>
#include "MString.h"
#include "Mall.h"
#include "MList.h"

typedef Number<int>     Integer;
typedef Number<double>  Double;

// Forward declaration
template <class T> class Sum;
template <class T> class Product;

// Declaration and Definition of class Symbol
template <class T>
class Symbol
{
    protected:
        // Constructor
        Symbol();
        Symbol(int);
        Symbol(int,T*,int*);
        Symbol(int,Symbol<T>**,T*,int*);
    public:
        // Data Fields
        int branches;
        Symbol<T> **ep;
        T *fac_exp;
        int *next_var;

        // virtual destructor
        virtual ~Symbol();

        // Non-virtual member functions
        void remove(int);               // remove unwanted node
        void removeB();                 // remove variables of type()=='B'
        int simpFuncs(Symbol<T>**,char); // simplify constant functions
        Symbol<T> * zero() const;       // special function for constant 0
        Symbol<T> * one() const;        // special function for constant 1
        void Shrink();                  // Simplify an expression
```

```
        // Pure virtual member functions
        virtual char   type() const = 0;
        virtual Sum<T> val()  const = 0;
        virtual T      nval() const = 0;

        virtual int  replace1(Symbol<T>*,Symbol<T>*) = 0;
        virtual int  equal_index(Symbol<T>*) = 0;
        virtual int  is_equal(Symbol<T>*) const = 0;
        virtual int  gather() = 0;
        virtual void copy(Symbol<T>**) const = 0;
        virtual void oprint(ostream&) const = 0;
        virtual void deleteOneNode() = 0;  // remove this node only
        virtual void deleteAllNodes() = 0; // remove all nodes at this node

        virtual int  mxpand(Symbol<T>**) = 0;
        virtual int  dxpand(Symbol<T>**) = 0; // Distributive Law
        virtual void diff(Symbol<T>**,Symbol<T>*) const = 0;
        virtual void integrate(Symbol<T>**,Symbol<T>*) const = 0;
        virtual int  funcSimp(int&,int&) = 0;    // Functions Simplification

        virtual int  shrink_1() = 0;  // removes ones from products
        virtual int  shrink_2() = 0;  // removes too many Sums and products
    // removes zero products and summands
        virtual int  shrink_3(Symbol<T>**) = 0;
        virtual int  shrink_4() = 0;  // sums ones from sums
        virtual int  shrink_5() = 0;  // removes sums with only one summand
};


// Declaration class Product
template <class T> class Product : public Symbol<T>
{
    private:
        // Data Field
        static int Comm;

        // Private Constructor
        Product(int, Symbol<T> **, T *, int *);

        // Private virtual functions
        virtual char   type() const;
        virtual Sum<T> val() const;
        virtual T      nval() const;
        virtual int    replace1(Symbol<T> *, Symbol<T> *);
        virtual int    equal_index(Symbol<T>*);
        virtual int    is_equal(Symbol<T> *) const;
        virtual int    gather();
```

```
            virtual void    copy(Symbol<T> **) const;
            virtual void    oprint(ostream &) const;
            virtual void    deleteOneNode();
            virtual void    deleteAllNodes();

            virtual int mxpand(Symbol<T> **);
            virtual int dxpand(Symbol<T> **);
            virtual void diff(Symbol<T> **, Symbol<T> *) const;
            virtual void integrate(Symbol<T> **, Symbol<T> *) const;
            virtual int funcSimp(int &, int &);

            virtual int shrink_1();
            virtual int shrink_2();
            virtual int shrink_3(Symbol<T> **);
            virtual int shrink_4();
            virtual int shrink_5();

    public:
        // Constructors
        Product(int);
        ~Product();

        friend void Commutative(T, int);
};

// Declaration of class Sum
template <class T> class Sum : public Symbol<T>
{
    private:
        // Private constructor
        Sum(const T,char);
        Sum(char,char);
        Sum(int,Symbol<T>**,T*,int*,String);

        // Private member functions
        void diffRules(Symbol<T>**,char) const;
        void IntRules(Symbol<T>**,char) const;

        // Private Virtual functions
        virtual char    type() const;
        virtual Sum<T> val() const;
        virtual T       nval() const;
        virtual int     replace1(Symbol<T>*,Symbol<T>*);
        virtual int     equal_index(Symbol<T>*);
        virtual int     is_equal(Symbol<T>*) const;
        virtual int     gather();
        virtual void    copy(Symbol<T>**) const ;
```

```
       virtual void    oprint(ostream&) const;
       virtual void    deleteOneNode();
       virtual void    deleteAllNodes();

       virtual int   mxpand(Symbol<T>**);
       virtual int   dxpand(Symbol<T>**);
       virtual void diff(Symbol<T>**,Symbol<T>*) const;
       virtual void integrate(Symbol<T>**,Symbol<T>*) const;
       virtual int   funcSimp(int&,int&);

       virtual int   shrink_1();
       virtual int   shrink_2();
       virtual int   shrink_3(Symbol<T>**);
       virtual int   shrink_4();
       virtual int   shrink_5();

       // Data fields
       int is_Bound, is_Set;
       Terms<T> *data;
       List<Sum<T>*> deplist;

 public:
       // Constructors
       Sum();
       Sum(const T);
       Sum(int,int);
       Sum(String,int);
       Sum(const Sum<T>&);
       ~Sum();

       // Member function available only to Sum<T>
       void clear();              // Renew a used variable
       void expand();             // Expand an expression
       void set(const T);         // Assign a value to a variable
       int put(const Sum<T>&,const Sum<T>&);
                                  // Assign an expression to another
       T       nvalue() const;    // Return the numeric value of a variable
       Sum<T> value() const;      // Return the value of an expression
       T       coeff(int) const;  // Return coefficient of constant term
       Sum<T> coeff(const Sum<T>&) const;     // Return coefficient of a term
       Sum<T> coeff(const Sum<T>&,int) const; // Return coefficient of a term
                                              // with degree int n.
       void depend(const Sum<T>&);   // declare dependency of variables
       void nodepend(const Sum<T>&); // declare no-dependency of variables
       int isdepend(const Sum<T>&) const;  // check dependency of variables
       int is_Number() const;        // check if a variable is pure number
```

```
        // Arithmetic operators
        Sum<T> & operator = (const Sum<T>&);
        Sum<T> operator + () const;
        Sum<T> operator - () const;
        Sum<T> operator += (const Sum<T>&);
        Sum<T> operator -= (const Sum<T>&);
        Sum<T> operator *= (const Sum<T>&);
        Sum<T> operator /= (const Sum<T>&);

        friend Sum<T> operator + (const Sum<T>&,const Sum<T>&);
        friend Sum<T> operator - (const Sum<T>&,const Sum<T>&);
        friend Sum<T> operator * (const Sum<T>&,const Sum<T>&);
        friend Sum<T> operator / (const Sum<T>&,const Sum<T>&);

        friend int operator == (const Sum<T>&,const Sum<T>&);
        friend int operator != (const Sum<T>&,const Sum<T>&);

        friend ostream & operator << (ostream&,const Sum<T>&);
        friend istream & operator >> (istream&,Sum<T>&);

        friend Sum<T> exp(const Sum<T>&);
        friend Sum<T> sinh(const Sum<T>&);
        friend Sum<T> cosh(const Sum<T>&);
        friend Sum<T> sin(const Sum<T>&);
        friend Sum<T> cos(const Sum<T>&);
        friend Sum<T> ln(const Sum<T>&);
        friend Sum<T> sqrt(const Sum<T>&);
        friend Sum<T> power(const Sum<T>&,T);
        friend Sum<T> df(const Sum<T>&,const Sum<T>&);
        friend Sum<T> df(const Sum<T>&,const Sum<T>&,int);
        friend Sum<T> Int(const Sum<T>&,const Sum<T>&);
        friend Sum<T> Int(const Sum<T>&,const Sum<T>&,int);

        friend Symbol<T>* Symbol<T>::zero() const;
        friend Symbol<T>* Symbol<T>::one() const;
        friend int Product<T>::mxpand(Symbol<T>**);
        friend void Product<T>::integrate(Symbol<T>**,Symbol<T>*) const;
};

// Definition for Symbol class
template <class T> Symbol<T>::Symbol()
    : branches(0), ep(NULL), fac_exp(NULL), next_var(NULL)
{ cerr << "Symbol: Default constructor invoked!" << endl;}

template <class T> Symbol<T>::Symbol(int num)
    : branches(num), ep(NULL), fac_exp(NULL), next_var(NULL)
{
```

```cpp
    // if num is zero, do nothing
    if(num)
    {
        ep = new Symbol<T>*[num]; assert(ep != NULL);
        fac_exp = new T[num];      assert(fac_exp != NULL);
        next_var = new int[num];   assert(next_var != NULL);

        // initialize ep[] to NULL, fac_exp[] to 1 and next_var[] to 0
        for(int i=0; i<num; i++)
        {
            ep[i] = NULL;
            fac_exp[i] = T(1);
            next_var[i] = 0;
        }
    }
}


template <class T> Symbol<T>::Symbol(int num, T *new_fe, int *new_nx)
    : branches(num), ep(NULL), fac_exp(NULL), next_var(NULL)
{
    // if num is zero, do nothing
    if(num)
    {
        ep = new Symbol<T>*[num]; assert(ep != NULL);
        fac_exp = new T[num];      assert(fac_exp != NULL);
        next_var = new int[num];   assert(next_var != NULL);
        // initialize ep[] to NULL, fac_exp[] and next_var[]
    // to respective values
        for(int i=0; i<num; i++)
        {
            ep[i] = NULL;
            fac_exp[i] = new_fe[i];
            next_var[i] = new_nx[i];
        }
    }
}


template <class T>
Symbol<T>::Symbol(int num, Symbol<T> **newep, T *new_fe, int *new_nx)
    : branches(num), ep(NULL), fac_exp(NULL), next_var(NULL)
{
    // if num is zero, do nothing
    if(num)
    {
        ep = new Symbol<T>*[num]; assert(ep != NULL);
        fac_exp = new T[num];      assert(fac_exp != NULL);
        next_var = new int[num];   assert(next_var != NULL);
```

```
        // initialize ep[] and fac_exp[] to repective values
        for(int i=0; i<num; i++)
        {
            ep[i] = newep[i];
            fac_exp[i] = new_fe[i];
            next_var[i] = new_nx[i];
        }
    }
}

// Destructor
template <class T> Symbol<T>::~Symbol() {}

// Remove the link and useless memory
// Note: branches will decrease by 1
//       index is the position of the node to be removed
template <class T> void Symbol<T>::remove(int index)
{
    int k, m;
    Symbol<T> **ept;
    T *fac_expt;
    int *next_vart;
    // delete node will result in branches decrease by 1
    branches--;
    ept = new Symbol<T>*[branches]; assert(ept != NULL);
    fac_expt = new T[branches];     assert(fac_expt != NULL);
    next_vart = new int[branches];  assert(next_vart != NULL);
    // copy node that are not to be deleted
    for(k=0; k<index; k++)
    {
        ept[k] = ep[k];
        fac_expt[k] = fac_exp[k];
        next_vart[k] = next_var[k];
    }

    for(k=index, m=index+1; k<branches; k++, m++)
    {
        ept[k] = ep[m];
        fac_expt[k] = fac_exp[m];
        next_vart[k] = next_var[m];
    }

    // delete old memory location
    ep[index]->deleteAllNodes();
    delete [] ep; delete [] fac_exp; delete [] next_var;

    // assign new memory location
```

```
   ep = ept;
   fac_exp = fac_expt;
   next_var = next_vart;
}


// remove variables of type()=='B' (defined variables)
template <class T> void Symbol<T>::removeB()
{
   int i;
   for(i=0; i<branches; i++)
   {
      if(ep[i]->type() == 'B')
      {
         ep[i]->ep[0]->copy(ep+i);
         if(ep[i]->type() == 'V') next_var[i] = 1;
      }
      ep[i]->removeB();

      if(!ep[i]->branches)
      {
         if((ep[i]->type() == 'd')||(ep[i]->type() == 'i'))
          // it is a differentiation or integration operator
          {
             if(ep[i]->ep[0]->type() == 'B')
               ep[i]->ep[0]->ep[0]->copy(ep[i]->ep);   // remove 'B' node

             if(ep[i]->ep[1]->type() == 'B')
               ep[i]->ep[1]->ep[0]->copy(ep[i]->ep+1); // remove 'B' node

             if(ep[i]->ep[0]->type() == 'V') ep[i]->next_var[0] = 1;
             if(ep[i]->ep[1]->type() == 'V') ep[i]->next_var[1] = 1;

             if(ep[i]->type() == 'd')
               ep[i]->ep[0]->diff(ep+i,ep[i]->ep[1]);
             else // type == 'i'
               ep[i]->ep[0]->integrate(ep+i,ep[i]->ep[1]);
          }
         else if(islower(ep[i]->type())) // it is a function
          {
             if(ep[i]->ep[0]->type() == 'B')
                 ep[i]->ep[0]->ep[0]->copy(ep[i]->ep); // remove 'B' node
             if(ep[i]->ep[0]->type() == 'V') ep[i]->next_var[0] = 1;
             ep[i]->ep[0]->Shrink();
             ep[i]->simpFuncs(ep+i,ep[i]->type());
          }
      }
   }
}
```

```
}

// Simplify constant functions
template <class T> int Symbol<T>::simpFuncs(Symbol<T> **s,char t)
{
    switch (t)
    {
        case 'e':
            if(ep[0]->branches==1 && ep[0]->ep[0]==zero())
              { (*s) = one(); return 1; }
            break;
        case 'j':
            if(ep[0]->branches==1 && ep[0]->ep[0]==zero())
               { (*s) = zero(); return 1; }
            break;
        case 'w':
            if(ep[0]->branches==1 && ep[0]->ep[0]==zero())
               { (*s) = one(); return 1; }
            break;
        case 'z':
            if(ep[0]->branches==1 && ep[0]->ep[0]==zero())
               { (*s) = zero(); return 1; }
            break;
        case 'c':
            if(ep[0]->branches==1 && ep[0]->ep[0]==zero())
               { (*s) = one(); return 1; }
            break;
        case 'l':
            if(ep[0]->branches==1 && ep[0]->ep[0]==one())
               { (*s) = zero(); return 1; }
            break;
        case 'q':
            if(ep[0]->branches==1)
            {
                if(ep[0]->ep[0]==zero())      { (*s) = zero(); return 1; }
                else if(ep[0]->ep[0]==one() && ep[0]->fac_exp[0] == T(1))
                   { (*s) = one(); return 1; }
            }
            break;
        default: ;
    }
    return 0;
}

template <class T> Symbol<T>* Symbol<T>::zero() const
{
    static Symbol<T> *addr = new Sum<T>(T(0),char());
```

```
    assert(addr != NULL);
    return addr;
}

template <class T> Symbol<T>* Symbol<T>::one() const
{
    static Symbol<T> *addr = new Sum<T>(T(1),char());
    assert(addr != NULL);
    return addr;
}


// Definition class Sum
// Create a Sum node that contain no data
template <class T> Sum<T>::Sum()
    : Symbol<T>(0), is_Bound(0), is_Set(0), data(new Variable<T>)
{ assert(data != NULL); }

// Create a Sum node that has n branches
template <class T> Sum<T>::Sum(int n, int)
    : Symbol<T>(n), is_Bound(0), is_Set(0), data(new Variable<T>)
{ assert(data != NULL); }

// Create a Number node with value num
template <class T> Sum<T>::Sum(const T num)
    : Symbol<T>(1), is_Bound(0), is_Set(0), data(new Variable<T>("", num))
{
    assert(data != NULL);
    if(num == T(0)) ep[0] = zero();
    else
    {
        ep[0] = one();
        fac_exp[0] = num;
    }
    next_var[0] = 1;
}

template <class T> Sum<T>::Sum(const T num, char)
    : Symbol<T>(0), is_Bound(0), is_Set(0), data(new Number<T>(num))
{ assert(data != NULL); }

template <class T> Sum<T>::Sum(char ftype, char)
    : Symbol<T>(2), is_Bound(0), is_Set(0)
{
    switch(ftype)
    {
        case 'd': data = new Fdf<T>;  break;
        case 'i': data = new FInt<T>;  break;
```

```
      case 'e': data = new Fexp<T>; break;
      case 'j': data = new Fsinh<T>; break;
      case 'w': data = new Fcosh<T>; break;
      case 'z': data = new Fsin<T>; break;
      case 'c': data = new Fcos<T>; break;
      case 'l': data = new Fln<T>; break;
      case 'q': data = new Fsqrt<T>; break;
      default : ;
   }
   assert(data != NULL);
   branches = 0;    // to indicate this is a leaf node
}


template <class T>
Sum<T>::Sum(int i,Symbol<T> **newep,T *new_ef, int *new_nx,String nm)
   : Symbol<T>(i, newep, new_ef, new_nx), is_Bound(0), is_Set(0),
     data(new Variable<T>(nm))
{ assert(data != NULL); }

template <class T> Sum<T>::Sum(String new_name, int)
   : Symbol<T>(0), is_Bound(0), is_Set(0), data(new Variable<T>(new_name))
{ assert(data != NULL); }

template <class T> Sum<T>::Sum(const Sum<T> &s)
   : Symbol<T>(1), is_Bound(0), is_Set(0), data(new Variable<T>)
{
   assert(data != NULL);
   if(s.type() == 'B') s.ep[0]->copy(ep);
   else                s.copy(ep);
   if (s.type() == 'V') next_var[0] = 1;
}


template <class T> Sum<T>::~Sum()
{
   if(!branches)
   {
      if(islower(type())) ep[0]->deleteAllNodes();
      deleteOneNode();
   }
   else deleteAllNodes();
   deplist.deleteAllNodes();
}


template <class T> Sum<T> & Sum<T>::operator = (const Sum<T> &s)
{
   if(this == &s) return *this;
   ((Sum<T>&)s).Shrink();
```

```
// delete the old value of *this if it has been assigned
// The removeB() is important, because if it is not included,
// y = a+b; a = x; y = c;
// The y = c will first free all the old memory of y which includes
// the memory of a
// y -> ""->a->x
//          ->b
if (is_Bound) { removeB(); ep[0]->deleteAllNodes(); }
delete [] ep; delete [] fac_exp; delete [] next_var;
branches = 1;
ep = new Symbol<T>*[1]; assert(ep != NULL);
fac_exp = new T[1];      assert(fac_exp != NULL);
next_var = new int[1];   assert(next_var != NULL);
is_Bound = 1; is_Set = 0;
next_var[0] = 0;

if(s.type() == 'B') s.ep[0]->copy(ep);
else if(s.branches) s.copy(ep);
else
{
    ep[0] = new Sum<T>(1,0); assert(ep[0] != NULL);
    s.copy(ep[0]->ep);
    ep[0]->next_var[0] = 1;
}

fac_exp[0] = T(1);
expand();

return *this;
}

// return the type of node pointed by the pointer
template <class T> char Sum<T>::type() const
{
    if(branches)
    {
        if(is_Bound) return 'B';
        return 'S';
    }
    return data->type();
}

// print the numerical value of the expression
template <class T> Sum<T> Sum<T>::val() const
{
    int i;
```

```
    Sum<T> s(T(0));

    // print the value, if this is a leaf node
    if(!branches)
    {
       if(islower(type())) // it is a function
       {
          if(((Sum<T>*)ep[0])->is_Number())
          return Sum<T>(((Function<T>*)data)->f(((Sum<T>*)ep[0])->nvalue()));
          else
          {
            Sum<T> r(type(),char()), p;
            p = ep[0]->val();
            p.copy(r.ep);
            if(r.ep[0]->type() == 'V') r.next_var[0] = 1;
               return r;
          }
       }
       else
       {
          if(is_Set) return Sum<T>(((Magnitude<T>*)data)->val());
          else       return *this;
       }
    }
    else
       for(i=0; i<branches; i++)
          s += Sum<T>(fac_exp[i]) * ep[i]->val();

    return s;
}

// replace expression 1 by expression 2
template <class T> int Sum<T>::replace1(Symbol<T> *s1, Symbol<T> *s2)
{
    int i, index, flag=0;
    for(i=0; i<branches; i++) flag += ep[i]->replace1(s1,s2);

    if(islower(type()))
       flag +=((Sum<T>*)ep[0])->put(*((Sum<T>*)s1),*((Sum<T>*)s2));

    if(s1->branches >= 1)
    {
      index = equal_index(s1);
      if(index >= 0)
      {
         s2->copy(ep+index);
         fac_exp[index] = T(1);
```

```
            flag++;
        }
    }
    else
    {
        for(i=0; i<branches; i++)
            if(ep[i]->is_equal(s1))
            {   s2->copy(ep+i); flag++; }
    }
    return flag;
}


// return the index of equal component
// return (-1) if no such component is found
template <class T> int Sum<T>::equal_index(Symbol<T> *)
{ return (-1); }

// check if 2 expressions are equal
template <class T> int Sum<T>::is_equal(Symbol<T> *s) const
{
    int i,k;
    int *ip;

    // if s is a product node, return 0
    if(s->type() == 'P')   return 0;

    // if s is equivalent to this return 1
    if(s == (Symbol<T>*)this) return 1;

    // so s not equal to this, if both are a leaf node, check for function
    if(!s->branches && !branches)
    {
        if(islower(type()) && s->type() == type())
        {
            if((type() == 'd')||(type() == 'i'))
            return (ep[0]->is_equal(s->ep[0]) && ep[1]->is_equal(s->ep[1]));
            return ep[0]->is_equal(s->ep[0]);
        }
        else return 0;
    }

    // if branches are not equal, return 0
    if(branches != s->branches) return 0;

    ip = new int[branches]; assert(ip != NULL);

    for(i=0; i<branches; i++) ip[i] = 0;
```

```
   // check if all the branches are equal
   for(i=0; i<branches; i++)
   {
      for(k=0; k<branches; k++)
      {
         if(ep[i]->is_equal(s->ep[k])   &&
            fac_exp[i] == s->fac_exp[k] && ip[k] == 0)
         { ip[k] = 1; break; }
      }
      // if there is no match terms, return 0
      if(k == branches) return 0;
   }
   delete [] ip;
   return 1;
}


// Collect/combine terms that have the same variable
template <class T> int Sum<T>::gather()
{
   int i, j, flag=0;

   for(i=0; i<branches; i++)    flag += ep[i]->gather();

   // comparing different subtrees, if equal, group/combine them
   for(i=0; i<branches; i++)
      for(j=i+1; j<branches; j++)
      {
         if(ep[i]->is_equal(ep[j]))
         {
            flag++;
            fac_exp[i] += fac_exp[j];    // sum factors

            // if fac_exp[] == 0 after added
            if(fac_exp[i] == T(0))
              if(branches > 2)
                { remove(j); remove(i); j=i;}
              else // remove one term and create a zero() node
              {
                 remove(j); j=i;
                 ep[0] = zero();
                 fac_exp[0] = T(1);
              }
            else { remove(j); j=i; }
         }
      }
   return flag;
```

```
}

// copy subtree from this to sp
template <class T> void Sum<T>::copy(Symbol<T> **sp) const
{
    if(branches)
    {
        int i;
        *sp = new Sum<T>(branches,0); assert(*sp != NULL);
        ((Sum<T>*)(*sp))->is_Bound = is_Bound;
        ((Sum<T>*)(*sp))->is_Set = is_Set;

        for(i=0; i<branches; i++) (*sp)->fac_exp[i] = fac_exp[i];
        for(i=0; i<branches; i++) (*sp)->next_var[i] = next_var[i];
        for(i=0; i<branches; i++) ep[i]->copy((*sp)->ep+i);
    }
    else
    {
        if(islower(data->type()))
        {
            *sp = new Sum<T>(data->type(),char()); assert(*sp != NULL);
            ep[0]->copy((*sp)->ep);
            if((data->type() == 'd')||(data->type() == 'i'))
              ep[1]->copy((*sp)->ep+1);
        }
        else *sp = (Symbol<T> *)this;
    }
}

// print the algebraic expression
template <class T> void Sum<T>::oprint(ostream &os) const
{
    // print the content, if this is a leaf node
    if(!branches)
    {
        data->oprint(os);
        if(islower(data->type()))
        {
            if((data->type() == 'd')||(data->type() == 'i'))
            {
                os << "(";
                if(ep[0]->branches==1 && ep[0]->fac_exp[0] == T(1))
                    ep[0]->ep[0]->oprint(os);
                else ep[0]->oprint(os);
                os << ",";
                ep[1]->oprint(os); os << ")";
            }
```

```
                else ep[0]->oprint(os);
        }
    }
    else
    {
        os << "(";

        // do not print + as this is the first term
        if(fac_exp[0] == T(-1))          { os << "-"; ep[0]->oprint(os); }
        else if(fac_exp[0] != T(1))
        {
            os << fac_exp[0];
            if(ep[0] != one())            { os << "*"; ep[0]->oprint(os); }
        }
        else ep[0]->oprint(os);

        // print the rest of the terms
        for(int i=1; i<branches; i++)
        {
            if(fac_exp[i] > T(0))
            {
                os << "+";
                if(fac_exp[i] == T(1))   ep[i]->oprint(os);
                else
                {
                    os << fac_exp[i];
                    if(ep[i] != one())    { os << "*"; ep[i]->oprint(os); }
                }
            }
            else // fac_exp[i] < T(0)
            {
                if(fac_exp[i] == T(-1)) { os << "-"; ep[i]->oprint(os); }
                else
                {
                    os << fac_exp[i];
                    if(ep[i] != one())    { os << "*"; ep[i]->oprint(os); }
                }
            }
        }
        os << ")";
    }
}

// Delete this node only
template <class T> void Sum<T>::deleteOneNode()
{
    delete data;
```

```cpp
   delete [] ep; delete [] fac_exp; delete [] next_var;
}


// Delete all the nodes rooted at this
template <class T> void Sum<T>::deleteAllNodes()
{
   if(branches)
   {
      for(int i=0; i<branches; i++)
         if(!next_var[i]) ep[i]->deleteAllNodes();
      deleteOneNode();
   }
   else if(islower(type()))
   {
      ep[0]->deleteAllNodes();
      deleteOneNode();
   }
}


// Binomial and Multinomial expansion
template <class T> int Sum<T>::mxpand(Symbol<T> **)
{
   int i, flag = 0;
   for(i=0; i<branches; i++) flag += ep[i]->mxpand(ep+i);
   return flag;

}


// Distributive law
template <class T> int Sum<T>::dxpand(Symbol<T> **)
{
   int i, flag = 0;
   for(i=0; i<branches; i++) flag += ep[i]->dxpand(ep+i);
   return flag;
}


// Differentiation
template <class T> void Sum<T>::diff(Symbol<T> **s, Symbol<T> *t) const
{
   int i;
   if(branches)
   {
      *s = new Sum<T>(branches,0); assert(*s != NULL);

      // copy all the coefficients
      for(i=0; i<branches; i++) (*s)->fac_exp[i] = fac_exp[i];
      // differentiate all the terms
```

```
       for(i=0; i<branches; i++)
       {
          ep[i]->diff((*s)->ep+i, t);

          if((*s)->ep[i]->type() == 'V' || (*s)->ep[i]->type() == 'N')
              (*s)->next_var[i] = 1;
          else (*s)->next_var[i] = 0;
       }
    }
    else // type()=='V' or 'N' or Functions dx/dx = 1, dy/dx = 0, dC/dx = 0
    {
       if(type() == 'd') // it is a differentiation operator
       {
          // (d/dx) df(y,x) = df(df(y,x),x)
          *s = new Sum<T>('d',char());    assert(*s != NULL);
          copy((*s)->ep);
          t->copy((*s)->ep+1);
       }
       else if(type() == 'i')
       {
        if(ep[1] == t) //d/dx Int(y(x),x) = y(x);
        {
            ep[0]->copy(s);
        }
        else if(isdepend(*((Sum<T>*)t)))
        {
        *s = new Sum<T>('d',char()); assert(*s != NULL);
        copy((*s)->ep);
        t->copy((*s)->ep+1);
        }
        else *s = zero();
       }
       else if(islower(type())) // it is functions
       {
          *s = new Sum<T>(1,0);               assert((*s) != NULL);
          (*s)->ep[0] = new Product<T>(2); assert((*s)->ep[0] != NULL);
          ep[0]->diff((*s)->ep[0]->ep,t);

          diffRules((*s)->ep[0]->ep+1, type());
       }
       else if(type() == 'N') *s = zero();
       else // type() == 'V'
       {
          if((Symbol<T>*)this == t) *s = one();
          else if(isdepend(*((Sum<T>*)t)))
          {
             *s = new Sum<T>('d',char()); assert(*s != NULL);
```

```
             copy((*s)->ep);
             t->copy((*s)->ep+1);
        }
        else *s = zero();
    }
  }
}


// Integration
template <class T> void Sum<T>::integrate(Symbol<T> **s,Symbol<T> *t) const
{
    int i;
    // make expression as simple as possible for integration of products
    ((Sum<T>*)this)->expand();

    if(branches)
    {
      *s = new Sum<T>(branches,0); assert(*s != NULL);
      // copy all the coefficients
      for(i=0; i<branches; i++) (*s)->fac_exp[i] = fac_exp[i];

      // integrate all the terms
      for(i=0; i<branches; i++)
       {
           ep[i]->integrate((*s)->ep+i, t);

           if ((*s)->ep[i]->type() == 'V' || (*s)->ep[i]->type() == 'N')
               (*s)->next_var[i] = 1;
           else (*s)->next_var[i] = 0;
       }
    }
    else   // type() == 'V' or 'N' or Functions
    {
        if(type() == 'i') // it is an integration operator
        {
          // Int(Int(y,x),x)
            *s = new Sum<T>('i',char()); assert(*s != NULL);
            copy((*s)->ep);
            t->copy((*s)->ep+1);
        }
        else if(type() == 'd') // it is a differentiation operator
        {
           // Int(df(y,x),x) = y
           if(ep[1] == t)
           {
             *s=new Sum<T>(1,0);
             ep[0]->copy((*s)->ep);
```

```
    }
    // Int(y(x),x)
    else if(isdepend(*((Sum<T>*)t)))
    {
     *s = new Sum<T>('i',char()); assert(*s != NULL);
     copy((*s)->ep);
     t->copy((*s)->ep+1);
    }
    else // Int(y,x)=yx
    {
     *s = new Sum<T>(1,0);                    assert((*s) != NULL);
     (*s)->ep[0] = new Product<T>(2);         assert((*s) != NULL);
     copy((*s)->ep[0]->ep);
     t->copy((*s)->ep[0]->ep+1);
    }
}
else if(islower(type())) // it is functions
{
    Symbol<T> *p=ep[0];
    while((p->branches==1)&&(p->type()!='V'))
     p=p->ep[0];
    if(p == t)
      IntRules(s,type());
    else if(isdepend(*((Sum<T>*)t)))
    {
     *s = new Sum<T>('i',char()); assert(*s != NULL);
     copy((*s)->ep);
     t->copy((*s)->ep+1);
    }
    else //Int(y,x)=yx
    {
     *s = new Sum<T>(1,0);                    assert((*s) != NULL);
     (*s)->ep[0] = new Product<T>(2);         assert((*s) != NULL);
     copy((*s)->ep[0]->ep);
     t->copy((*s)->ep[0]->ep+1);
    }
}
else if(type() == 'N')
{
 // int(0,x)=0
 if(((Number<T>*)data)->val()==T(0)) *s = zero();
 else    //int(C,x)=Cx
 {
  *s = new Sum<T>(1,0);   assert((*s) != NULL);
  t->copy((*s)->ep);
 }
}
```

```
        else // type() == 'V'
        {
           if((Symbol<T>*)this == t)   //Int(x,x)=(x^2)/2
           {
            *s = new Sum<T>(1,0);                assert ((*s) != NULL);
            (*s)->ep[0]=new Product<T>(1); assert ((*s)->ep[0] != NULL);
            (*s)->fac_exp[0] = T(1)/T(2);
            (*s)->ep[0]->fac_exp[0] = T(2);
            t->copy((*s)->ep[0]->ep);
           }
           else if(isdepend(*((Sum<T>*)t)))
           {
              *s = new Sum<T>('i',char()); assert(*s != NULL);
              copy((*s)->ep);
              t->copy((*s)->ep+1);
           }
           else // Int(y,x)=yx, y is a variable other than x
           {
            *s = new Sum<T>(1,0);                assert((*s) != NULL);
            (*s)->ep[0] = new Product<T>(2);     assert((*s)->ep[0] != NULL);
            copy((*s)->ep[0]->ep);
            t->copy((*s)->ep[0]->ep+1);
           }
        }
    }
  }
}


// Simplify Special Functions
template <class T> int Sum<T>::funcSimp(int&,int&)
{ return 0; }

// Simplify algebraic into simplest possible form
template <class T> void Symbol<T>::Shrink()
{
   int i, flag;
   if(type() == 'B') ep[0]->Shrink();
   else
   {
      removeB();
      do
      {
         do
         {
            flag = 0;
            for(i=0; i<branches; i++) flag += shrink_2();
            for(i=0; i<branches; i++) flag += shrink_5();
            for(i=0; i<branches; i++) flag += shrink_3(ep+i);
```

```
                 for(i=0; i<branches; i++) flag += shrink_1();
                 for(i=0; i<branches; i++) flag += shrink_4();
             } while(flag);  // repeat loop if there is any simplification
         } while(gather()); // repeat loop if there are some terms combined
     }
}

// return coefficient of s
template <class T> Sum<T> Sum<T>::coeff(const Sum<T> &s) const
{
    int i, j, k, m, found;
    Symbol<T> *ptr, *sptr, *mptr;
    Sum<T> result(T(0));

    // simplify expressions first before extract a coefficient
    ((Sum<T>*)this)->Shrink();
    ((Sum<T>&)s).Shrink();

    // x.coeff(x) -> 1
    if(!branches &&
        (&s==this ||
        (s.type() == 'B' && s.ep[0]->type() == 'S' &&
         s.ep[0]->branches == 1 && s.ep[0]->ep[0] == (Symbol<T>*)this)))
      { result = T(1); return result; }

    // initializing various pointers: ptr, sptr, mptr
    // depends on different situation
    if(type() == 'B')
    {
        ptr = ep[0];
        if(s.type() == 'B')
        {
          mptr = s.ep[0];
          if(mptr->type() == 'S' && mptr->branches == 1
              && mptr->fac_exp[0] == T(1) && !mptr->ep[0]->branches)
              mptr = mptr->ep[0];
          if(mptr->branches == 1) sptr = s.ep[0]->ep[0];
        }
        else
        {
          mptr = (Symbol<T>*)(&s);
          if(mptr->type() == 'S' && mptr->branches == 1
              && mptr->fac_exp[0] == T(1) && !mptr->ep[0]->branches)
              mptr = mptr->ep[0];
          if(mptr->branches == 1) sptr = s.ep[0];
        }
    }
```

```
    else
    {
        ptr = (Symbol<T>*)this;
if(s.type() == 'B')
        {
            mptr = s.ep[0];
            if(mptr->type() == 'S' && mptr->branches == 1
                && mptr->fac_exp[0] == T(1) && !mptr->ep[0]->branches)
                mptr = mptr->ep[0];
            if(mptr->branches == 1) sptr = s.ep[0]->ep[0];
        }
        else
        {
            mptr = (Symbol<T>*)(&s);
            if(mptr->type() == 'S' && mptr->branches == 1
                && mptr->fac_exp[0] == T(1) && !mptr->ep[0]->branches)
                mptr = mptr->ep[0];
            if(mptr->branches == 1) sptr = s.ep[0];
        }
    }

    // if s is a products of 2 variables and above
    if(mptr->branches)
    {
        if(mptr->branches == 1)      // avoid z = a+b
        {                            // assumed p->branches==1
            for(i=0; i<ptr->branches; i++)
            {
                if(ptr->ep[i]->type() == 'P')
                {
                    // allocate memory to store indices of matched-terms
                    int *index = new int[sptr->branches]; assert(index != NULL);
                    // check if terms in this & s match, if so record the index
                    for(m=0; m<sptr->branches; m++)
                    {
                        for(j=0; j<ptr->ep[i]->branches; j++)
                        {
                            if(ptr->ep[i]->ep[j]->is_equal(sptr->ep[m])
                                && ptr->ep[i]->fac_exp[j] == sptr->fac_exp[m])
                            { index[m] = j; break; }
                        }
                        // term not match (use index[0] = -1 to indicate)
                        if(j == ptr->ep[i]->branches) { index[0] = -1; break; }
                    }

                    // if the required term is matched
                    if(index[0] >= 0)
```

```
                {
                    // if every term matches, then the coefficient must be
                    // a numeric number
                    if(ptr->ep[i]->branches == sptr->branches)
                    { result += Sum<T>(ptr->fac_exp[i]); }
                    else    // the coefficient could be symbolic variables
                    {
                        // allocate memory for coefficients
                        Sum<T> tempsum(1,0);
                        Product<T> *p;
                        p = new Product<T>(ptr->ep[i]->branches-sptr->branches);
                        tempsum.ep[0] = p;

                        // extract terms that does not belong to the comparison
                        // list. These terms are the coefficients
                        found = 0;
                        for(m=0, k=0; k<ptr->ep[i]->branches; k++)
                        {
                            for(j=0; j<sptr->branches; j++)
                            {
                                if(index[j] == k) { found = 1; break; }
                            }
                            if(found) { found = 0; continue; }

                            ptr->ep[i]->ep[k]->copy(tempsum.ep[0]->ep+m);
                            tempsum.ep[0]->fac_exp[m] = ptr->ep[i]->fac_exp[k];
                            tempsum.ep[0]->next_var[m] = ptr->ep[i]->next_var[k];

                            m++;
                        }

                        // add this term to the final coefficient
                        result += Sum<T>(ptr->fac_exp[i]) * tempsum;
                    }
                }
                delete [] index;
            }
        }
    }
}
else // if s is of type 'V' or 'N'
{
    if(! ptr->branches && ptr->is_equal(mptr)) result += T(1);
    for(i=0; i<ptr->branches; i++)
    {
        // if this is a one term expression
        if(ptr->ep[i]->type() != 'P')
```

```
            {
            if(ptr->ep[i]->is_equal(mptr)) result += Sum<T>(ptr->fac_exp[i]);
            }
            else
            {
               for(j=0; j<ptr->ep[i]->branches; j++)
               {
                  // check if there exists a match term
                  if(ptr->ep[i]->ep[j]->is_equal(mptr) &&
                     (ptr->ep[i]->fac_exp[j] == T(1)))
                  {
                     Sum<T> tempsum(1,0);
                     Product<T> *p = new Product<T>(ptr->ep[i]->branches-1);
                     tempsum.ep[0] = p;
                     for(k=0; k<j; k++)
                     {
                        ptr->ep[i]->ep[k]->copy(tempsum.ep[0]->ep+k);
                        tempsum.ep[0]->fac_exp[k] = ptr->ep[i]->fac_exp[k];
                        tempsum.ep[0]->next_var[k] = ptr->ep[i]->next_var[k];
                     }

                     for(m=j, k=j+1; k<ptr->ep[i]->branches; k++, m++)
                     {
                        ptr->ep[i]->ep[k]->copy(tempsum.ep[0]->ep+m);
                        tempsum.ep[0]->fac_exp[m] = ptr->ep[i]->fac_exp[k];
                        tempsum.ep[0]->next_var[m] = ptr->ep[i]->next_var[k];
                     }
                     result += Sum<T>(ptr->fac_exp[i]) * tempsum;
                     break;
                  }
               }
            }
         }
   }
   return result;
}


// return coefficient of constant term
template <class T> T Sum<T>::coeff(int) const
{
   int i;
   T result;
   ((Sum<T>*)this)->Shrink();
   if(type() == 'B')
   {
      for(i=0; i<ep[0]->branches; i++)
         if(ep[0]->ep[i] == one()) return ep[0]->fac_exp[i];
```

```
   }
   else
   {
      for(i=0; i<branches; i++)
         if(ep[i] == one())
         {
            result = fac_exp[i];
            return result;
         }
   }
   return T(0);
}


// return coefficient of s^n
// note that s could only be of type() == 'V'
template <class T> Sum<T> Sum<T>::coeff(const Sum<T> &s,int n) const
{
   if(n == 1) return coeff(s);
   else if(n)                    // if n != 0,1
   {
      Sum<T> news(1,0);
      Product<T> *p = new Product<T>(1); assert(p != NULL);
      news.ep[0] = p;
      s.copy(news.ep[0]->ep);
      news.ep[0]->fac_exp[0] = T(n);
      if(s.type() == 'V') news.ep[0]->next_var[0] = 1;
      return coeff(news);
   }
   else   // n == 0
   {
      int i, j;
      Symbol<T> *ptr, *sptr;
      Sum<T> *temp;
      Sum<T> result(T(0));

      // initializing various pointers: ptr, sptr
      if(type() == 'B')
      {
         ptr = ep[0];
         if(s.type() == 'B') sptr = s.ep[0];
         else                sptr = (Symbol<T>*)(&s);

         if(sptr->type() == 'S' && sptr->branches == 1
            && sptr->fac_exp[0] == T(1)) sptr = sptr->ep[0];
      }
      else
      {
```

```
        ptr = (Symbol<T>*)this;
        if(s.type() == 'B') sptr = s.ep[0];
        else                      sptr = (Symbol<T>*)(&s);
        if(sptr->type() == 'S' && sptr->branches == 1
            && sptr->fac_exp[0] == T(1)) sptr = sptr->ep[0];
    }


    if(!(ptr->branches || ptr->is_equal(sptr)))
    {
      temp = (Sum<T>*)ptr;
      result += *temp;
    }


    for(i=0; i<ptr->branches; i++)
    {
      if(ptr->ep[i]->type() != 'P')
      {
        if(! ptr->ep[i]->is_equal(sptr))
        {
          temp = (Sum<T>*)(ptr->ep[i]);
          result += Sum<T>(ptr->fac_exp[i]) * *temp;
        }
      }
      else
      {
        // to check if s exists in the expression
        for(j=0; j<ptr->ep[i]->branches; j++)
          if (ptr->ep[i]->ep[j]->is_equal(sptr)) break;
    // if it does not exist,
// then this term is part of the coefficient
        if(j == ptr->ep[i]->branches)
        {
          Sum<T> tempsum(1,0);
          ptr->ep[i]->copy(tempsum.ep);
          result += Sum<T>(ptr->fac_exp[i]) * tempsum;
        }
      }
    }
    return result;
  }
}


// declare dependency of variables
template <class T> void Sum<T>::depend(const Sum<T> &x)
{ if(!deplist.is_Include((Sum<T>*)&x)) deplist.add((Sum<T>*)&x); }

// declare no-dependency of variables
```

```
template <class T> void Sum<T>::nodepend(const Sum<T>& x)
{
   ListIterator<Sum<T>*> m((List<Sum<T>*>&)deplist);
   for(m.init(); !m; ++m)
      if(m() == &x)
      {
         m.delete_Current();
         return;
      }
}


// check implicit dependency of variables
template <class T> int implicit_depend(Symbol<T> **s,Symbol<T> *x,int n)
{
 int i;
 for(i=0;i<n;i++)
 {
  if(s[i] == x) return 1;
 }
 for(i=0;i<n;i++)
 {
  //check for explicit and implicit dependency of variables and sums
  if((s[i]->type() == 'S')||(s[i]->type() == 'B')||(s[i]->type() == 'V'))
  {
   if(((Sum<T>*)s[i])->isdepend(*((Sum<T>*)x))) return 1;
  }
  else if(s[i]->type() != 'N')
  {
   //check for implicit dependency - or implicit dependency
   // of function parameters
   if(implicit_depend(s[i]->ep,x,s[i]->branches+(s[i]->branches==0)))
    return 1;
  }
 }
 return 0;
}


// check dependency of variables
template <class T> int Sum<T>::isdepend(const Sum<T> &x) const
{
   if(type() == 'N') return 0;
   if(this == &x) return 1;
   if(deplist.is_Include((Sum<T>*)&x)) return 1;

   ListIterator<Sum<T>*> m((List<Sum<T>*>&)deplist);
   for(m.init(); !m; ++m)
      if(m()->isdepend(x)) return 1;
```

```
    // check for implicit dependency
    // or implicit dependency of function parameters
    if((branches)||((branches==0)&&islower(type())))
    return implicit_depend(ep,(Symbol<T>*)&x,branches+(branches==0));
    return 0;
}


// check if a variable is pure number
template <class T> int Sum<T>::is_Number() const
{
    Symbol<T> *ptr;
    if(type() == 'B') ptr = ep[0];
    else ptr = (Symbol<T>*)this;
    if(ptr->branches == 1 && ptr->ep[0]->type() == 'N') return 1;
    return 0;
}


// Exponential Function
template <class T> Sum<T> exp(const Sum<T> &s)
{
    Sum<T> r('e',char());
    ((Sum<T>&)s).Shrink();
    if(s.type() == 'B') s.ep[0]->copy(r.ep);
    else if(s.branches) s.copy(r.ep);
    else
    {
        r.ep[0] = new Sum<T>(1,0); assert(r.ep[0] != NULL);
        r.ep[0]->ep[0] = (Symbol<T>*)&s;
        r.ep[0]->next_var[0] = 1;
    }
    return r;
}


// Hyperbolic Sine Function
template <class T> Sum<T> sinh(const Sum<T> &s)
{
    Sum<T> r('j',char());
    ((Sum<T>&)s).Shrink();
    if(s.type() == 'B') s.ep[0]->copy(r.ep);
    else if(s.branches) s.copy(r.ep);
    else
    {
        r.ep[0] = new Sum<T>(1,0); assert(r.ep[0] != NULL);
        r.ep[0]->ep[0] = (Symbol<T>*)&s;
        r.ep[0]->next_var[0] = 1;
    }
```

```
    return r;
}


// Hyperbolic Cosine Function
template <class T> Sum<T> cosh(const Sum<T> &s)
{
    Sum<T> r('w',char());
    ((Sum<T>&)s).Shrink();
    if(s.type() == 'B') s.ep[0]->copy(r.ep);
    else if(s.branches) s.copy(r.ep);
    else
    {
        r.ep[0] = new Sum<T>(1,0); assert(r.ep[0] != NULL);
        r.ep[0]->ep[0] = (Symbol<T>*)&s;
        r.ep[0]->next_var[0] = 1;
    }
    return r;
}


// Sine Function
template <class T> Sum<T> sin(const Sum<T> &s)
{
    Sum<T> r('z',char());
    ((Sum<T>&)s).Shrink();
    if(s.type() == 'B') s.ep[0]->copy(r.ep);
    else if(s.branches) s.copy(r.ep);
    else
    {
        r.ep[0] = new Sum<T>(1,0); assert(r.ep[0] != NULL);
        r.ep[0]->ep[0] = (Symbol<T>*)&s;
        r.ep[0]->next_var[0] = 1;
    }
    return r;
}


// Cosine Function
template <class T> Sum<T> cos(const Sum<T> &s)
{
    Sum<T> r('c',char());
    ((Sum<T>&)s).Shrink();
    if(s.type() == 'B') s.ep[0]->copy(r.ep);
    else if(s.branches) s.copy(r.ep);
    else
    {
        r.ep[0] = new Sum<T>(1,0); assert(r.ep[0] != NULL);
        r.ep[0]->ep[0] = (Symbol<T>*)&s;
        r.ep[0]->next_var[0] = 1;
```

```
      }
      return r;
}


// Logarithm Function
template <class T> Sum<T> ln(const Sum<T> &s)
{
    Sum<T> r('l',char());
    ((Sum<T>&)s).Shrink();
    if(s.type() == 'B') s.ep[0]->copy(r.ep);
    else if(s.branches) s.copy(r.ep);
    else
    {
       r.ep[0] = new Sum<T>(1,0); assert(r.ep[0] != NULL);
       r.ep[0]->ep[0] = (Symbol<T>*)&s;
       r.ep[0]->next_var[0] = 1;
    }
    return r;
}


// Square Root Function
template <class T> Sum<T> sqrt(const Sum<T> &s)
{
    Sum<T> r('q',char());
    ((Sum<T>&)s).Shrink();
    if(s.type() == 'B') s.ep[0]->copy(r.ep);
    else if(s.branches) s.copy(r.ep);
    else
    {
       r.ep[0] = new Sum<T>(1,0); assert(r.ep[0] != NULL);
       r.ep[0]->ep[0] = (Symbol<T>*)&s;
       r.ep[0]->next_var[0] = 1;
    }
    return r;
}


// raise Sum s to a power n
template <class T> Sum<T> power(const Sum<T> &s, T n)
{
    static T zero(0);
    Sum<T> r(1,0);
    if(n != zero)
    {
       Product<T> *p = new Product<T>(1);
       r.ep[0] = p;

       // simplify expression first before rise to a power n
```

```
        ((Sum<T>&)s).Shrink();

        if(s.type() == 'B')    s.ep[0]->copy(r.ep[0]->ep);
        else                   s.copy(r.ep[0]->ep);

        if(s.type() == 'V')    r.ep[0]->next_var[0] = 1;

        r.ep[0]->fac_exp[0] = n;
    }
    else
    {
        r.ep[0] = s.one();
        r.fac_exp[0] = T(1);
    }
    return r;
}

// rise T x to a power n
template <class T> T power(T x,T n)
{
    int N;
    T Y(1);
    if(n < T(0))
    {
        x = T(1)/x;
        N = int(-n);
    }
    else
        N = int(n);
    if(N==0) return T(1);
    while(1)
    {
        if(N%2)
        {
            Y = Y * x;
            N /= 2;
            if(N==0) return Y;
        }
        else N /= 2;
        x = x * x;
    }
}

// Partial derivative of s with respect to t
template <class T> Sum<T> df(const Sum<T> &s,const Sum<T> &t)
{
    Sum<T> result(1,0);
```

```
      Symbol<T> *tt;
      ((Sum<T>&)s).Shrink();
      if(t.type() == 'B') tt = t.ep[0];
      else                  tt = (Symbol<T>*)&t;

      if(tt->type() == 'S' && tt->branches == 1 && tt->fac_exp[0] == T(1))
        tt = tt->ep[0];

      s.diff(result.ep,tt);
      result.Shrink();
      return result;
}


// The n-th partial derivative of s with respect to t
template <class T> Sum<T> df(const Sum<T> &s, const Sum<T> &t, int n)
{
    Sum<T> result(s);
    for(int i=0; i<n; i++) result = df(result,t);
    return result;
}


// Partial integral of s with respect to t
template <class T> Sum<T> Int(const Sum<T> &s, const Sum<T> &t)
{
    Sum<T> result(1,0);
    Symbol<T> *tt;
    ((Sum<T>&)s).Shrink();
    if(t.type() == 'B') tt = t.ep[0];
    else                  tt = (Symbol<T>*)&t;

    if(tt->type() == 'S' && tt->branches == 1 && tt->fac_exp[0] == T(1))
      tt = tt->ep[0];
    s.integrate(result.ep,tt);
    result.Shrink();
    return result;
}


// The n-th partial integral of s with respect to t
template <class T> Sum<T> Int(const Sum<T> &s, const Sum<T> &t, int n)
{
    Sum<T> result(s);
    for(int i=0; i<n; i++)
        result = Int(result,t);
    return result;
}

// set a numerical value to a variable
```

```
template <class T> void Sum<T>::set(const T num)
{
    if(type() == 'B') { (*(Sum<T>*)ep[0]).set(num); return; }
    if(type() == 'S' && branches==1 && fac_exp[0]==T(1))
        { (*(Sum<T>*)ep[0]).set(num); return; }
    is_Set = 1;
    ((Variable<T>*)data)->set(num);
}


// Assign an expression to another
template <class T> int Sum<T>::put(const Sum<T> &s1,const Sum<T> &s2)
{
    int i, flag=0;
    Symbol<T> *ptr;
    if(s1.type() == 'B') return put(*(Sum<T>*)(s1.ep[0]),s2);
    if(type() == 'B') ptr = ep[0];
    else              ptr = (Symbol<T> *)this;
    ((Sum<T>&)s1).Shrink();

    if(s1.type() == 'S' && s1.branches == 1)
    {
        for(i=0; i<ptr->branches; i++)
        {
            flag += ptr->ep[i]->replace1(s1.ep[0],(Symbol<T>*)(&s2));
            if(ptr->ep[i]->is_equal(s1.ep[0]))
            { s2.copy(ptr->ep+i); flag++; }
        }
    }
    else // (s1.branches == 0)
    {
        for(i=0; i<ptr->branches; i++)
        {
            flag += ptr->ep[i]->replace1((Symbol<T>*)(&s1),(Symbol<T>*)(&s2));
            if(ptr->ep[i]->is_equal((Symbol<T>*)(&s1)))
            { s2.copy(ptr->ep+i); flag++; }
        }
    }
    Shrink();
    return flag;
}


// Return the numeric value of a variable
template <class T> T Sum<T>::nvalue() const
{
    return nval();
}
```

```cpp
template <class T> T Sum<T>::nval() const
{
   if(type() == 'B') return (*(Sum<T>*)ep[0]).nval();
   if(type() == 'S' && branches==1 && fac_exp[0]==T(1))
     return (*(Sum<T>*)ep[0]).nval();
   int i;
   T s(0);
   if(!branches)
   {
     if(islower(type())) // it is a function
        return ((Function<T>*)data)->f(((Sum<T>*)ep[0])->nval());
      else
        return ((Magnitude<T>*)data)->val();
   }
   else
      for(i=0; i<branches; i++)
         s += fac_exp[i] * ((Sum<T>*)ep[i])->nval();
   return s;
}


// Return the value of an expression
template <class T> Sum<T> Sum<T>::value() const
{
   return val();
}


// Renew a used variable
template <class T> void Sum<T>::clear()
{
   if(ep) { removeB(); ep[0]->deleteAllNodes(); }
   delete [] ep; delete [] fac_exp;
   delete [] next_var; deplist.deleteAllNodes();

   branches = 0;
   ep = NULL; fac_exp = NULL; next_var = NULL;
   is_Bound = 0; is_Set = 0;
}


// Expand an expression
template <class T> void Sum<T>::expand()
{
   int i, flag;
   Shrink();
   do {
      do {
         flag = 0;
         for(i=0; i<branches; i++)  flag += ep[i]->mxpand(ep+i);
```

```
            if(!flag) for(i=0; i<branches; i++)  flag += ep[i]->dxpand(ep+i);
            Shrink();
        } while(flag);
    } while(gather());
    Shrink();
}


// unary plus
template <class T> Sum<T> Sum<T>::operator + () const
{ return *this; }

// unary minus
template <class T> Sum<T> Sum<T>::operator - () const
{ return *this * T(-1); }

template <class T> Sum<T> Sum<T>::operator += (const Sum<T> &s)
{ return *this = *this + s; }

template <class T> Sum<T> Sum<T>::operator -= (const Sum<T> &s)
{ return *this = *this - s; }

template <class T> Sum<T> Sum<T>::operator *= (const Sum<T> &s)
{ return *this = *this * s; }

template <class T> Sum<T> Sum<T>::operator /= (const Sum<T> &s)
{ return *this = *this / s; }

template <class T> Sum<T> operator + (const Sum<T> &s1,const Sum<T> &s2)
{
    Sum<T> r(2,0);
    if(s1.type() == 'B') s1.ep[0]->copy(r.ep);
    else                 s1.copy(r.ep);

    if(s2.type() == 'B') s2.ep[0]->copy(r.ep+1);
    else                 s2.copy(r.ep+1);

    if(s1.type() == 'V') r.next_var[0] = 1;
    if(s2.type() == 'V') r.next_var[1] = 1;
    return r;
}

template <class T> Sum<T> operator - (const Sum<T>& s1,const Sum<T>& s2)
{
    Sum<T> r(2,0);
    if(s1.type() == 'B') s1.ep[0]->copy(r.ep);
    else                 s1.copy(r.ep);
    if(s2.type() == 'B') s2.ep[0]->copy(r.ep+1);
```

```
      else                    s2.copy(r.ep+1);
    if(s1.type() == 'V') r.next_var[0] = 1;
    if(s2.type() == 'V') r.next_var[1] = 1;
    r.fac_exp[1] = T(-1);
    return r;
}


template <class T> Sum<T> operator * (const Sum<T> &s1,const Sum<T> &s2)
{
    Sum<T> r(1,0);
    Product<T> *p;
    p = new Product<T>(2); assert(p != NULL);
    r.ep[0] = p;
    if(s1.type() == 'B') s1.ep[0]->copy(r.ep[0]->ep);
    else                    s1.copy(r.ep[0]->ep);

    if(s2.type() == 'B') s2.ep[0]->copy(r.ep[0]->ep+1);
    else                    s2.copy(r.ep[0]->ep+1);

    if(s1.type() == 'V') r.ep[0]->next_var[0] = 1;
    if(s2.type() == 'V') r.ep[0]->next_var[1] = 1;
    return r;
}


template <class T> Sum<T> operator / (const Sum<T> &s1,const Sum<T> &s2)
{
    ((Sum<T>&)s2).Shrink();
    if(s2.branches == 1 && s2.ep[0] == s2.zero())
    {
        cerr << "Division by zero" << endl;
        exit(1);
    }

    Sum<T> r(1,0);
    Product<T> *p = new Product<T>(2);
    r.ep[0] = p;
    if(s1.type() == 'B') s1.ep[0]->copy(r.ep[0]->ep);
    else                    s1.copy(r.ep[0]->ep);

    if(s2.type() == 'B') s2.ep[0]->copy(r.ep[0]->ep+1);
    else                    s2.copy(r.ep[0]->ep+1);

    if(s1.type() == 'V') r.ep[0]->next_var[0] = 1;
    if(s2.type() == 'V') r.ep[0]->next_var[1] = 1;

    r.ep[0]->fac_exp[1] = T(-1);
    return r;
```

```
}


template <class T> int operator == (const Sum<T> &s1,const Sum<T> &s2)
{
   ((Sum<T>&)s1).Shrink();
   ((Sum<T>&)s2).Shrink();
   return s1.is_equal((Symbol<T>*)&s2);
}

template <class T> int operator != (const Sum<T> &s1,const Sum<T> &s2)
{ return !(s1==s2); }

template <class T> ostream & operator << (ostream &os,const Sum<T> &s)
{
   Symbol<T> *ptr;
   ((Sum<T>&)s).Shrink();
   if(s.type() == 'B') ptr = s.ep[0];
   else                ptr = (Symbol<T>*)&s;

   // print the content, if this is a leaf node
   if(!ptr->branches)
   {
     ((Sum<T>*)ptr)->data->oprint(os);
     if(islower(((Sum<T>*)ptr)->data->type()))
       {
          if(ptr->ep[0]->branches) ptr->ep[0]->oprint(os);
          else if((((Sum<T>*)ptr)->data->type() == 'd')||
                  (((Sum<T>*)ptr)->data->type() == 'i'))
          {
             os << "("; ptr->ep[0]->oprint(os); os << ",";
             ptr->ep[1]->oprint(os); os << ")";
          }
          else
          { os << "("; ptr->ep[0]->oprint(os); os << ")"; }
       }
   }
   else
   {
      // do not print + as this is the first term
      if(ptr->fac_exp[0] == T(-1)) { os << "-"; ptr->ep[0]->oprint(os); }
      else if(ptr->fac_exp[0] != T(1))
      {
         os << ptr->fac_exp[0];
         if(ptr->ep[0] != ptr->one())
           { os << "*"; ptr->ep[0]->oprint(os); }
      }
      else ptr->ep[0]->oprint(os);
```

```
// print the rest of the terms
for(int i=1; i<ptr->branches; i++)
{
   if(ptr->fac_exp[i] > T(0))
   {
      os << "+";
      if(ptr->fac_exp[i] == T(1)) ptr->ep[i]->oprint(os);
      else
      {
         os << ptr->fac_exp[i];
         if(ptr->ep[i] != ptr->one())
            { os << "*"; ptr->ep[i]->oprint(os); }
      }
   }
   else
   {
      if(ptr->fac_exp[i] == T(-1))
      { os << "-"; ptr->ep[i]->oprint(os); }
      else
      {
         os << ptr->fac_exp[i];
         if(ptr->ep[i] != ptr->one())
            { os << "*"; ptr->ep[i]->oprint(os); }
      }
   }
}
   }
   return os;
}

template <class T> istream & operator >> (istream &s,Sum<T> &)
{ return s; }

// Shift factor out to outer level, rised to appropriate power
// in order to preserve the canonical form representation
template <class T> int Sum<T>::shrink_1()
{
   int i, flag=0, k;
   Symbol<T> *ptr;
   for(i=0; i<branches; i++)
   {
      if(ep[i]->type() == 'P')
      {
         if(ep[i]->branches == 1 && ep[i]->fac_exp[0] == T(1))
         {
            ptr = ep[i]->ep[0];
```

```
            next_var[i] = ep[i]->next_var[0];
            ep[i]->deleteOneNode();
            ep[i] = ptr;
            flag++;
          }
          else
          for(k=0; k < ep[i]->branches; k++)
          {
            if(ep[i]->ep[k]->type() == 'S' &&
               ep[i]->ep[k]->branches == 1 &&
               ep[i]->ep[k]->fac_exp[0] != T(1))
            {
            flag++;
            // Shift factor out to outer level
            fac_exp[i] *= power(ep[i]->ep[k]->fac_exp[0],ep[i]->fac_exp[k]);
            ep[i]->ep[k]->fac_exp[0] = T(1);
            }
          }
       }
    }
    for(i=0; i<branches; i++) flag += ep[i]->shrink_1();
    return flag;
}


// removes extra Sum nodes: moves the lower level of a consecutive level
// of Sum node to the upper level
template <class T> int Sum<T>::shrink_2()
{
    int i, k, j, n, flag=0;
    Symbol<T> **ept;
    T *fac_expt;
    int *next_vart;

    for(i=0; i<branches; i++)
    {
       if(ep[i]->type() == 'S')
       {
          n = branches - 1 + ep[i]->branches;
          ept = new Symbol<T>*[n]; assert(ept != NULL);
          fac_expt = new T[n];       assert(fac_expt != NULL);
          next_vart = new int[n];   assert(next_vart != NULL);

          // copy nodes that are not affected
          for(k=0; k<i; k++)
          {
             ept[k] = ep[k];
             fac_expt[k] = fac_exp[k];
```

```
            next_vart[k] = next_var[k];
        }

        // moves Sum node to one level up
        for(j=i, k=0; k < ep[i]->branches; k++, j++)
        {
            ept[j] = ep[i]->ep[k];
            fac_expt[j] = ep[i]->fac_exp[k] * fac_exp[i];
            next_vart[j] = ep[i]->next_var[k];
        }

        // copy nodes that are not affected
        for(k=i+1; k<branches; k++, j++)
        {
            ept[j] = ep[k];
            fac_expt[j] = fac_exp[k];
            next_vart[j] = next_var[k];
        }
        branches = n;

        ep[i]->deleteOneNode();
        delete [] ep; delete [] fac_exp; delete [] next_var;
        ep = ept;
        fac_exp = fac_expt;
        next_var = next_vart;
        flag++;
        }
    }
    for(i=0; i<branches; i++) flag += ep[i]->shrink_2();
    return flag;
}

// remove zero Summands
template <class T> int Sum<T>::shrink_3(Symbol<T>**)
{
    int i, flag = 0;
    // for-loop is in reverse order since remove(j) decreases branches by 1
    for(i=branches-1; i>=0; i--)
    {
        if(fac_exp[i] == T(0))
        {
            ep[i]->deleteAllNodes();
            ep[i] = zero();  fac_exp[i] = T(1);
            next_var[i] = 1; flag++;
        }

        if(ep[i] == zero())
```

```
        {
            if(branches > 1)
              { remove(i); flag++; }
            else if(branches == 1 && fac_exp[i] != T(1))
              { fac_exp[i] = T(1); flag++; }
        }
    }
    for(i=0; i<branches; i++) flag += ep[i]->shrink_3(ep+i);
    return flag;
}


// sums Numbers (type()=='N') from Sums
template <class T> int Sum<T>::shrink_4()
{
    int i, k, flag=0;
    for(i=0; i<branches; i++) flag += ep[i]->shrink_4();

    // find the first 'N' node
    for(i=0; i<branches; i++)
      if(ep[i]->type() == 'N') break;

    // check for the rest of the node, if there exists any 'N' node, add to
    // the first node
    for(k=i+1; k<branches; k++)
      if(ep[k]->type() == 'N')
        {
        flag++;

        fac_exp[i] = fac_exp[i]*((Number<T>*)((Sum<T>*)ep[i])->data)->val()
                   + fac_exp[k]*((Number<T>*)((Sum<T>*)ep[k])->data)->val();
        ep[i] = one();
        remove(k); k--;
        }
    return flag;
}


template <class T> int Sum<T>::shrink_5()
{
    int i, flag=0;
    for(i=0; i<branches; i++) flag += ep[i]->shrink_5();
    return flag;
}


// Private member functions
template <class T> void Sum<T>::diffRules(Symbol<T> **s,char t) const
{
    switch(t)
```

```
      {
      case 'e': // (d/dx)exp(u) = exp(u) (du/dx)
                (*s) = new Sum<T>('e',char()); assert((*s) != NULL);
                ep[0]->copy((*s)->ep);
                break;
      case 'j': // (d/dx)sinh(u) = cosh(u) (du/dx)
                (*s) = new Sum<T>('w',char()); assert((*s) != NULL);
                ep[0]->copy((*s)->ep);
                break;
      case 'w': // (d/dx)cosh(u) = sinh(u) (du/dx)
                (*s) = new Sum<T>('j',char()); assert((*s) != NULL);
                ep[0]->copy((*s)->ep);
                break;
      case 'z': // (d/dx)sin(u) = cos(u) (du/dx)
                (*s) = new Sum<T>('c',char()); assert((*s) != NULL);
                ep[0]->copy((*s)->ep);
                break;
      case 'c': // (d/dx)cos(u) = -sin(u) (du/dx)
                (*s) = new Sum<T>(1,0); assert((*s) != NULL);
                (*s)->ep[0] = new Sum<T>('z',char());
                (*s)->fac_exp[0] = T(-1);
                ep[0]->copy((*s)->ep[0]->ep);
                break;
      case 'l': // (d/dx)ln(u) = 1/(u) (du/dx)
                (*s) = new Sum<T>(1,0); assert((*s) != NULL);
                (*s)->ep[0] = new Product<T>(1); assert((*s)->ep[0] != NULL);
                (*s)->ep[0]->fac_exp[0] = T(-1);
                ep[0]->copy((*s)->ep[0]->ep);
                break;
      case 'q': // (d/dx)sqrt(u) = 1/(2*sqrt(u)) (du/dx)
                (*s) = new Sum<T>(1,0); assert((*s) != NULL);
                (*s)->ep[0] = new Product<T>(1); assert((*s)->ep[0] != NULL);
                (*s)->ep[0]->ep[0] = new Sum<T>(1,0);
                assert((*s)->ep[0]->ep[0] != NULL);
                (*s)->ep[0]->ep[0]->ep[0] = new Sum<T>('q',char());
                assert((*s)->ep[0]->ep[0]->ep[0] != NULL);
                (*s)->ep[0]->fac_exp[0] = T(-1);
                (*s)->ep[0]->ep[0]->fac_exp[0] = T(2);
                ep[0]->copy((*s)->ep[0]->ep[0]->ep[0]->ep);
                break;
      default : ;
      }
}

template <class T> void Sum<T>::IntRules(Symbol<T> **s,char t) const
{
   switch (t)
```

```
    {
    case 'e': // Int(exp(u),y) = exp(u)
            copy(s);
            break;
    case 'j': // Int(sinh(u),u) = cosh(u)
            (*s) = new Sum<T>('w',char()); assert((*s) != NULL);
            ep[0]->copy((*s)->ep);
            break;
    case 'w': // Int(cosh(u),u = sinh(u)
            (*s) = new Sum<T>('j',char()); assert((*s) != NULL);
            ep[0]->copy((*s)->ep);
            break;
    case 'z': // Int(sin(u),u) = -cos(u)
            (*s) = new Sum<T>(1,0); assert((*s) != NULL);
            (*s)->ep[0] = new Sum<T>('c',char());
            (*s)->fac_exp[0] = T(-1);
            ep[0]->copy((*s)->ep[0]->ep);
            break;
    case 'c': // Int(cos(u),u) = sin(u)
            (*s) = new Sum<T>('z',char()); assert((*s) != NULL);
            ep[0]->copy((*s)->ep);
            break;
    case 'l': // Int(ln(u),u) = u*ln(u)-u
            (*s) = new Sum<T>(2,0);              assert((*s) != NULL);
            (*s)->ep[0] = new Product<T>(2); assert((*s)->ep[0] != NULL);
            (*s)->ep[0]->ep[1] = new Sum<T>('l',char());
            assert((*s)->ep[0] != NULL);
            (*s)->fac_exp[1] = T(-1);
            ep[0]->copy((*s)->ep[0]->ep);
            ep[0]->copy((*s)->ep[0]->ep[1]->ep);
            ep[0]->copy((*s)->ep+1);
            break;
    case 'q': // Int(sqrt(u),u) = (2/3)*u^(3/2) = (2/3)*(sqrt(u))^3
            (*s) = new Sum<T>(1,0); assert((*s) != NULL);
            (*s)->ep[0] = new Product<T>(1); assert((*s)->ep[0] != NULL);

            (*s)->fac_exp[0]=T(2)/T(3);
            (*s)->ep[0]->fac_exp[0] = T(3)/T(2);
            ep[0]->copy((*s)->ep[0]->ep);
            break;
    default : ;
    }
}

// Definition of class Product
// operator * is commutative by default
template <class T> int Product<T>::Comm = 1;
```

```
template <class T> Product<T>::Product(int i) : Symbol<T>(i) {}

template <class T>
Product<T>::Product(int i, Symbol<T> **newep, T *new_fe, int *new_nx)
   : Symbol<T>(i, newep, new_fe, new_nx) {}

template <class T> Product<T>::~Product()
{ deleteOneNode(); }

// return the type Product
template <class T> char Product<T>::type() const
{ return 'P'; }

// print the numerical value of the expression (return type Sum<T>)
template <class T> Sum<T> Product<T>::val() const
{
   Sum<T> s(T(1));
   for(int i=0; i<branches; i++)
     s *= power(ep[i]->val(),fac_exp[i]);
   return s;
}

// print the numerical value of the expression (return type T)
template <class T> T Product<T>::nval() const
{
   T s(1);
   for (int i=0; i<branches; i++)
      s *= power(ep[i]->nval(), fac_exp[i]);
   return s;
}

// replace expression 1 by expression 2
template <class T> int Product<T>::replace1(Symbol<T> *s1, Symbol<T> *s2)
{
   int i, index, flag=0;
   for(i=0; i<branches; i++) flag += ep[i]->replace1(s1,s2);
   if(s1->branches >= 1)
   {
      index = equal_index(s1);
      if(index >= 0)
      {
         s2->copy(ep+index);
         fac_exp[index] = T(1);
         flag++;
      }
   }
```

```
        else
        {
           for(i=0; i<branches; i++)
              if(ep[i]->is_equal(s1))
              { s2->copy(ep+i); flag++; }
        }
        return flag;
}

// return the index of equal component
// return (-1) if no such component is found
template <class T> int Product<T>::equal_index(Symbol<T> *s)
{
    int i,j,k;
    if(Comm) // Commutative Algebra
    {
        if(branches >= s->branches)
        {
            int *ip,        // ip[] is used as an indicator for the visited node
                count=0; // number of match components
            ip = new int[branches]; assert(ip != NULL);
            for(i=0; i<branches; i++) ip[i] = 0;
            for(j=0; j<branches; j++)
            {
                for(i=0; i<s->branches; i++)
                    if(s->ep[i]->is_equal(ep[j]) && s->fac_exp[i] == fac_exp[j])
                    {  ip[count] = j; ++count; break; }
            }

            if(count == s->branches)
            {
                for(count--; count>0; count--) remove(ip[count]);
                return ip[0];
            }
        }
        return (-1);
    }
    else // Non-commutative Algebra
    {
        for(j=0; j<branches-s->branches+1; j++)
        {
            for(i=0, k=j; i<s->branches; i++, k++)
                if(!(s->ep[i]->is_equal(ep[k]) && s->fac_exp[i] == fac_exp[k]))
                    break;
            if(i == s->branches)
            {
                for(i--, k--; i>0; i--, k--) remove(k);
```

```
            return k;
        }
    }
    return (-1);
    }
}

// check if 2 expressions are equal
template <class T> int Product<T>::is_equal(Symbol<T> *s) const
{
  if(Comm) // for commutative algebra
  {
    int i,k;
    int *ip;  // ip[] is used as an indicator for the visited node

    // a product node must equals to a product node
    if(s->type() != 'P')  return 0;

    // if s is equivalent to this return 1
    if(s == (Symbol<T>*)this) return 1;

    // if either s or this is a leaf node, return 0
    if(!s->branches || !branches) return 0;

    // if branches are not equal, return 0
    if(branches != s->branches) return 0;

    // branches are equal
    ip = new int[branches]; assert(ip != NULL);

    for(i=0; i<branches; i++) ip[i] = 0;

    // check if all the subtree are equal
    for(i=0; i<branches; i++)
    {
        for(k=0; k<branches; k++)
        {
            if(ep[i]->is_equal(s->ep[k]) &&
                fac_exp[i] == s->fac_exp[k] && ip[k] == 0)
            { ip[k] = 1; break; }
        }

        // if there is no match terms, return 0
        if(k == branches) return 0;
    }
    delete [] ip;
    return 1;
```

```
    }
    else // for non-commutative algebra
    {
        int i;

        // a product node must equals to a product node
        if(s->type() != 'P')  return 0;

        // if s is equivalent to this return 1
        if(s == (Symbol<T>*)this) return 1;

        // if either s or this is a leaf node, return 0
        if(!s->branches || !branches) return 0;

        // if branches are not equal, return 0
        if(branches != s->branches) return 0;

        // branches are equal
        // check if all the subtree are equal
        for(i=0; i<branches; i++)
        {
            if(!(ep[i]->is_equal(s->ep[i])) || fac_exp[i] != s->fac_exp[i])
                return 0;
        }
        return 1;
    }
}

// Collect/combine terms that have the same variable
template <class T> int Product<T>::gather()
{
    if(Comm) // for commutative algebra
    {
        int i, j, flag;
        flag = 0;
        for(i=0; i<branches; i++) flag += ep[i]->gather();

        // comparing different subtrees, if equal, group/combine them
        for(i=0; i<branches; i++)
            for(j=i+1; j<branches; j++)
            {
                if(ep[i]->is_equal(ep[j]))
                {
                    flag++;
                    fac_exp[i] += fac_exp[j];   // sum factors

                    // if fac_exp[] == 0 after added
```

```
        if(fac_exp[i] == T(0))
           if(branches > 2)
           { remove(j); remove(i); j=i;}
           else // remove one term and create a one() node
           {
               remove(j); j=i;
               ep[0] = one();
               fac_exp[0] = T(1);
           }
        else { remove(j); j=i; }
    }
    else if(!ep[i]->branches && !ep[j]->branches)
    {
        if(islower(ep[i]->type()) && ep[i]->type() == ep[j]->type())
           flag += funcSimp(i,j);
    }
  }
  return flag;
}
else // for non-commutative algebra
{
  int i, j, flag;
  flag = 0;
  for(i=0; i<branches; i++) flag += ep[i]->gather();

  // comparing different subtrees, if equal, group/combine them
  for(i=0; i<branches; i++)
     if((j=i+1) < branches)
     {
        if(ep[i]->is_equal(ep[j]))
        {
           flag++;
           fac_exp[i] += fac_exp[j];   // sum factors

           // if fac_exp[] == 0 after added
           if(fac_exp[i] == T(0))
              if(branches > 2)
              { remove(j); remove(i); }
              else // remove one term and create a one() node
              {
                  remove(j);
                  ep[0] = one(); assert(ep[0] != NULL);
                  fac_exp[0] = T(1);
              }
           else remove(j);
        }
        else if(!ep[i]->branches && !ep[j]->branches)
```

```
            {
                if(islower(ep[i]->type()) && ep[i]->type() == ep[j]->type())
                    flag += funcSimp(i,j);
            }
        }
        return flag;
    }
}


// copy subtree from this to sp
template <class T> void Product<T>::copy(Symbol<T> **sp) const
{
    int i;
    *sp = new Product<T>(branches); assert(*sp != NULL);
    for(i=0;i<branches;i++) (*sp)->fac_exp[i] = fac_exp[i];
    for(i=0;i<branches;i++) (*sp)->next_var[i] = next_var[i];
    for(i=0;i<branches;i++) ep[i]->copy((*sp)->ep+i);
}


// print the algebraic expression
template <class T> void Product<T>::oprint(ostream &os) const
{
    // do not print * as this is the first terms
    ep[0]->oprint(os);
    if(fac_exp[0] != T(1)) os << "^(" << fac_exp[0] << ")";

    // print the rest of the terms
    for(int i=1; i<branches; i++)
    {
        os << "*";
        ep[i]->oprint(os);
        if(fac_exp[i] != T(1)) os << "^(" << fac_exp[i] << ")";
    }
}


// Delete this node only
template <class T> void Product<T>::deleteOneNode()
{
    delete [] ep; delete [] fac_exp; delete [] next_var;
}


// Delete all the nodes rooted at this
template <class T> void Product<T>::deleteAllNodes()
{
    if(branches)
    {
        for(int i=0; i<branches; i++)
```

```
            if(!next_var[i]) ep[i]->deleteAllNodes();
        deleteOneNode();
    }
}


// Binomial/Multinomial Expansion
template <class T> int Product<T>::mxpand(Symbol<T> **f)
{
    if(Comm) // for commutative algebra
    {
        int i, flag=0, m;
        (*f) = new Product<T>(branches, ep, fac_exp, next_var);
        assert(*f != NULL);
        for(i=0; i<branches; i++)
        {
            if(ep[i]->type() == 'S' && ep[i]->branches > 1
                && fac_exp[i] > T(1)
                && double(fac_exp[i]) == int(double(fac_exp[i])))
            {
                flag++;
                int k, n,
                    newbranches = int(fac_exp[i]) + 1,
                    lastindex = newbranches - 1,
                    bcenter = int(fac_exp[i]) >> 1;
                T *bcoeff = new T[newbranches-2]; assert(bcoeff != NULL);

                // Calculate the binomial coefficient
                bcoeff[0] = fac_exp[i];

                for(k=(int)fac_exp[i]-1, n=2, m=1; m<bcenter; m++, k--, n++)
                    bcoeff[m] = bcoeff[m-1]*T(k)/T(n);

                for(k=0, m=newbranches-3; m>=bcenter; m--, k++)
                    bcoeff[m] = bcoeff[k];

                // Create new nodes to store expanded expression
                (*f)->ep[i] = new Sum<T>(newbranches,0);
                assert((*f)->ep[i] != NULL);
                (*f)->fac_exp[i] = T(1);

                Sum<T> *sum1 = new Sum<T>(1,0);
                Sum<T> *sum2 = new Sum<T>(ep[i]->branches-1,0);
                assert(sum1 != NULL); assert(sum2 != NULL);

                // if the expression to be expanded contains more than 2 terms
                // then group front/rear expressions into 1 term in order to
                // apply binomial expansion
```

```
if(ep[i]->branches > 2)
{
    sum1->ep[0] = ep[i]->ep[0];
    sum1->fac_exp[0] = ep[i]->fac_exp[0];

    for(m=1; m<ep[i]->branches; m++)
    {
        k = m-1;
        sum2->ep[k] = ep[i]->ep[m];
        sum2->fac_exp[k] = ep[i]->fac_exp[m];
    }
}
else    // front/rear are clear
{
    sum1->ep[0] = ep[i]->ep[0];
    sum1->fac_exp[0] = ep[i]->fac_exp[0];
    sum2->ep[0] = ep[i]->ep[1];
    sum2->fac_exp[0] = ep[i]->fac_exp[1];
}


// Binomial Expansion
// For term 0th
(*f)->ep[i]->fac_exp[0] = T(1);
(*f)->ep[i]->ep[0] = new Product<T>(1);
assert((*f)->ep[i]->ep[0] != NULL);
sum1->copy((*f)->ep[i]->ep[0]->ep);
(*f)->ep[i]->ep[0]->fac_exp[0] = fac_exp[i];

// For term (lastindex)th
(*f)->ep[i]->fac_exp[lastindex] = T(1);
(*f)->ep[i]->ep[lastindex] = new Product<T>(1);
assert((*f)->ep[i]->ep[lastindex] != NULL);
sum2->copy((*f)->ep[i]->ep[lastindex]->ep);
(*f)->ep[i]->ep[lastindex]->fac_exp[0] = fac_exp[i];

// All other terms
for(m=1; m<lastindex; m++)
{
    // coefficient of each term
    (*f)->ep[i]->fac_exp[m] = bcoeff[m-1];
    (*f)->ep[i]->ep[m] = new Product<T>(2);
    assert((*f)->ep[i]->ep[m] != NULL);
    sum1->copy((*f)->ep[i]->ep[m]->ep);
    sum2->copy((*f)->ep[i]->ep[m]->ep+1);

    // degree of each term
    (*f)->ep[i]->ep[m]->fac_exp[0] = fac_exp[i]-T(m);
```

```
                    (*f)->ep[i]->ep[m]->fac_exp[1] = T(m);
                }
                delete [] bcoeff;
                sum1->deleteOneNode();
                sum2->deleteOneNode();
                ep[i]->deleteAllNodes();

                for(m=0; m<newbranches; m++)
                    flag += (*f)->ep[i]->ep[m]->mxpand((*f)->ep[i]->ep+m);
            }
        }
        return flag;
    }
    else // for non-commutative algebra
    {
        int i, flag=0, flag1;
        (*f) = new Product<T>(branches, ep, fac_exp, next_var);
        assert(*f != NULL);

        for(i=0; i<branches; i++)
        {
            if(ep[i]->type() == 'S' && ep[i]->branches > 1
               && fac_exp[i] > T(1)
               && double(fac_exp[i]) == int(double(fac_exp[i])))
            {
                int m, newbranches = int(fac_exp[i]);
                flag++;
                (*f)->ep[i] = new Product<T>(newbranches);
                assert((*f)->ep[i] != NULL);
                (*f)->fac_exp[i] = T(1);

                for(m=0; m<newbranches; m++) ep[i]->copy((*f)->ep[i]->ep+m);

                ep[i]->deleteAllNodes();
                do {
                    flag1 = 0;
                    flag1 = (*f)->ep[i]->dxpand((*f)->ep+i);
                } while(flag1);
            }
        }
        return flag;
    }
}

// Distributive law
template <class T> int Product<T>::dxpand(Symbol<T> **f)
{
```

```
    int i, flag=0, k, m;
    for(i=0; i<(*f)->branches; i++) flag += (*f)->ep[i]->dxpand((*f)->ep+i);

    // find the index of the term that could apply distributive law
    for(i=0; i<branches; i++)
        if(ep[i]->type() == 'S' && ep[i]->branches > 1) break;

    // check for fac_exp[]==1 or else we use binomial/multinomial expansion
    if(i < branches && fac_exp[i] == T(1))
    {
        flag++;
        *f = new Sum<T>(ep[i]->branches,0); assert(*f != NULL);
        for(k=0; k < ep[i]->branches; k++)
        {
            (*f)->ep[k] = new Product<T>(branches); assert((*f)->ep[k] != NULL);
            (*f)->fac_exp[k] = ep[i]->fac_exp[k];

            for(m=0; m<i; m++)
            {
                ep[m] -> copy((*f)->ep[k]->ep+m);
                (*f)->ep[k]->fac_exp[m] = fac_exp[m];
                (*f)->ep[k]->next_var[m] = next_var[m];
            }

            // Do not assign fac_exp[] because is preset to be 1 by constructor
            // it is set to be 1 because of distributive law
            ep[i]->ep[k] -> copy((*f)->ep[k]->ep+i);
            (*f)->ep[k]->next_var[i] = ep[i]->next_var[k];

            for(m=i+1; m<branches; m++)
            {
                ep[m]->copy((*f)->ep[k]->ep+m);
                (*f)->ep[k]->fac_exp[m] = fac_exp[m];
                (*f)->ep[k]->next_var[m] = next_var[m];
            }
        }
        deleteAllNodes();
    }
    return flag;
}

// Differentiation
template <class T> void Product<T>::diff(Symbol<T> **s,Symbol<T> *t) const
{
    int i, j, k, newbranches = branches + 1;
    *s = new Sum<T>(branches,0); assert(*s != NULL);
```

```
// d/dx (a1 a2 ... aN) = da1/dx * a2 .. aN + a1 * da2/dx .. aN + ...
for(i=0; i<branches; i++)
{
    if(fac_exp[i] == T(1))
    {
        (*s)->ep[i] = new Product<T>(branches);
        assert((*s)->ep[i] != NULL);

        for(j=0; j<i; j++)
        {
            (*s)->ep[i]->fac_exp[j] = fac_exp[j];
            (*s)->ep[i]->next_var[j] = next_var[j];
            ep[j]->copy((*s)->ep[i]->ep+j);
        }

        (*s)->ep[i]->fac_exp[i] = fac_exp[i];
        ep[i]->diff((*s)->ep[i]->ep+i,t);

        if((*s)->ep[i]->ep[i]->type() == 'V' ||
            (*s)->ep[i]->ep[i]->type() == 'N')
            (*s)->ep[i]->next_var[i] = 1;
        else (*s)->ep[i]->next_var[i] = 0;

        for(j=i+1; j<branches; j++)
        {
            (*s)->ep[i]->fac_exp[j] = fac_exp[j];
            (*s)->ep[i]->next_var[j] = next_var[j];
            ep[j]->copy((*s)->ep[i]->ep+j);
        }
    }
    else    // fac_exp[i] != T(1)
    {
        (*s)->ep[i] = new Product<T>(newbranches);
        assert((*s)->ep[i] != NULL);
        for(j=0; j<=i; j++)
        {
            (*s)->ep[i]->fac_exp[j] = fac_exp[j];
            (*s)->ep[i]->next_var[j] = next_var[j];
            ep[j]->copy((*s)->ep[i]->ep+j);
        }

        for(k=i+1, j=i+2; j<newbranches; j++, k++)
        {
            (*s)->ep[i]->fac_exp[j] = fac_exp[k];
            (*s)->ep[i]->next_var[j] = next_var[k];
            ep[k]->copy((*s)->ep[i]->ep+j);
        }
```

```
            // d/dx (u^n) = n u^(n-1) du/dx
            (*s)->ep[i]->fac_exp[i] = fac_exp[i] - T(1);
            (*s)->fac_exp[i] *= fac_exp[i];
            ep[i]->diff((*s)->ep[i]->ep+i+1, t);
      }
   }
}


// Integration
template <class T>
void Product<T>::integrate(Symbol<T> **s,Symbol<T> *t) const
{
 if(branches == 1)
 {
  if(ep[0] == t)
  {
   if(fac_exp[0]!=T(-1))    // Int(x^n,x)=x^(n+1)/(n+1)
   {
    *s = new Sum<T>(1,0);                    assert(*s != NULL);
    (*s)->ep[0] = new Product<T>(1);         assert((*s)->ep[0] != NULL);
    t->copy((*s)->ep[0]->ep);
    (*s)->fac_exp[0]=T(1)/(fac_exp[0]+T(1));
    (*s)->ep[0]->fac_exp[0]=(fac_exp[0]+T(1));
    (*s)->ep[0]->next_var[0]=0;
   }
   else                     // Int(x^(-1),x)=ln(x);
   {
    *s = new Sum<T>('l',char());             assert(*s != NULL);
    (*s)->ep[0] = new Sum<T>(1,0);           assert((*s)->ep[0] != NULL);
    t->copy((*s)->ep[0]->ep);
   }
  }
  else if(implicit_depend(ep,t,branches))
  {
   *s = new Sum<T>('i',char());  assert(*s != NULL);
   copy((*s)->ep);
   t->copy((*s)->ep+1);
  }
  else
  {
   *s = new Product<T>(2);  assert(*s != NULL);
   copy((*s)->ep);
   t->copy((*s)->ep+1);
  }
 }
 else //branches>1
```

```
{
  int i,j,constant_found=-1;

  //search for constants with respect to integration
  if(Comm == 1)
  for(i=0;(i<branches)&&(constant_found<0);i++)
  {
    if((ep[i]->type() == 'S')||(ep[i]->type()=='B')||(ep[i]->type()=='V'))
    {
      if(!((Sum<T>*)ep[i])->isdepend(*((Sum<T>*)t))) constant_found=i;
    }
    else
    {
      if(ep[i]->type() == 'N') constant_found=i;
      else
        if(!implicit_depend(ep[i]->ep,t,ep[i]->branches+(ep[i]->branches==0)))
          constant_found=i;
    }
  }

  if(constant_found<0)
  {
    *s = new Sum<T>('i',char());  assert(*s != NULL);
    copy((*s)->ep);
    t->copy((*s)->ep+1);
  }
  else
  {
    *s = new Product<T>(branches);              assert(*s != NULL);
    Product<T> *p=new Product<T>(branches-1); assert(p != NULL);

    ep[constant_found]->copy((*s)->ep);
    for(i=0,j=0;i<branches;i++)
    if(i!=constant_found)
    {
      p->fac_exp[j]=fac_exp[i];
      p->next_var[j]=next_var[i];
      ep[i]->copy(p->ep+(j++));
    }
    p->integrate((*s)->ep+1,t);
    delete p;
  }
}
}

// Simplify Special Functions
template <class T> int Product<T>::funcSimp(int& i,int& j)
```

```
{
   switch(ep[i]->type())
   {
      case 'e':
         Symbol<T> *ptr;

         ptr = ep[i]->ep[0];
         ep[i]->ep[0] = new Sum<T>(2,0); assert(ep[i] != NULL);
         ep[i]->ep[0]->fac_exp[0] = fac_exp[i];
         ep[i]->ep[0]->fac_exp[1] = fac_exp[j];
         ep[i]->ep[0]->ep[0] = ptr;
         ep[j]->ep[0]->copy(ep[i]->ep[0]->ep+1);
         fac_exp[i] = fac_exp[j] = T(1);
         ep[i]->ep[0]->Shrink();
         remove(j); j=i;
         return 1;
      default: return 0;
   }
}

// removes ones from products
template <class T> int Product<T>::shrink_1()
{
   int i, flag=0;

   for(i=0; i<branches; i++)
   {
      if(ep[i] == one())
      {
         if(branches > 1)
         {
            remove(i); i--;
            flag++;
         }
         // (branches==1), put 1^(n) = 1
         else if(fac_exp[i] != T(1))  fac_exp[i] = T(1);
      }
   }
   for(i=0; i<branches; i++) flag += ep[i]->shrink_1();
   return flag;
}

// removes extra Product nodes: moves the lower level of a consecutive
// level of Product node to the upper level
template <class T> int Product<T>::shrink_2()
{
   int i, j, k, n, flag=0;
```

```
Symbol<T> **ept;
T *fac_expt;
int *next_vart;

for(i=0; i<branches; i++)
{
   if(ep[i]->type() == 'P')
   {
      n = branches - 1 + ep[i]->branches;
      ept = new Symbol<T>*[n]; assert(ept != NULL);
      fac_expt = new T[n];     assert(fac_expt != NULL);
      next_vart = new int[n];  assert(next_vart != NULL);

      // copy nodes that are not affected
      for(k=0; k<i; k++)
      {
         ept[k] = ep[k];
         fac_expt[k] = fac_exp[k];
         next_vart[k] = next_var[k];
      }

      // moves Product node to one level up
      for(j=i, k=0; k < ep[i]->branches; k++, j++)
      {
         ept[j] = ep[i]->ep[k];
         fac_expt[j] = ep[i]->fac_exp[k] * fac_exp[i];
         next_vart[j] = ep[i]->next_var[k];
      }

      // copy nodes that are not affected
      for(k=i+1; k<branches; k++, j++)
      {
         ept[j] = ep[k];
         fac_expt[j] = fac_exp[k];
         next_vart[j] = next_var[k];
      }

      branches = n;
      ep[i]->deleteOneNode();
      delete [] ep; delete [] fac_exp; delete [] next_var;
      ep = ept;
      fac_exp = fac_expt;
      next_var = next_vart;
      flag++;
   }
}
for(i=0; i<branches; i++) flag += ep[i]->shrink_2();
```

```
      return flag;
}


// remove zero Products
template <class T> int Product<T>::shrink_3(Symbol<T> **f)
{
   int i, flag=0;
   for(i=0; i<branches; i++)
      if(ep[i] == zero())
      {
         *f = zero();
         deleteAllNodes();
         return 1;
      }

   for(i=0; i<branches; i++) flag += ep[i]->shrink_3(ep+i);
   return flag;
}


// sums Numbers (type()=='N') from Sums
template <class T> int Product<T>::shrink_4()
{
   int i, flag=0;
   for(i=0; i<branches; i++) flag += ep[i]->shrink_4();
   return flag;
}


// remove Sums with only one summand
template <class T> int Product<T>::shrink_5()
{
   int i, flag=0;
   Symbol<T> *ptr;
   for(i=0; i<branches; i++)
      if(ep[i]->type() == 'S' &&
         ep[i]->branches == 1 &&
         ep[i]->fac_exp[0] == T(1))
      {
         ptr = ep[i]->ep[0];
         next_var[i] = ep[i]->next_var[0];
         ep[i]->deleteOneNode();
         ep[i] = ptr;
         flag++;
      }

   for(i=0; i<branches; i++) flag += ep[i]->shrink_5();
   return flag;
}
```

```
// declare a commutative/non-commutative operator *
template <class T> void Commutative(T,int flag)
{
    if(flag) Product<T>::Comm = 1;
    else     Product<T>::Comm = 0;
}
#endif
```

Chapter 11

PVM and Abstract Data Types

Parallel Virtual Machine (PVM) [16] is a software system that permits a network of heterogeneous Unix computers to be used as a single large parallel computer. Thus large computational problems can be solved by using the aggregate power of many computers. Applications, which may be written in Fortran 77, C or C++, can be parallelized by using message-passing constructs common to most distributed-memory computers. By sending and receiving messages, multiple tasks of an application can cooperate to solve a problem in parallel.

The PVM system is composed of two parts. The first part is a daemon which resides on all the computers making up the virtual machine. The second part of the system is a library of PVM interface routines. This library contains C routines for message passing, spawning processes, coordinating tasks, and modifying the virtual machine. All basic data types in C and strings (arrays of characters) can be packed, sent and unpacked. Data encoding is a binary representation for data objects (e.g., integers, floating point numbers) such as XDR or the native format of a microprocessor. PVM can contain data in XDR or native format. XDR stands for eXternal Data Representation. This is an Internet standard data encoding (essentially big-endian integers and IEEE format floating point numbers). PVM converts data to XDR format to allow communication between hosts with different native data formats.

Typical applications of PVM are when the problems under consideration can be parallelized or the divide-and-conquer method can be applied. For example consider the *logistic map*

$$x_{t+1} = 4x_t(1 - x_t), \qquad t = 0, 1, 2, \ldots$$

with x_0 (initial value) given. The map can be split up into two maps given by

$$x_{t+2} = 16x_t(1 - x_t)(1 - 4x_t(1 - x_t))$$
$$x_{t+3} = 16x_{t+1}(1 - x_{t+1})(1 - 4x_{t+1}(1 - x_{t+1})).$$

The two series are independent and can be calculated separately. Two tasks are spawned to calculate the two series. These are then finally combined. Since the two maps are identical (except for a offset of one in their indices) two identical slaves can be spawned.

Next we show how the PVM may be used together with Abstract Data Types (ADT). In the following program, we calculate the addition of two rational numbers A and B, where

$$A = \frac{41152}{218107} \quad \text{and} \quad B = \frac{123234}{333445} \quad \text{then} \quad A + B = \frac{40600126678}{72726688615}.$$

The program consists of two parts. The master program and the slave program.

- The master program specifies the values for A and B, packs the numbers and then sends them to the slave program across the network. The slave program does the calculation $A+B$ and returns the result of the calculation to the master program.

- The slave program which resides on the remote machine first receives the packet sent by the master program, unpacks it, performs the addition and finally returns the result of the calculation back to the master program.

Note that the library routines of the PVM do not support the packing of abstract data types. Therefore, we have to supply our own packing and unpacking functions:

- pvm_pkvlong() packs the Verylong data type.

- pvm_upkvlong() unpacks the Verylong data type.

- pvm_pkrat() packs the Rational data type.

- pvm_upkrat() unpacks the Rational data type.

```
// master.cc (The master program)
// In this example we show how to send abstract data types.
// Note that only the data members needed to reconstruct the object
// are packed and sent.

#include <iostream.h>
#include <pvm3.h>
```

```cpp
#include "Verylong.h"
#include "Rational.h"

// We write our own versions of the PVM packing functions
// Pack and unpack a Verylong
void pvm_pkvlong(const Verylong *l)
{
    char *s = (char*)(*l);    // type conversion : Verylong -> char*
    int length = strlen(s);
    pvm_pkint(&length,1,1);
    pvm_pkstr(s);
}


void pvm_upkvlong(Verylong *l)
{
    int length;
    pvm_upkint(&length,1,1);
    char *s = new char[length+1];
    pvm_upkstr(s);
    Verylong v(s);
    *l = v;
}


// Pack and unpack a Rational
void pvm_pkrat(const Rational<Verylong> *r)
{
    Verylong v = r->num();
    pvm_pkvlong(&v);
    v = r->den();
    pvm_pkvlong(&v);
}


void pvm_upkrat(Rational<Verylong> *r)
{
    Verylong num, den;
    pvm_upkvlong(&num);
    pvm_upkvlong(&den);
    Rational<Verylong> a(num,den);
    *r = a;
}


void main()
{
    // Enroll task into PVM by querying task ID
```

```
    int tid = pvm_mytid();

    // Create two rational numbers:
    Rational<Verylong> A("41152","218107"), B("123234","333445");

    // Spawn the slave task
    int slaveid;
    int error = pvm_spawn("slave", (char**)NULL,
                          PvmTaskDefault, "", 1, &slaveid);

    if(error > 0)  // If error < 0, the slave could not be spawned
    {
        // Send the two rational numbers
        pvm_initsend(PvmDataDefault);

        pvm_pkrat(&A);
        pvm_pkrat(&B);
        pvm_send(slaveid, 1);

        // Receive the answer
        pvm_recv(slaveid, 2);
        Rational<Verylong> C;
        pvm_upkrat(&C);

        cout << A << " + " << B << " = " << C << endl;
    }
    // Unhook current task from PVM
    pvm_exit();
}


// slave.cc (The slave program)

#include <iostream.h>
#include <pvm3.h>
#include "Verylong.h"
#include "Rational.h"

// We write our own versions of the PVM packing functions
// Pack and unpack a Verylong
void pvm_pkvlong(const Verylong *l)
{
    char *s = (char*)(*l);    // type conversion : Verylong -> char*
    int length = strlen(s);
    pvm_pkint(&length,1,1);
```

```
   pvm_pkstr(s);
}

void pvm_upkvlong(Verylong *l)
{
   int length;
   pvm_upkint(&length,1,1);
   char *s = new char[length+1];
   pvm_upkstr(s);
   Verylong v(s);
   *l = v;
}

// Pack and unpack a Rational
void pvm_pkrat(const Rational<Verylong> *r)
{
   Verylong v = r->num();
   pvm_pkvlong(&v);
   v = r->den();
   pvm_pkvlong(&v);
}

void pvm_upkrat(Rational<Verylong> *r)
{
   Verylong num, den;
   pvm_upkvlong(&num);
   pvm_upkvlong(&den);
   Rational<Verylong> a(num,den);
   *r = a;
}

void main()
{
   // Enroll task into PVM by querying parent task ID
   int ptid = pvm_parent();

   // Receive the two rational numbers
   Rational<Verylong> A, B;

   pvm_recv(ptid,1);
   pvm_upkrat(&A); pvm_upkrat(&B);

   // Do the calculation
   Rational<Verylong> C = A + B;
```

```
    // Send the answer back
    pvm_initsend(PvmDataDefault);
    pvm_pkrat(&C);
    pvm_send(ptid, 2);

    // Task completed successfully
    pvm_exit();
}
```

The output of the program is

41152/218107 + 123234/333445 = 40600126678/72726688615

Note that in the packing function pvm_pkvlong(), the statement

```
    char *s = (char*)(*l);
```

converts a Verylong number to the built-in data type char*. This step is necessary because PVM supports only the packing of built-in data types. The conversion operator has been implemented in the Verylong class (see Chapter 6).

Chapter 12

Error Handling Techniques

An exception is an error that occurs at run-time. A typical example is division by zero. Thus it may be a good idea to classify a program into distinct subsystems that either execute successfully or fail. Thus, local error checking should be implemented throughout the system for a sound program. It is, therefore, important to identify the possible source of errors. They could be caused by the programmer himself. This case includes the detection of an internal logic error, such as an assertion failure. Sometimes, special attention has to be paid to the preconditions of calling a function. For example, an attempt to get the seventh character of a three-character String would cause an error. However, some problems are not logic errors, but a failure to get some resource during run-time. These errors might include running out of memory, a write failure due to disk full, etc.

A fault-tolerant system is usually designed hierarchically, with each level coping with as many errors as possible. However, some errors cannot be handled locally. Thus, we need a global error communication mechanism, which would be one of the following:

- Error state, i.e. the different return values of a function or class method represent the different status of the function execution.

- Exception handling.

In C we can use the function void assert(int test) to test a condition and possibly abort. This function is a macro that expands to an if statement; if test evaluates to zero, assert prints a message on stderr and aborts the program by calling abort. Using C++'s exception handling subsystem we can in a structured and controlled manner handle run-time errors. C++ exception handling is built upon three keywords: try, catch and throw.

12.1   Error State

One can implement *error states* via class variables. Typically, each object has an error flag (error state variable). If an error occurs, the error state variable is set to a value indicating the error type. For example, the <iostream.h> library of C++ has an integer state variable called state. The different bits of this variable indicate different error states. The error state can be queried via the class methods bad(), eof(), fail(). The user can then implement his own error-handling mechanism.

12.2   Exception Handling

A more sophisticated method that controls error detection and error handling is called *exception handling*. It is a non-local mechanism which provides a means of communicating errors between the classes/functions and programs that make use of these classes/functions. It also lets the programmers write their code without worrying about errors at every function call.

In the exception handling mechanism in C++, the function that detects an error raises an exception using the keyword throw. The handler function then catches and handles the exception. The function that raises the exception must come within a try-block. The handlers, which are declared using the keyword catch, are placed at the end of the try-block.

As an example, consider the following Rational class, together with the main() program:

```
// except.cpp

#include <iostream.h>
#include "MString.h"

class Rational
{
 private:
   long int num, den;

 public:
   Rational(const long int = 0, const long int = 1);
};

class Check_Error
{
 private:
   String reason;
```

```
  public:
    Check_Error(const String &s) : reason(s) {}
    String diagnostic() const { return reason; }
};

Rational::Rational(const long int N, const long int D)
    : num(N), den(D)
{
    if(D==0) throw Check_Error("Division by zero !");
}

int main()
{
    try
    {
        long int a, b;

        a = 5; b = 0;

        Rational R1(a,b);
    }

    catch(Check_Error diag)
    {
        cerr << "Internal error : " << diag.diagnostic() << endl;
        return 1;
    }
    return 0;
}
```

```
Result
======
Internal error : Division by zero !
```

The constructor of the Rational class checks if the denominator of the number is zero. It throws an exception when this error occurs. The thrown object contains the diagnostic message, which is

```
    Division by zero !
```

in this case.

The class Check_Error handles all the exceptions that might happen within the program. It contains a private data member which stores a string containing the reason for the exception. The member function diagnostic() prints the diagnostic message of the failure. It is used in the catch function of the program.

Note that in the main() function the whole program is placed in the try-block. Immediately after the block is the catch function, which is the actual exception handler. The catch function specifies the type of exception it handles.

Chapter 13

Gnuplot and PostScript

Gnuplot is a command-driven interactive function-plotting program. It can be used to plot ordinary or user-defined mathematical functions and data points in both two- and three-dimensional space. The original software was developed by Thomas Williams and Colin Kelley. It is available over a number of platforms, such as Unix, Atari, VMS, MS-DOS and OS/2. It accommodates many of the needs of today's scientists for graphic data representation.

A wide-range of commands and functions are available in Gnuplot. It handles both curves (two-dimensional) and surfaces (three-dimensional). Surfaces can be plotted as a mesh, floating in three-dimensional coordinate space, or as a contour plot on the $x - y$ plane. For two-dimensional plots, the plot styles includes lines, points and error bars, etc. Graphs may be labelled with arbitrary labels and arrows, axes labels, a title, date and time, and a key.

The **plot** and **splot** commands are two primary commands in Gnuplot. The command **plot** is used to plot two-dimensional functions and data, whereas **splot** plots three-dimensional surfaces and data. For example,

    gnuplot> plot cos(x)

plots a simple cosine curve, and

    gnuplot> splot x*x*y

plots a three-dimensional surface. To plot a data file called **result.dat**, we type

    gnuplot> plot 'result.txt'

Gnuplot has an on-line help, which can be invoked by the `help` command, with the following syntax:

    gnuplot> help <*topic*>

where <*topic*> refers to a particular topic in the context. For other gnuplot commands, we refer to the user's manual [59].

PostScript was designed and developed by Adobe Systems Incorporated. Basically, it is a page-description language that is designed specifically to provide a device-independent description of a printable document from a computer system to a printer. It is also a general-purpose programming language that contains a wide range of graphics operations. It supports data types such as reals, booleans, arrays and strings. Moreover, it contains variables and allows the construction of more complex procedures and functions by combining different operators.

PostScript is embedded in an interpreter program that generally runs in an independent device, such as a laser printer. The interpreter program translates PostScript operations and data into device-specific codes and generates the graphics being described on the page. One of the greatest strengths of PostScript is that it is independent of the output device, such as a typesetter or a high dot-density publishing device.

A PostScript file should start with the `%!` character. It is a magic number to indicate a PostScript file rather than some kind of text file. PostScript is a *postfix* language, where the operators follow their arguments. For example, the statement

    x y moveto

moves the current pointer to the coordinate (x, y). In general, the syntax of a given command is

    arg_1 arg_2 ... arg_n operator

The following PostScript program draws a triangle and displays the text message "SymbolicC++" at a 50° angle to the horizontal.

```
%!PostScript                  % PostScipt "Magic Number"
70 140 moveto                 % set starting point
300 620 lineto                % add line segment
530 140 lineto                % add line segment
closepath                     % close the shape
stroke                        % draw the path
/Palatino-Roman findfont      % find the required font
65 scalefont                  % scale the font size
setfont                       % select the current font
140 148 moveto                % set current point
50 rotate                     % rotate coordinate system
(SymbolicC++) show            % specific the text message
showpage                      % display page
```

For other commands in the PostScript language, we refer to [21], [33].

Suppose we intend to visualize the time evolution of the logistic map

$$x_{t+1} = 4x_t(1 - x_t), \qquad t = 0, 1, 2, \ldots, \quad x_0 \in [0, 1]$$

with $x_0 = 1/3$. We first generate the data set (timeev.dat) using the following program:

```
// logis.cpp

#include <fstream.h>    // for ofstream, close()
#include "Verylong.h"
#include "Rational.h"

void main()
{
    ofstream data("timeev.dat");
    Rational<Verylong> c1("1"), c2("4"),
                       x0("1","3");            // initial value x0 = 1/3

    data << 0 << " " << double(x0) << endl;
    int i;
    for(i=1; i<=10; i++)
    {
        x0 = c2*x0*(c1 - x0);
        data << i << " " << double(x0) << endl;
    }
    data.close();
}
```

The time evolution of the map (`timeev.dat`) can now be viewed using Gnuplot. In the Gnuplot environment, we type the following command:

```
gnuplot> plot [0:10] 'timeev.dat'
```

This command plots the first eleven points of the time evolution. We can create a postscript file using the commands:

```
gnuplot> set term postscript default
gnuplot> set output "timeev.ps"
gnuplot> plot 'timeev.dat'
```

The postscript file (`timeev.ps`) may be sent to the printer using the command:

```
copy timeev.ps lpt1
```

for MS-DOS, OS/2, Windows NT and

```
lp timeev.ps
```

for Unix systems.

Bibliography

[1] Ammeraal Leen, *STL for C++ Programmers*, John Wiley, Chichester (1997).

[2] Anderson J. R., Corbett A. T. and Reiser B. J., *Essential LISP*, Addison-Wesley (1987).

[3] Aslaksen H., *Multiple-valued Complex Functions and Computer Algebra*, SIGSAM Bull., **30**, 12-20 (1996).

[4] Ayres F., *Modern Algebra*, Schaum's Outline Series, McGraw-Hill, New York (1965).

[5] Barnsley M. F., *Fractals Everywhere*, 2nd ed., Academic Press Professional, Boston (1993).

[6] Berry J. T., *The Waite Group's C++ Programming*, 2nd ed., SAMS, Carmel, Indiana (1992).

[7] Birtwistle G. M., Dahl O. J., Myhrhaug B. and Nygaard K., *SIMULA BEGIN*, Studentlitteratur, Sweden, Auerbach, Philadelphia (1973).

[8] Brackx F., *Computer Algebra with LISP and REDUCE : An Introduction to Computer-aided Pure Mathematics*, Kluwer Academic, Boston (1991).

[9] Budd T. A., *Classic Data Structures in C++*, Addison-Wesley (1994).

[10] Char B. W., *First Leaves - A Tutorial Introduction to MAPLE V*, Springer-Verlag, New York (1991).

[11] Dautcourt G., Jann K. P., Riemer E. and Riemer M., *Astronomische Nachrichten*, **102**, 1 (1981).

[12] Davenport J. H., Siret Y. and Tournier E., *Computer Algebra: Systems and Algorithms for Algebraic Computation*, 2 ed., Academic Press, London (1993).

[13] Ellis M. A. and Stroustrup B., *The Annotated C++ Reference Manual*, Addison-Wesley, (1990).

[14] Epstein R. L. and Carnielli W. A., *Computability*, Wadsworth & Brooks/Cole, Pacific Grove, California (1989).

[15] Fröberg C. E., *Numerical Mathematics, Theory and Computer Applications*, Benjamin-Cummings, Menlo Park (1985).

[16] Geist A., Beguelin A., Dongarra J., Jiang Weicheng, Manchek R. and Sunderam V., *PVM: Parallel Virtual Machine, A User's Guide and Tutorial for Networked Parallel Computing*, MIT Press, Cambridge, MA (1994).

[17] Gray P. D. and Mohamed R., *Smalltalk-80: A Practical Introduction*, Pitman, London (1990).

[18] Hearn A., *REDUCE User's Manuel, Version 3.5*, RAND publication CP78 (Rev. 10/93).

[19] Hehl F. W., Winkelmann V. and Meyer H., *REDUCE*, 2 ed., Springer-Verlag (1993).

[20] Hekmatpour S., *An Introduction to LISP and Symbol Manipulation*, Prentice Hall, New York (1988).

[21] Holzgang D. A., *Understanding PostScript Programming*, second edition, SYBEX Inc., Alameda (1987).

[22] Howson A. G., *A Handbook of Terms used in Algebra and Analysis*, Cambridge University Press, Cambridge (1972).

[23] Jamsa K., *Success with C++*, Jamsa Press, Las Vegas (1994).

[24] Jenks R. D. and Sutor R. S., *Axiom: The Scientific Computation System*, Springer-Verlag, New York (1992).

[25] Kim W., *Introduction to Object-Oriented Databases*, MIT Press, Cambridge, MA (1990).

[26] Knuth D. E., *Fundamental Algorithms*, vol. 1 of *The Art of Computer Programming*, Addison-Wesley, Reading, MA (1968).

[27] Knuth D. E., *Seminumerical Algorithms*, 2nd ed., vol. 2 of *The Art of Computer Programming*, Addison-Wesley, Reading, MA (1981).

[28] Knuth D. E., *Sorting and Searching*, vol. 3 of *The Art of Computer Programming*, Addison-Wesley, Reading, MA (1973).

[29] Lafore R., *Object-Oriented Programming in TURBO C++*, Waite Group Press, Mill Valley (1991).

[30] Lang S., *Linear Algebra*, Addison-Wesley, Reading, MA (1968).

[31] Lippman S. B., *C++ Primer*, 2nd ed., Addison-Wesley, Reading, MA (1991).

[32] McCarthy J. et al., *LISP 1.5 Programmer's Manual*, 2 ed., MIT Press (1965).

[33] McGilton H. and Campione M., *PostScript by Example*, Addison-Wesley (1992).

[34] MacCallum M. A. H. and Wright F. J., *Algebraic Computing with Reduce*, Clarendon Press, Oxford (1991).

[35] Mével A. and Guéguen T., *Smalltalk-80*, Macmillan Education, London (1987).

[36] Meyer B., *Object-Oriented Software Construction*, Prentice-Hall International (1988).

[37] Meyers S., *Effective C++*, Addison-Wesley, Reading, MA (1992).

[38] Norvig P., *Paradigms of Artificial Intelligence Programming: Case Studies in Common Lisp*, Morgan Kaufman, San Mateo (1991).

[39] Oevel W., Postel F., Rüscher G., and Wehmeier S., *Das MuPAD Tutorium*, SciFace Software, Paderborn (1998).

[40] Press W. H., Teukolsky S. A., Vetterling W. T. and Flannery B. P., *Numerical Recipes in C, The Art of Scientific Computing*, Cambridge University Press (1992).

[41] Rayna G., *REDUCE: Software for Algebraic Computation*, Springer-Verlag, New York (1987).

[42] Reiser M., *The Oberon System : user guide and programmer's manual*, Addison-Wesley (1991).

[43] Reiser M. and Wirth N., *Programming in Oberon*, Addison-Wesley (1992).

[44] Risch R. H., *Transactions of the American Mathematical Society*, **139**, 167–189 (1969).

[45] Smith M. A., *Object-Oriented Software in C++*, Chapman and Hall, London (1993).

[46] Steeb W. H. and Euler N., *Nonlinear Evolution Equations and Painlevé Test*, World Scientific, Singapore (1988).

[47] Steeb W. H. and Lewien D., *Algorithms and Computation with Reduce*, BI-Wissenschaftsverlag, Mannheim (1992).

[48] Steeb W. H., Lewien D. and Boine-Frankenhein O., *Object-Oriented Programming in Science with C++*, BI-Wissenschaftsverlag, Mannheim (1993).

[49] Steeb W. H., *Quantum Mechanics using Computer Algebra*, World Scientific, (1994).

[50] Steeb W. H. and Euler N., *Continuous Symmetries, Lie Algebras and Differential Equations*, World-Scientific, Singapore (1996).

[51] Steeb W. H., *Problems and Solutions in Theoretical and Mathematical Physics*, World Scientific, Singapore (1996).

[52] Steeb W. H., *The Nonlinear Workbook: Chaos, Fractals, Cellular Automata, Neural Networks, Genetic Algorithms, Fuzzy Logic with C++, Java, SymbolicC++ and Reduce Programs*, World Scientific, Singapore (1999).

[53] Stephani H., *General Relativity*, Cambridge University Press, Cambridge (1985).

[54] Stoutemeyer D. R., *Notices of the American Mathematical Society*, Volume 38, pp. 778–785, (1991).

[55] Stroustrup B., *The C++ Programming Language*, 2nd ed., Addison-Wesley, Reading, MA (1991).

[56] Tennent R. D., *Principles of Programming Languages*, Prentice-Hall, Englewood Cliffs (1981).

[57] Thomas P. and Weedon R., *Object-Oriented Programming in Eiffel*, Addison-Wesley, Wokingham (1985).

[58] Touretzky D. S., *Common Lisp: A Gentle Introduction to Symbolic Computation*, Benjamin/Cummings, Redwood City (1990).

[59] Williams T., Kelley C. et al., *Gnuplo: An Interactive Plotting Program*, http://www.cs.dartmouth.edu/gnuplot-info.html.

[60] Winston P. H., *Lisp*, Addison-Wesley, (1989).

[61] Wirth N. and Gutknecht J., *Project Oberon: The Design of an Operating System and Compilers*, Addison-Wesley (1992).

[62] Wolfram S., *Mathematica: A System for Doing Mathematics by Computer*, 2nd ed., Addison-Wesley (1992).

[63] Yan S. S., Collofello J. S. and MacGergor T., *Ripple Effect Analysis of Software Maintenance*, In Proc. COMPSAC'78, pp.60–65 (1978).

Index